Microplastic Pollutants

Microplastic Pollutants

Christopher Blair Crawford

Brian Quinn

ELSEVIER

AMSTERDAM • BOSTON • HEIDELBERG • LONDON
NEW YORK • OXFORD • PARIS • SAN DIEGO
SAN FRANCISCO • SINGAPORE • SYDNEY • TOKYO

Elsevier
Radarweg 29, PO Box 211, 1000 AE Amsterdam, Netherlands
The Boulevard, Langford Lane, Kidlington, Oxford OX5 1GB, United Kingdom
50 Hampshire Street, 5th Floor, Cambridge, MA 02139, United States

Notices
Knowledge and best practice in this field are constantly changing. As new research and experience
broaden our understanding, changes in research methods, professional practices, or medical treatment
may become necessary.

Practitioners and researchers must always rely on their own experience and knowledge in evaluating and
using any information, methods, compounds, or experiments described herein. In using such information
or methods they should be mindful of their own safety and the safety of others, including parties for
whom they have a professional responsibility.

To the fullest extent of the law, neither the Publisher nor the authors, contributors, or editors, assume any
liability for any injury and/or damage to persons or property as a matter of products liability, negligence
or otherwise, or from any use or operation of any methods, products, instructions, or ideas contained in
the material herein.

Library of Congress Cataloging-in-Publication Data
A catalog record for this book is available from the Library of Congress

British Library Cataloguing-in-Publication Data
A catalogue record for this book is available from the British Library

ISBN: 978-0-12-809406-8

For information on all Elsevier publications
visit our website at https://www.elsevier.com/

Working together
to grow libraries in
developing countries

www.elsevier.com • www.bookaid.org

Publisher: Candice G. Janco
Acquisition Editor: Laura S. Kelleher
Editorial Project Manager: Emily Thomson
Production Project Manager: Poulouse Joseph
Designer: Mark Rogers

Typeset by TNQ Books and Journals

Contents

About the Author, Christopher Blair Crawford

Christopher Blair Crawford, PhD researcher in Chemistry and award-winning Scientist. After receiving a distinction in his previous scientific studies, and having achieved a degree in Forensic Science with First Class Honours, Christopher continued his studies in the area of Chemistry. Following his research alongside an industrial collaborator, he achieved his degree of Master of Science in Drug Design and Discovery with Distinction. In addition, Christopher was awarded two distinguished University Medals, in both Forensic Science and in Drug Design and Discovery. His other studies and interests have included Forensic Medical Science and Business Management, as well as Applied Science and Technology. Christopher works in partnership with the government at his research institution, actively investigating the chemical and toxicological interactions of microplastic pollutants, as well as identifying their potential as vectors for the accumulation and trophic transfer of contaminants in the aquatic environments of the world.

About the Author, Brian Quinn

Brian Quinn, PhD, reader in Ecotoxicology at the University of the West of Scotland and award winning Scientist specialising in the impact of contaminants of emerging concern in the aquatic environment. Extensive experience in the government, academic and industry sectors with a prominent academic record in scientific publication, peer review and regularly presents at international conferences. He is currently leading the MASTS Scottish Microplastics Research Group and is an active member of leading national and international scientific review panels.

Foreword

We are familiar with the ocean's garbage patches where areas the size of Texas are littered with discarded plastics. They contain items which were thrown away by our grandparents and which will not degrade in our lifetime. Fading buoys, plastic cartons, toys, flip flops, sheeting – items that once had a purpose and made our lives easier – but which now trap the flotsam and jetsam of the seas in their tangled midst, as well as fish, marine birds and mammals.

Plastics are versatile, ubiquitous and cheap to produce. They allow us to make technological advances unimaginable even a century ago. But they are almost everlasting, and therein lies a major environmental problem of the future.

We are beginning to realise the impact of these durable materials and are actively starting to think of ways to clear up the garbage patches in our oceans with a variety of projects proposed, which would slowly collect the debris for proper destruction. What we are less familiar with are microplastics – tiny unseen plastic fragments. Their impact is not visible, yet likely to be as great. And if ridding the ocean of plastic garbage is difficult, doing the same for microplastics is near impossible, meaning that the only solution is to make us aware of their impact and think how we can prevent their production or their potential for damage. This is why *Microplastic Pollutants* is such a timely and important book.

Microplastics can be produced in several ways. They can be manufactured to be intentionally small such as those used in place of sand in sand blasting. And we have become aware of the use of microbeads in face and body scrubs, which are typically washed down our drains with the shower water. Because of their small size, these microplastics escape water treatment plants finding their way into lakes, rivers and oceans. We may congratulate ourselves on not using such products, or in campaigning to have them removed. Yet we are each unintentionally the source of another tsunami of microplastics – those produced every time we wash our clothes in a washing machine. Up to 19,000 fibres are produced from a single wash with the favourite fabric of lovers of the outdoors, fleece, being one of the most egregious sources.

Microplastics can also come from the breakdown of larger items of plastic degraded by sunlight and then, in their newly brittle and discoloured state, broken down into thousands of tiny specks of plastic under the constant pounding of waves.

Such microplastics are a cause of immediate danger to animals. Assumed to be food, they are snapped up by a variety of aquatic creatures or drawn into their body along with currents of water as they filter them for food. Mussels and other filter feeding bivalves are particularly at risk. They clog the feeding tubes of aquatic creatures, causing them to starve or develop abnormally. But they also cause a more insidious form of long-term harm.

It has recently been discovered that microplastics concentrate the pollutants found in aquatic environments which include well-known carcinogens, such as polychlorinated biphenyls. This is because these types of pollutants are chemically attracted to microplastics, which essentially pick them up like a sponge. The level of pollutants associated with microplastics becomes much higher than in the surrounding environment. Thus, aquatic creatures in waters containing microplastics are likely to ingest more concentrated levels of pollutants than they would in waters free from these micropollutants. The authors have been investigating the ability of pollutants to leach from the plastic once ingested and become absorbed by the animal. If their hypothesis is correct, microplastics are the vector by which pollutants in the aquatic environment can become concentrated, travelling up the food chain to bigger and bigger animals and perhaps even to humans.

This book then is timely, shining a bright light on a hitherto little known area of research. It lays out in great detail the history of plastics, which have become such a fundamental part of our modern society. The fate of plastic objects, first as waste and then when we assume them to be no longer of concern, as microplastics, is elegantly described.

Proper data sets are an essential first step in tackling a problem and the authors offer important elements in acquiring these, showing how to collect, separate and identify these insidious pollutants.

Above all this book is important. It is a clarion call to action and should be a required reading for anyone who values the environment. I congratulate Christopher Blair Crawford and Brian Quinn on bringing microplastics into the spotlight.

Vivienne Parry, OBE
Science writer and broadcaster

Preface

Throughout the millennia, we have used the raw materials in our environment to create tools, build shelter and improve upon the efficiency of processes. From our use of flint to light fires to the creation of artefacts and cookware from bronze and iron, nature has provided us with an abundance of useful materials. Yet seeking to improve upon these materials, we created plastics, a uniquely human creation that simply does not exist in nature. In doing so, we furnished ourselves with the ability to break away from the confines and limitations of ceramics, metals and wood. Since then, plastic materials have been at the forefront of our modern evolution, propelling us from our humble beginnings to the grandeurs of technological innovation. At the centre of this amazing progression, human curiosity, ingenuity and our adeptness for symbolic communication has been the main driving force at play in this evolutionary success. Indeed, in comparison to the countless other extant species with whom we share this world, we possess the unique and powerful ability to substantially manipulate phenomena and the environment around us. Yet, while our advancement is unsurpassed, it is important to remember that such an influential capability begets an obligation of responsibility upon us all to mitigate the adverse effects of the pollution created from our activities.

There was once a time when colourful plastic items, bearing the scars of years at sea, would occasionally arrive upon our shores from distant lands. Perhaps to the intrigue of the beach comber, the rare appearance of a plastic artefact on the shoreline would instil a natural sense of curiosity, while leaving the finder blissfully unaware that its arrival quietly signalled the beginning of a new synthetic world. In the 1950s, we regarded plastics as the wonder materials of the future, destined to fulfil our hopes and dreams of a better tomorrow in which our lives would be made easier. To a great extent this did transpire and we found evermore uses for these materials. Indeed, our world was rapidly transformed and as time passed by, the production of plastics increased exponentially to keep up with the demand. Yet while we went to and fro about our daily lives disposing of plastics as we saw fit, vast accumulations of these materials began to build up in our seas and oceans. Brought ashore by the ebb and flow of the tide, what was once glass bottles and natural rope had turned to vast swathes of plastic bottles, synthetic ropes, fishing line and everyday household items. For many years, this plastic litter was simply regarded as an inert substance that was an unsightly blemish upon the world's landscapes. Nowadays, and after years of extensive research, this view has changed considerably and we now know that plastic litter is causing significant harm to the aquatic environment and affecting the things we value. In fact, we may now be experiencing the effects of this directly, in the form of plastics and toxic chemicals entering our food chain.

The large amount of plastic materials that entered the aquatic environment in the past initiated an environmental catastrophe that may span thousands of years yet to come. Today, vast amounts of land-based discarded plastics are still finding their way into the world's oceans with estimates projecting up to 32 million tons entering the oceans each year by 2050. Instead of simply dissolving away, plastic materials are generally insoluble in water. However, the natural environment is dynamic. Consequently, some plastics are degraded by sunlight and become brittle and fracture, while others are thrown around by waves, ultimately breaking apart into countless tiny colourful pieces, termed microplastics. Some of these tiny pieces of plastic float on the surface water, while others sink to the seabed or are carried along by currents, meandering and swirling as they go. At the same time, we purposely manufacture microplastics as tiny spheres for use in cosmetics and personal care products, such as face scrubs. These microscopic, and almost undetectable, microplastics are washed down our drains and out into the oceans where they join the trillions of others already there. Yet their colourful appearance, small size and induced motion mean that they are easily mistaken for food by predators, who eagerly consume them. Therein lies the danger.

We are now detecting microplastics within the ocean's aquatic life. The same oceans provide an abundant source of food for much of our world's population. Simultaneously, thousands of different toxic chemicals pollute our oceans, some of which persist for several decades before they breakdown. Many of these toxic chemicals are readily attracted to microplastics, where they can concentrate to levels up to one million times higher than that of the surrounding water. If these contaminated microplastics are subsequently consumed by an organism, the contaminants may leach off the microplastics and deliver a dose of toxic chemicals into that organism. If this contaminated organism is then consumed by a predator, the toxic chemicals are also consumed. In the process of biomagnification, the toxic chemicals may potentially be transferred to successively higher levels in the food chain, with organisms at higher levels tending to accumulate a substantially greater concentration of toxic chemicals in their tissues. Importantly, as humans we regularly consume predators residing in the top regions of many food chains. Yet we still have much to learn about the potential dangers posed to aquatic life, and even ourselves. Thus, the research carried out by myself, my co-author and others seeks to ascertain the risks and effects of microplastics and their consequences upon our world. As the first ever book dedicated exclusively to the subject of microplastics, it is my hope that by reading this book you will gain a greater awareness of these implications and ultimately, help bring the world's attention upon the issue of microplastic pollutants.

Christopher Blair Crawford

Acknowledgments

To all our family and friends, and our colleagues and publishers, who have supported us throughout the creation of this book, we express our deepest gratitude. Thank you all.

And to you, dear reader, for choosing to read our book from the many others that merit reading. We recognise that an enriching endeavour like reading requires your time and for that, we thank you greatly.

The emergence of plastics

1

Introduction

Since the dawn of the modern human race, and throughout our 200,000-year history, never before has the world seen materials like plastics. Indeed, plastics are such a new phenomenon in this world, that to date, practically no biological organism in the environment has sufficiently evolved to readily consume them. For this reason, plastics represent an unprecedented turning-point, not only in our own evolutionary history, but in the evolutionary history of the Earth. These remarkable, versatile and yet ubiquitous substances have fundamentally changed how we live and revolutionised the modern world. Unfortunately, these same substances which have allowed us to make great leaps and technological advancements may ultimately lead to significant environmental problems in the near future. Unless we are able to develop new technologies, processes or approaches to deal with the persistence of plastics in the environment, we will continue to observe increasing accumulations of these somewhat everlasting substances. But what about biodegradable plastics, you might ask? Indeed, fairly recent advancements have resulted in the creation of plastics produced from natural substances, such as soybeans and corn starch, which can be biologically broken down. However, most plastics in use today are simply not biodegradable and are in fact, highly resistant to degradation. Indeed, the billions of tonnes of plastics already released into the environment, since the origin of their creation, remain with us to the present day in one form or another and may take thousands of years to completely degrade.

Remarkably, the persistence of plastic and its effects on the aquatic environment were demonstrated in 2005, when a single piece of white plastic was found among numerous other pieces of plastic recovered from the stomach of a Laysan Albatross chick carcass. It is likely that the bird died due to starvation from consistently ingesting plastic, which has no nutritional content and can cause intestinal blockages. Examination of the piece of white plastic revealed that it was imprinted with a serial number. Cross-checking and verification of that number revealed that, astonishingly, the plastic had originated from an American Navy seaplane which was shot down thousands of miles away near Japan in 1944 during the Second World War. Subsequent computational modelling and simulations revealed that the piece of plastic had undergone a 60-year odyssey within the North Pacific subtropical oceanic gyre. Bounded by the gyre are two distinct swirling vortices of plastic garbage. One vortex is situated off the coast of Japan and is termed the Western Garbage Patch. The other vortex is situated off the coast of America and is termed the Eastern Garbage Patch. Together, these two patches form the Great Pacific garbage patch, a colossal swirling area of significant plastic accumulation (see Chapter 3). Amazingly, it was calculated that the piece of plastic from the seaplane had spent approximately 10 years travelling around the Western Garbage Patch. Eventually it escaped from this vortex and was carried thousands of miles by the North Pacific Gyre to the Eastern Garbage Patch. It

Microplastic Pollutants. http://dx.doi.org/10.1016/B978-0-12-809406-8.00001-3

then spent the next 50 years circling around this region, until it was eventually eaten by a Laysan Albatross chick in 2005. The carcass of that bird was then discovered by researchers at Midway Atoll, a Hawaiian Island in the middle of the North Pacific Ocean. Importantly, since the plastic was still intact, had it not have been collected by the researchers then it would have been freely available to be consumed again by another organism, after which this deadly cycle of plastic ingestion could potentially be repeated for thousands of years.

Thus, although plastics could quite rightly be considered as one of our greatest ever achievements, and the pinnacle of technological innovation, unfortunately plastics are also increasingly becoming recognised as representing one of the greatest environmental challenges that our species has ever known. In the face of their groundbreaking versatility and contemporary marvel, plastic litter has profoundly negative consequences. In this book, we shall explore in great detail, the way in which plastics behave in the environment and their effects upon the natural world. However, before we do so, it is important to examine the remarkable history of the development of these materials in order that we can appreciate precisely what plastics are and how this diverse class of unique materials gradually became intertwined within our modern lives to form a fundamental part of our contemporary society.

What are plastics?

The term 'plastic' first appeared in the 1630s in which it was used to describe a substance that could be moulded or shaped. The term derives from the Ancient Greek term *plastikos*, which refers to something that is suitable for moulding, and the Latin term *plasticus* which pertains to moulding or shaping. The modern use of the term plastic was first coined by Leo Hendrick baekeland in 1909 and today, plastic is a generic term used to describe a vast array of materials. But what do we mean when we talk about plastics, and what exactly are these strange and yet familiar substances?

As an integral part of our daily lives, we tend to use the word plastic on a regular basis without paying much attention to what plastics actually are. From plastic bags and writing pens to pipes and electrical equipment, different types of plastic are everywhere. However, they all have something in common; all plastic substances are composed of large chain-like molecules, termed macromolecules. These large molecules are composed of many recurring smaller molecules connected together in a sequence. We term a substance with this kind of molecular arrangement as a 'polymer'. The existence of macromolecules, and their characterisation as polymers, was first demonstrated by German chemist Hermann Staudinger in the 1920s. Thereafter, he formed the first polymer journal in 1940 and received the Noble Prize in Chemistry in 1953. Parenthetically, the word polymer is a combination of the Ancient Greek words *poly* (meaning many) and *meres* (meaning parts). Each individual molecule in a polymer chain is considered to be a single unit and we call each of these a 'monomer'. In this case, the prefix 'mono' is used to mean single. Thus, monomers are small molecules that have the ability to bond together to form long chains. We can visualise these large chains by thinking about a polymer as being akin to a pearl necklace. If we can imagine the entire necklace to be the polymer, then each individual pearl would be considered as a monomer (Fig. 1.1).

The process of connecting monomers together to form a polymer is called 'polymerisation'. In Fig. 1.2, we have the monomer ethylene. When ethylene is polymerised, it forms the common plastic polyethylene (Fig. 1.3). These large molecular chains can then be moulded and shaped to form solid objects. In Fig. 1.4, we can see that a polyethylene bag is actually composed of an amorphous mass of tangled branched polymer chains. Each of these chains is, in turn, composed of many repeating ethylene monomers.

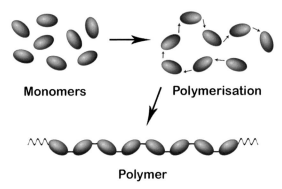

Figure 1.1 Small molecules (monomers) connect together in a repeating sequence (polymerise) to form a large chain-like molecule (polymer).

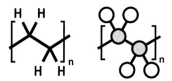

Figure 1.2 Ethylene (monomer).

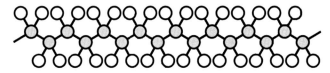

Figure 1.3 Polyethylene (polymer).

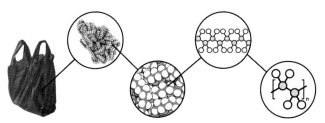

Figure 1.4 A polyethylene bag is composed of a mass of large polymer chains, which are in turn composed of many repeating ethylene monomers.

Thus, all plastics are polymers, and polymers are simply long chains of many repeating monomers. Typically, in commercial plastics, there will be between 10,000 and 100,000,000 monomers per chain, depending on the type of plastic. Often, each monomer of a polymer is the same as the next monomer in the sequence and this is termed as a 'homopolymer'. However, some polymers are composed of different alternating monomers, and this is termed as a 'copolymer'. In other cases, polymers can be composed of branched chains in a variety of architectures, thus exhibiting a deviation from a simple linear polymer chain. Furthermore, two polymers can be mixed together to form a plastic blend that simultaneously displays the characteristics of each individual polymer, thereby providing the benefits of both. Alternatively, mixing of two polymers may form a blend with enhanced properties, in comparison to each individual polymer. Accordingly, we shall explore the properties and characteristics of plastics in more detail within Chapter 4. The vast majority of polymers in today's world are the synthetic plastics created by humans. However, there are also many natural polymers in existence. For example, our DNA is a polysaccharide, composed of many repeating sugar units (monosaccharides). Similarly, our hair is a polypeptide protein filament which is composed of long chains of amino acids, connected together by peptide bonds. Thus, it is important to note that although all plastics are polymers, not all polymers are plastics.

The advent of plastics

The exact point at which plastic entered our world has no fixed point in time and rather, we should think of the generic term 'plastic' as referring to a large assortment of different polymers created as a result of our attempts to improve upon the raw materials provided by nature. Consequently, the advent of plastics is quite indistinct, and indeed, one could contend that the start of our path to a plastic society first began with the creation of rubber balls by the Mesoamericans around 1600 BC or perhaps with the introduction of natural rubber by French explorer Charles Marie de La Condamine in 1736. Incidentally, while on an expedition in South America, tensions within the expedition group resulted in La Condamine breaking off from the others and travelling alone to Ecuador. Along the way, he stumbled across the Pará rubber tree (*Hevea brasiliensis*) in 1736. Later that year, La Condamine brought samples of this natural rubber back to France where he subsequently studied the substance and presented a scientific paper purporting the beneficial properties of natural rubber in 1751.

Alternatively, perhaps the Grandfathers of plastic are British manufacturing engineer Thomas Hancock, who patented the vulcanization of natural rubber in 1843 in the United Kingdom, and American chemist Charles Goodyear, who patented vulcanization of natural rubber in the United States 8 weeks later in 1844. Considerable controversy ensues as to who was the true inventor of the vulcanization process, and indeed a lengthy and bitter court battle ensued when Goodyear attempted to patent the vulcanization process in the United Kingdom. Ultimately, to Goodyear's dismay, the judge ruled in favour of Hancock. Nonetheless, both men are considered as rubber pioneers who made significant contributions to the development of vulcanized rubber.

In the case of Hancock, the process was discovered after many years of experimenting with natural rubber and sulphur. In the case of Goodyear, the process was discovered as a result of many years spent trying to improve the characteristics of natural rubber. Goodyear recognised that natural rubber had an inherent problem due to its unstable thermal characteristics. In the heat, natural rubber has a tendency to melt and become sticky, producing a particularly pungent odour in the process. Conversely, in the cold the substance becomes hard and brittle. In his attempts to address this problem, Goodyear became obsessed with trying to improve the properties of natural rubber. He subsequently mixed the substance with all manner of things, including turpentine, witch-hazel and even cream cheese. Often struggling with debt from overspending on materials and the complaints of his neighbours regarding the unpleasant odours emanating from his property, Goodyear continued his trial-and-error experiments heedlessly, but nothing appeared to work. Eventually good fortune arose when on one occasion he mixed natural rubber with sulphur and lead carbonate. The exact nature as to what happened next is shrouded in myth and legend. Indeed, one account claims that during an argument with his business partner Nathaniel Hayward, Goodyear threw the rubber onto a hot stove. However, instead of catching fire, they noticed that the rubber charred and became solid. Another account states that Goodyear absentmindedly left the rubber on the stove and later noticed the charring effect. Irrespective of the events surrounding the discovery, experimentation with the mixture over the next few years ultimately resulted in Goodyear perfecting the vulcanization process.

On the other hand, perhaps we could consider that the development of the first fully-synthetic polymeric compound known as Bakelite, by Belgian chemist Leo Hendrick Baekeland in 1907, was the moment that heralded the birth of plastic. Certainly, Bakelite transformed the world at that time with an assortment of new products. Indeed, by the end of the 1930s more than 200,000 tonnes of Bakelite had been made into a vast array of household items. Nevertheless, for the purposes of this book, we shall meet in the middle and consider the starting point of plastic as situated somewhere between the vulcanization of rubber and the creation of the first fully-synthetic polymer. Accordingly, the advent of the plastic age was the development of the first semi-synthetic polymer.

The early history of plastics

1862–98

Our story of plastic begins within the United Kingdom in 1862. That year, the second of the decennial Great International Exhibition was taking place with 6.1 million attendees and 29,000 exhibitors from 36 different countries. The first exhibition had taken place in 1851, with considerable support from Prince Albert, and was heralded as a momentous success. Consequently, the Royal Society of Arts decided upon a second exhibition in 1861. However, delays with the assigned committee, and the outbreak of war amongst France and Austria, resulted in postponement of the second exhibition until 1862. A further setback came with the death of Prince Albert in December 1861, thereby preventing opening of the event by the Queen. Even so, the Queen ensured that everything possible was undertaken to ensure the event was a grand national ceremony. Accordingly, the exhibition went ahead on 1 May 1862 at a 90 square metre site in South Kensington, London. Parenthetically, the same site today houses the Natural History Museum. The exhibitions were organised into 36 classes, ranging from the latest engineering advancements, such as the creation of ice by refrigeration; to the latest scientific advancements, such as the very first safety match. However, amongst all 29,000 exhibits was Exhibit 1112 displaying a peculiar substance termed Parkesine. The substance was created by, and named after, its inventor Alexander Parkes. Parkesine was demonstrated at the exhibition as being a hard but flexible substance capable of being cast, carved and painted. Thus, the substance attracted great attention as being a suitable replacement for ivory and consequently, Parkes was awarded a bronze medal at the exhibition for the outstanding quality of his creation. Today, we know Parkesine as a celluloid and it is regarded in the historical literature as the very first thermoplastic. Incidentally, Parkes had created Parkesine by combining vegetable oil or camphor (plasticiser); pyroxylin (nitrocellulose with fewer nitrated groups) and nitrobenzole or analine (solvent).

In 1863 in the Unites States, inventor John Wesley Hyatt had recognised the current trend in declining ivory. In response to an advertisement offering a reward of $10,000 to anyone who could find a suitable substance that could replace ivory, Hyatt started research on creating an alternative nitrocellulose-based material for the production of billiard balls. Meanwhile in the United Kingdom, there was considerable sensation surrounding Parkesine. Consequently, Parkes entered into a partnership with rubber textile manufacturer Daniel Spill and set up the Parkesine Co. Ltd in 1866 to mass-produce Parkesine. The company mainly produced domestic items, such as handles, hair combs and jewellery. Though, in 1867, Parkesine Co. Ltd also started the production of Parkesine-based billiard balls. Unfortunately, as a result of Parkes attempts to reduce expenditures and satisfy investors, Parkesine Co. Ltd took a significant downturn. To save production costs, Parkes was using inferior cellulose in the production process and consequently, the resulting products failed to meet the impressive standards previously set at the International Exhibition in 1862. As a result, customers disparaged the products and Parkesine Co. Ltd collapsed in 1868. Interestingly, Parkes largely blamed the failure of the company on the inherent flammability of their products. On the other hand, Spill cited difficulties in creating a pure product that was suitably ivory-like in colour and would appeal to customers. Following the collapse of Parkesine Co. Ltd, Spill then formed the Xylonite Company in 1869 in an effort to reignite demand for Parkesine and make a second attempt at the market. Around the same time, French chemist Paul Schützenberger first introduced cellulose acetate after discovering that the substance is formed when cellulose is reacted with acetic anhydride. Ultimately, cellulose acetate would later go on to be used in the production of film for the motion picture industry, as well as being used to produce fibres.

Back in the United States, Hyatt and his brother Isaiah had successfully invented a technique for creating billiard balls, based on Parkes discovery of Parkesine, which utilised camphor and nitrocellulose. The brothers then patented their technique in the United States in 1869 before they established the Albany Billiard Ball Company and filing further patents thereafter. The billiard balls they produced sold successfully.

However, to guarantee sufficient durability of the balls, the Hyatt's only added very minute amounts of pigment to the balls highly-flammable nitrocellulose coating. This ensured that the coating was as pure as possible and the colours were vibrant. Unfortunately, overenthusiastic players were occasionally startled when violent striking of the billiard balls resulted in a small explosive bang. As a consequence, one owner of a billiard club wrote a letter of complaint to the Hyatt's stating that upon hearing the explosion of colliding billiard balls, every armed man in the room drew their gun.

Back in the United Kingdom, and upon learning of the patents published by the Hyatt brothers, Daniel Spill raised a court action citing infringement of his own 1870 United States patents detailing the combination of camphor and nitrocellulose and the bleaching of pyroxylin. A long bitter court battle then ensued between Spill and the Hyatt brothers. At the same time, demand for rubber could not be met and the resulting shortage was having a detrimental effect on the rubber-based dental plate industry. In response, the Hyatt's formed the Albany Dental Plate Company in 1870 in New Jersey, producing nitrocellulose-based dental plates. In 1872, they renamed themselves to the Celluloid Manufacturing Company, thereby coining the term 'celluloid' (a modification of the term cellulose) which they trademarked. The company enjoyed great prosperity and Hyatt developed several novel industrial processes for the development of celluloid.

Around the same time, Spill's efforts at reigniting demand for Parkesine proved to be in vain, and the Xylonite Company came to an end in 1874. Undeterred by the failure, Spill relocated to new premises and formed the Danielle Spill Company. As with Parkesine Co. Ltd, this new company produced mainly handles, ornaments and jewellery but of higher quality than the products produced by Parkesine Co. Ltd. This was achieved by using a modified version of Parkesine developed by Spill and was termed Xylonite. As such, the company drew the interest of three entrepreneurs; Alexander MacKay, Ernest Leigh Bennet and Herbert Leigh Bennet. In partnership with Spill, the four men formed the British Xylonite Company in 1875.

In 1880, the fierce court battle between Spill and the Hyatt brothers finally came to an end when the judge ruled that the Hyatt brothers had indeed infringed Spill's patents. Consequently, the processes that the Hyatt's had been using to manufacture their products were now forbidden by law. Spill then put on sale his United States patents for Parkesine and Xylonite, which were subsequently purchased by L. L. Brown. Accordingly, a rival company to the Celluloid Manufacturing Company was set up nearby by Brown in New Jersey, United States and named the Zylonite Company. However, in response to the court ruling, the Celluloid Manufacturing Company slightly modified their manufacturing process, replacing ethanol with methanol thereby allowing them to continue on producing celluloid unabated. Concerned by this, Spill and the Zylonite Company raised another court action, and a lengthy and bitter 4-year court battle ensued. In the end, the judge (who by a happenstance was the same judge that ruled in their previous court battle) decided in 1884 that the previous judgement should be reversed. The judge ruled that in point of fact, Alexander Parkes was the true inventor of the techniques utilising camphor and therefore, Spill's patents were not valid. Consequently, Spill's patented industrial processes became completely unrestricted, and the Celluloid Manufacturing Company quickly reinstated its original production methods. In a short time, the company rapidly expanded and ultimately absorbed the Zylonite Company.

Back in the United Kingdom, the British Xylonite Company was experiencing considerable financial instability. However, their fortunes changed in 1885 when the Hyatt brothers proposed an arrangement with them. At that time, the Hyatt's were involved in the production and sale of celluloid collars and cuffs in France with the French company Compagnie Française du Celluloide. These had proven to be quite popular and the Hyatt's, who were keen to become involved in the United Kingdom, proposed that the British Xylonite Company begin working alongside them and the French company to produce and sell these products in the United Kingdom. After some discussion, the British Xylonite Company agreed to the partnership and became engaged in the production of white celluloid collars and cuffs. To their delight, these proved to be exceptionally fashionable with the British middle class. The collars and cuffs became a means by which gentlemen, who were unable to afford expensive cloth collars, could differentiate themselves from the working class. Incidentally, at the time cloth collars were generally only worn by the upper class due to affordability issues and laundering requirements. However, the water-repellent characteristics of celluloid collars and cuffs meant that they could be easily rinsed with water, dried and reused immediately. For this reason, celluloid collars and cuffs became exceptionally popular and befitted gentlemen striving for better things. The popularity of celluloid collars and cuffs brought the British Xylonite Company enormous prosperity and despite the setback of a fire in 1885, the company went on to build new premises in 1886. Parenthetically, fires were a common occurrence in the celluloid industry, and this was mainly attributable to the production of impure and unstable nitrocellulose. By 1898, the British Xylonite Company had rapidly expanded and their range of products increased accordingly. The company greatly prospered, eventually changing its name to BX Plastics. They went on to produce a large assortment of everyday household items, such as hair combs, boot heels, handles and children's toys, as well as, many other products over the next 100 years. At the same time, the Hyatt brothers and the Celluloid Manufacturing Company flourished, eventually growing to become the American Celluloid and Chemical Corporation and adsorbing rival companies. In time, the American Celluloid and Chemical Corporation eventually became adsorbed itself by the Celanese Corporation.

1899–1931

The next major advent in the plastic story came with the discovery of casein plastics, produced using a protein (casein) found in cow's milk. Although some early work was undertaken by German inventor Wilhelm Krische, in which he explored the development of coatings utilising casein, it was not until 1899 that a considerable breakthrough was made. While working at home, German chemist Adolf Spitteler noticed the formation of a solid water-resistant substance when his cat spilled a bottle of formaldehyde onto its saucer of milk. Recognising the commercial potential of this new material, Spitteler and Krische established patents in Germany and the United

States during 1899. Following research and development, in 1900 they displayed objects fashioned from the material at the Plastics Universal Exhibition. A French and German company subsequently acquired the patents and invested a considerable amount of time in developing a more transparent product. The product was marketed as Galalith and in 1904, the companies combined to create the company International Galalith Gesellschaft Hoft and Co. and successfully manufactured casein plastic for use in the button industry.

Also in 1904, British electrical engineer Sir James Swinborne was concerned by the inferiority of current wiring insulation. By happenchance, he came across a phenolic resinous substance produced by Galician chemist Adolf Luft while visiting a patent office. Swinborne was amazed at the properties of this strange solid substance and subsequently initiated deliberations with Luft. Thereafter, he created the company Fireproof Celluloid Syndicate Ltd which was based in the United Kingdom. The purpose of this new company was to investigate the properties and possible commercialisation of the substance. Following a large amount of research, they successfully created some low quality flat and cylindrical pieces of plastic, but nevertheless continued their research. Meanwhile, in 1907 Belgian chemist Leo Hendrick Baekeland, while reacting formaldehyde and phenol, serendipitously created a cross-linked thermoset plastic termed polyoxybenzyl methylene glycolol anhydride. Baekeland decided to name this new substance after himself and thus, the world met Bakelite. Unlike casein, which is only considered to be a semi-synthetic polymer, Bakelite was the world's first fully-synthetic polymer.

Note: Baekeland is often credited as the inventor of phenolic plastics but surprisingly, this is not entirely accurate since it ignores the large amount of groundwork undertaken prior to Baekeland's invaluable commercial influence on the plastic industry. Indeed, 35 years earlier in 1872, a publication was released by German chemist Adolf von Baeyer whereby he described the production of an unworkable resinous product from the reaction of formaldehyde and phenol. Unfortunately, interest in the substance by the wider scientific community was practically non-existent at that point in time. However, in 1891 a student of Baeyer, Werner Kleeberg, was successful in creating a sticky workable resinous substance. Kleeberg utilised concentrated hydrochloric acid and a surplus of formaldehyde thereby creating a violent exothermic reaction. Shortly after, the substance cooled and formed an inflexible solid. Nevertheless, at that time Kleeberg was more interested in the production of pure crystalline solids and thus, did not take the discovery any further. However, in 1899 a British gentleman named Arthur Smith secured the first patent which described the formation of a resinous substance from the reaction of acetaldehyde and phenol. Smith had utilised various alcohols in an attempt to moderate the violent exothermic reaction but struggled with the new found technology.

By 1909, research at the Fireproof Celluloid Syndicate Ltd had managed to achieve the creation of a good quality hard lacquer. The lacquer proved to be their most promising product, and Swinborne then filed a patent application on 8 December 1909. It was at this point where Swinborne encountered Baekeland, and it was to the great disappointment of Swinborne when he discovered that Baekeland had already filed a similar patent the previous day, thereby rendering Swinborne's patent invalid.

Consequently, he withdrew his patent application and heedlessly, the Fireproof Celluloid Syndicate Ltd continued manufacture regardless, producing and selling a new lacquer to prevent brass from tarnishing. Meanwhile in the field of semi-synthetic plastics, Russian student Victor Schutze had created a method for creating a hard plastic from the curds of milk and patented the technique in 1909. Subsequently, the Syrolit Company was formed in the United Kingdom to begin producing a casein based semi-synthetic plastic substance, termed Syrolit. In 1910, following on from the success of the anti-tarnish lacquer, the Fireproof Celluloid Syndicate Ltd was wound up and Swineborne and colleagues immediately created The Damard Lacquer Company instead, solely focussing on producing the lacquer.

In 1912, German chemist Fritz Klatte was attempting to develop a protective coating that could be applied to the wings of an aircraft. Klatte serindipously discovered that he could create the monomer vinyl chloride (a compound first created by French chemist and physicist Henri Victor Regnault in 1835) by combining hydrogen chloride, acetylene and mercury using sunlight as a catalyst. Klatte had unwittingly left his mixture near a sunny window and noticed that over time, the mixture turned to a cloudy sludge before solidifying. Furthermore, Klatte created the monomer vinyl acetate by a similar process. Greisham Electron, the company at which Klatte worked, patented both processes at Klatte's request, but the company did not take the discovery any further. At the same time, Russian organic chemist Ivan Ostromislensky, while developing synthetic rubber, filed a patent describing a technique for the polymerisation of vinyl chloride using a solvent (such as benzene) or the action of sunlight to form polyvinyl chloride (PVC). Nonetheless, difficulties in making the brittle polymeric material more flexible hindered any further development.

By 1913, the production of casein-based Syrolit had proved to be uneconomical and troublesome and the Syrolit Company, facing imminent liquidation, was subsequently acquired by Erinoid Company in 1913, who then created a workable process to produce Syrolit. In 1914, the First World War broke out and over the ensuing 4 years, competition between German-based Galalith and United Kingdom-based Erinoid was non-existent, thereby allowing both companies to flourish. During the war, The Damard Lacquer Company supplied vast amounts of lacquer to the United Kingdom's Royal Navy for insulation of electrical wires. Parenthetically, Baekeland had approved the use of his patents during the time of war, but implicitly requested that the companies exploiting his patents should create a commercial partnership following cessation of the war.

In the wake of the First World War, many rival manufacturers began the production of casein plastic following the cessation of existing patents in Germany. In 1919 production started in the United States, where the substance was termed Aladdinite, and thereafter an increasingly large number of manufacturers began production of casein plastic over the next decade. In 1920, German chemist Hermann Staudinger and his colleagues noticed that when liquid formaldehyde was stored at $-80°C$ for a period of 60 min, the liquid formed a gel-like substance. Intrigued by this, they investigated further and found that storage of formaldehyde under the same conditions for 24 h resulted in a polymerisation reaction to form polyformaldehyde, otherwise known as polyoxymethylene (POM). Although Staudinger was able to form rudimentary films and fibres from POM, the poor thermal stability of the substance precluded it from commercial development at that time. In 1922, Staudinger and doctoral research student Jakob Fritschi, published a paper in which they proposed that polymeric materials consist of large chains composed of many smaller repeating molecules that are linked together. Although their proposition was met with great opposition at the time, it would eventually go on to provide the fundamental theoretical basis for polymer chemistry. In 1924, Staudinger coined and defined the term 'macromolecule'.

By 1925, Fritz Klatte's German patents concerning vinyl chloride and vinyl acetate had expired. At the same time, Belgian chemist Julius Nieuwland had created divinylacetylene by polymerising acetylene in the presence of the inorganic catalyst copper(I)oxide. However, American chemist Elmer Keiser Bolton noted that the polymeric substance exploded upon impact and sought to create a more stable version with the help of Nieuwland. Eventually, the pair successfully created a non-explosive polymer but unfortunately, exposure to light resulted in complete degradation of the substance. Also in 1925, the French microbiologist Maurice Lemoigne first discovered the biodegradable polymer polyhydroxybuterate (PHB), which was present in granular deposits within the cytoplasm of the rod-like aerobic spore forming bacterium, *Bacillus megaterium*.

In 1926, a significant advancement, based on the earlier work on polymerisation of vinyl chloride by Fritz Klatte and Ivan Ostromislensky, was made by American inventor Waldo Lonsbury Semon while working on a rubber substitute at the B.F. Goodrich Company in the United States. At the time, polyvinyl chloride was a hard unworkable substance that was of no commercial interest, partly due to the current recession at the time. However, Semon experimented with the material and managed to produce a PVC substance that was more flexible and pliable by plasticising it. This involved adding tricresyl phosphate, which has a boiling point of $255°C$, and subsequently heating the mixture to a high temperature. Around the same time, Baekeland was involved in a fierce court battle in the United States in an attempt to lodge his Bakelite patents there. His case proved to be successful and his patents were approved in 1927. Following this, Baekeland then negotiated a merger between The Damard Lacquer Company and two other United Kingdom manufacturers, thereby forming Bakelite Ltd. The purpose of this new company was to produce and sell Bakelite products in the United Kingdom.

In 1928, further progress with PVC was made with the development of copolymers by a variety of research groups. At that point, the main issue with PVC was that manufacture required heating the polymer to a high temperature of around $160°C$ to make it pliable. However, the safe upper temperature limit was only $20°C$ more and consequently, overheating past $180°C$ would result in complete thermal decomposition of the polymer. Furthermore, accurate temperature control technology was in its infancy, and thus, the researchers had to find a method by which they could reduce the lower temperature threshold at which the polymer became pliable. Eventually, this was achieved by creating a polymer composed of different monomers; termed a copolymer. This then resolved the manufacturing issues and large scale production of PVC began in Germany and the United States, thereby signalling the birth of the vinyl industry. At that time, vinyl represented an unprecedented alternative to mainstream rubber, and ultimately, the material became a huge commercial success.

Also in 1928, the American chemist Wallace Hume Carothers took up a managerial position at the Experimental Station of the American conglomerate DuPont. Carothers and his small team of organic chemists began working on polyesters in an attempt to develop a synthetic rubber material that possessed a molecular weight of more than 4200. At that time, 4200 was the highest weight that anyone had reached and had been achieved by German chemist Emil Fischer several years earlier. However, by 1929 the research team could only manage to achieve molecular weights of between 3000 and 4000. To their dismay, they tried many times but could not pass the threshold set by Fischer. Indeed, obtaining accurate molecular weights at that time was especially challenging, and frustratingly, although some materials appeared to marginally surpass the threshold, they could not be absolutely certain. In 1930, American chemist Elmer Keiser Bolton took up the position of research director at DuPont and was keen to get results from Carothers team. Based on the results of Bolton's previous work on acetylene alongside Belgian chemist Julius Nieuwland, Bolton instructed the team to take a new direction and examine the polymerisation of acetylene. The aim was to create a synthetic rubber that did not degrade when exposed to light. Finally, a breakthrough came later that year when a member of Carothers team, Arnold Collins, reacted hydrogen chloride gas and monovinyl acetylene to produce a substance they later termed chloroprene. Upon leaving the substance to stand, to Collins amazement, a polymerisation reaction occurred in which a solid rubber-like substance was formed, termed polychloroprene. Collins noted that if this elastomeric substance was thrown at a solid surface, it would bounce. DuPont then named the synthetic rubber material Duprene (later termed Neoprene). Meanwhile, another member of Carothers team, Julian Hill was attempting to synthesise polymers with a molecular weight of more than 5000, termed superpolymers. To their amazement, Hill successfully created a polyester polymer with a molecular weight of 12,000. After he allowed the hot substance to cool, he found that it could be pulled and drawn out in to thin flexible fibres. Unfortunately, the melting point of these fibres was too low to allow them to be suitable as a commercial fibre. However, Hill had discovered the cold-drawn process, and consequently, this set the stage for the synthesis of Nylon fibres in later years.

Around the same time in 1930, polystyrene was first commercially manufactured by German company IG Farben; whereby they produced styrene from the dehydrogenation of ethylbenzene and subsequently polymerised the styrene to polystyrene inside a reaction vessel. The substance was then passed through a pipe and sliced into pellets. Shortly after the creation of polystyrene, injection moulding tests were carried out with the substance. At the same time, German chemists Hans Fikentscher and Claus Heuck, while working at IG Farben, first synthesised polyacrylonitrile (PAN). However, their research was abandoned following difficulties working with the insoluble material when nothing could be found that was capable of dissolving it.

Also in 1930, postgraduate research student William Chalmers was studying at Mcgill University in Montreal, Canada. During his research, Chalmers collaborated with the head of Akron University rubber laboratory, George Stafford Whitby. Following Whitby's recommendations, Chalmers polymerised methyl methacrylate to form a transparent solid resinous substance that was very similar to glass, but far more flexible. Subsequently, on his own inventiveness, he also polymerised methacrylic nitrile to form a similar glass-like substance. Recognising the commercial importance of his discovery, Chalmers applied for patents in Canada and the United States in 1931. Thereafter, he contacted Imperial Chemical Industries (ICI) in the United Kingdom, and Röhm and Haas in Germany. Both companies were interested in commercially exploiting the new found discovery and a research team at ICI, headed by Rowland Hill, began investigating ways in which efficient moulding of poly(methyl methacrylate) (PMA) could be achieved. This resulted in ICI patenting the production of articles moulded from PMA in 1931. Simultaneously, German chemists Otto Röhm and Walter Baeur, while working at Röhm and Haas GmbH, began investigating ways in which PMA could be processed into large acrylic sheets to function as a direct replacement for glass.

Also in 1931, Dupont began production of their synthetic rubber compound Duprene. Unfortunately, sales were limited and unsuitable for end products due to the unpleasant smell of the substance. This disagreeable odour resulted from the by-products created during manufacture. Consequently, a more efficient production process was developed and the resultant product became odour-free. Dupont then experienced a rapid increase in interest from manufacturers keen to use Duprene in their end products and consequently, sales increased considerably. At the same time, German chemist Herbert Rein discovered that polyacrylonitrile (PAN), which was considered impervious to solvents, would dissolve in benzylpyridinium chloride. Meanwhile in the United States, casein plastic production had flourished and the many different manufacturers producing casein merged to form a single company named the American Plastics Corporation.

Note: Casein plastic cannot be regarded as a thermosetting plastic or thermoplastic, although it does possess some thermoplastic properties, as well as, exhibiting a hygroscopic nature. One particular advantage of casein plastic is the vast array of possible colours in which the product can be manufactured. Incidentally, this proved to be especially popular with the button industry, thereby establishing their prominence as the main consumer of casein.

1932–44

In 1932, Scottish chemist John William Crawford, while working at ICI in Scotland, developed a revolutionary commercially efficient method for the production of methyl methacrylate using the inexpensive, and easily available, raw materials sulphuric acid, acetone, methanol and hydrogen cyanide. This breakthrough allowed ICI to produce the monomer of poly(methyl methacrylate) (PMA) on a large scale at a considerably reduced cost and represented a significant leap forward in the large-scale production of PMA. The method was subsequently patented by ICI in 1932. Meanwhile, Carothers and his team at DuPont, created polylactic acid (PLA) after heating lactic acid under vacuum conditions.

Around the same time in 1932, British organic chemist Eric Fawcett and British physical chemist Reginald Gibson began high-pressure chemical research at ICI. After much experimentation, a breakthrough occurred in 1933. The pair of chemists had placed a mixture of benzaldehyde and ethylene inside a sealed reaction vessel and pressurised the container to 1400 atm while heating to 170°C. Despite the absence of any apparent reaction, subsequent inspection of the vessel revealed it to be internally lined with a white waxy coating. Further analysis as to the identification of compound revealed that Fawcett and Gibson had unwittingly created the most widespread plastic in use today; polyethylene. Fawcett and Gibson then undertook further experimentation by heating and pressuring pure ethylene alone. Unfortunately, the result was a powerful explosion which completely obliterated their equipment, demolished part of their laboratory and ultimately, resulted in postponement of their research by company officials for reasons of safety. Consequently, Fawcett moved to the United States and began working for the American conglomerate DuPont. However, Gibson stayed in the United Kingdom and secretly continued development of polyethylene at ICI in the evenings.

Also in 1933, German chemists Otto Röhm and Walter Baeur, while working at Röhm and Haas GmbH, had successfully managed to create an efficient method of producing poly(methyl methacrylate) (PMA) sheets. Consequently, production commenced later that year. Their glass substitute was sold under the trade name Plexiglas in Germany, while sheets produced at the American branch of Röhm and Haas were marketed under the name Oroglass. By 1934, ICI had also began large scale production of PMA moulding powder, as well as sheets, and sold the products under the trade names Diakon and Perspex, respectively. At the same time, Carothers and his team at Dupont reported that they had successfully polymerised ε-caprolactone, utilising potassium carbonate and heat, to form polycaprolactone (PCL). Also in 1934, the first fluorinated flexible polymers were discovered by German chemists Fritz Schloffer and Otto Scherer, while working at IG Farben. Their reporting of the research at the time described the synthesis of polychlorotrifluoroethylene (PCTFE), polychlorodifluoroethylene and polybromotrifluoroethylene. The thermoplastic PCTFE was then commercialised that same year by German company Hoechst AG under the trade name Hostaflon C2. Incidentally, Schloffer and Scherer also investigated the polymerisation of tetrafluoroethylene (TFE) to form polytetrafluoroethylene (PTFE), but the substance was deemed as unworkable owing to its inert and insoluble nature. Consequently, Hoechst AG rejected an application to manufacture the material, citing that the repellent properties of PTFE, and the difficulties obtaining sufficient quantities of TFE, would prohibit mass production.

In 1935, Reginald Gibson was still secretly working on the development of polyethylene in the evenings at ICI. Despite the previous explosion that had demolished his laboratory, Gibson again decided to heat 98% pure ethylene inside a sealed reaction vessel. However, this time he reduced the pressure to 200 atm while heating to 170°C. Trace levels of oxygen, present in the ethylene, catalysed the polymerisation of ethylene and a small yield of polyethylene (8 g), in the form of a white powder, was produced. On this occasion Gibson had been exceptionally fortunate as it was later determined that a leak, discovered in the reaction vessel, had allowed the release of just enough pressure to avert another explosion. Meanwhile, Wallace Carothers and his team were focussing their efforts at the research facility of DuPont, where they were attempting to create polyester and polyamide synthetic fibres. Although they did not achieve much success with polyester fibres, the breakthrough came when Carothers reacted adipic acid with hexamethylenediamine to produce the polymeric substance we now recognise as Nylon. Once the Nylon had cooled, Carothers used a cold-drawn technique, previously discovered by Julian hill in 1930, to stretch the Nylon into long fibres. These fibres appeared to be commercially viable, and consequently, DuPont then invested a significant amount of resources into bringing Nylon to the market.

In 1936, ICI were impressed by the polyethylene material Gibson had created and recognising the commercial potential, quickly initiated the development of a high-pressure polyethylene production method in which several major engineering challenges were addressed. These involved designing effective cooling of the highly exothermic polymerisation reaction and developing suitable methods in which to pressurise a vessel up to approximately 20,000 atm without creating an explosion. Their polyethylene product was then brought to market soon after under the trade name Polythene. Also in 1936, ICI granted DuPont a licence to use their United States patent to produce the methyl methacrylate polymer Perspex. Manufacture of Perspex by DuPont then commenced under the trade name Pontalite, after which the name was changed later that year to Lucite.

By 1937, PVC production was well underway in Germany and the United States. Meanwhile in the United Kingdom, ICI was conducting research on the production of vinyl chloride using hydrochloric acid and ethyne. Furthermore, new ways in which the compound could be polymerised were also studied. At the same time, DuPont decided to change the name of their synthetic rubber material Duprene to Neoprene to indicate that the substance was not an end-product, but rather a material used for the creation of end products. Ultimately, Neoprene generated significant sales for DuPont and eventually went on to become the material of choice in the manufacture of diver's wet suits. Also in 1937, German chemist Otto Bayer and his colleagues, while working at German company IG Farben, synthesised the first polyurethane from polyisocyanate and polyol. As a result of their discovery, polyurethanes would ultimately go on to become ubiquitous worldwide. In the present day, polyurethanes are utilised in a wide range of applications, ranging from shopping cart and skateboard wheels to hoses and carpet underlay.

In 1938, American chemist Roy Plunkett was working at DuPont attempting to develop an alternative refrigerant to the patented refrigerant tetrafluorodichloroethane. This particular refrigerant was the most commonly used at that time but was exclusively sold by DuPont to an American division of General Motors called Frigidaire. As many other refrigerator manufacturers were requesting a similar refrigerant, DuPont had to find an alternative that was not patented. Consequently, Plunkett was tasked with developing this new refrigerant and decided to create tetrafluoroethylene gas, which he would then react with hydrochloric acid. Consequently, Plunkett stored the gas in metal containers, which he then cooled to decrease the pressure inside. The following day, Plunkett and his associate Jack Rebok attempted to release the gas into a reaction chamber. However, to their surprise, no gas was released from the containers. Perplexed, and after much fiddling with their apparatus, the pair decided to cut open the containers to investigate. Inside, they found a silky white substance coating the container walls. The substance was polytetrafluoroethylene (PTFE). After some experimentation, Plunkett observed that PTFE was smooth and highly frictionless, an electrical insulator and resistant to heat, acids and ultraviolet light. Plunkett then subsequently patented the substance on behalf of DuPont. However, at that time, commercial development was hindered due to the explosion hazards of handling the monomer TFE, which formed dangerous peroxides. Thus, a significant amount of research and resources had to be invested to overcome this manufacturing challenge.

Also in 1938, the first full-scale production of Nylon began with the first commercial usage of the substance as bristles in a toothbrush (marketed under the name Dr West's Miracle-Tuft). At that time, toothbrush bristles consisted of horse and boar hair, and these were problematic. The animal hairs would often fall out, or retain moisture thereby allowing bacteria to multiply. Consequently, this new nylon bristled toothbrush proved to be exceptionally popular since these problems were overcome. Nylon then subsequently found great success in the production of ladies' stockings. Nevertheless, the commercial development of Nylon had been especially challenging for DuPont and ultimately required the combined work of 230 chemists and engineers. Furthermore, from the initial discovery of the substance 3 years beforehand, until they finally brought Nylon to the consumer market, DuPont claimed that they had spent approximately 27 million dollars.

In 1939, the Second World War began and Nylon, owing to its resilience, abrasion resistance and durability, was utilised to make parachutes and ropes. Furthermore, poly(methyl methacrylate) (PMA) with its durable, lightweight and shatter resistant properties found large-scale and almost exclusive usage, by both opposing sides, as window glazing in their military aircraft. At that time, the worldwide production of plastic materials had reached approximately 300,000 tonnes per annum. By 1940, ICI had developed a new polymerisation method for the efficient production of an emulsion PVC polymer which they marketed as Corvic. In 1941, the Japanese initiated an incursion of the British-owned peninsula of Malaya, an area of vast rubber plantations which supplied raw rubber resources to the West.

Around the same time, the Dow Chemical Company in the Unites States created the first extruded closed-cell polystyrene foam and marketed the product as Styrofoam. Meanwhile, the British were exploiting the electrical insulating properties of polyethylene in their radar equipment cables. As this was a time of war, demand for polyethylene was considerably high, and consequently, the British gave the Americans their undisclosed method of making polyethylene to increase production. Manufacture of polyethylene then promptly started in the United States, where it was mass-produced by the Union Carbide and Carbon Corporation and the conglomerate DuPont. At the same time, British chemists John Rex Whinfield and James Tennant Dixon, while working for the Calico Printers Association, created the first commercial polyester fibre by synthesising polyethylene terephthalate (PET). Whinfield and Dixon carried out an esterification reaction of ethylene glycol and terephthalic acid thereby yielding the monomer bis(2-hydroxyethyl)terephthalate. A condensation reaction of the monomers then produced the polymer PET. Whinfield and Dixon noted that the material was able to be pulled and stretched into fibres. They named the material Terylene and patented it that year, but as a result of the war, it would be several years later before it was granted.

In 1942, Malaya was successfully captured by the Japanese and this resulted in significant shortages of rubber in the West. Consequently, there was a significant demand for a rubber substitute, especially for insulating cables, and thus, ICI began production of their PVC product, Corvic. However, since the main use of Corvic was for electrical insulation, there was an increasing demand for a substance with improved insulating properties. Ultimately, this led to the creation of a modified version of Corvic in 1943, known as suspension grade PVC, and marketed as Corvic DQ. Meanwhile, rival companies, such as Bakelite Ltd also began production of their own PVC products. Furthermore, The Distillers Company, developed a PVC production technique utilising 1,2-dichloroethane and opened a rival manufacturing plant to produce a type of PVC known as a paste polymer. Incidentally, the rubber shortage also hindered the production of rubber-based dentures, and concern among military leaders of the effects of this on their soldiers spurred the creation and development of Perspex dentures. These new dentures proved to be highly popular and in light of their success, Perspex eventually became the main material of choice in denture production. Around 1944, and recognising the commercial possibilities of polytetrafluoroethylene (PTFE), DuPont undertook production of the material at a pilot plant and registered PTFE under the trademark Teflon. Ultimately, Teflon would go on to become a household name synonymous with the non-stick coatings used in cookware and bakeware.

The contemporary history of plastics

2

The modern history of plastics

1945–62

In 1945, the Second World War ended. At this point, it is important to note that while many of the popular plastics in use today had their beginnings in the early 1930s and 1940s, the Second World War played a significant part in providing the impetus for their advancement and development. For example, owing to its chemically resistant properties, polytetrafluoroethylene (PTFE) found usage during the war as a means to handle the highly toxic and corrosive uranium hexafluoride during the development of nuclear weapons. The substance was also discovered to be invisible to Doppler radar and consequently, PTFE was used to fabricate the antenna cap of proximity fuses. Additionally, and as a consequence of research undertaken near the end of the war to develop bullet-proof polymer sheets, the impact-resistant thermoplastic acrylonitrile butadiene styrene (ABS) was first created, and subsequently patented in 1946, by American chemist Lawrence Daly, while working at the United States Rubber Company. Ultimately, ABS would go on to be used in a multitude of applications, ranging from automobile parts to Lego bricks. At the same time, poly(methyl methacrylate) (PMA), following the wide scale use of the material during the Second World War, was introduced to the automobile industry by American company Chrysler as lenses for vehicle tail lights.

By 1946, the price of PVC had fallen considerably, and the UK-based Distillers Company and the US-based B.F. Goodrich Chemical Company established an agreement, resulting in the formation of PVC paste and suspension manufacturer British Geon Limited. Around the same time in America, inventor Earl Silas Tupper first introduced airtight polyethylene food containers under the trade name Tupperware and inventor Jules Montenier developed and marketed an antiperspirant that was dispensed by squeezing a plastic bottle. As a result of the huge popularity of these products, by 1947 the commercial use of plastic bottles and containers had risen exponentially and for the first time in history, plastic was competing with glass as packaging for foods and drinks.

By 1948, there were several varieties of PVC available (emulsion, suspension and paste), from a multitude of manufacturers, to suit a large array of purposes such as electrical insulation, waterproofing, upholstery and vinyl long-play music records. In 1949 German chemical engineer Fritz Stastny, while conducting research into developing effective cable insulation at German company BASF AG, serendipitously created and then patented, extruded closed-cell polystyrene foam. Stastny had absentmindedly incubated a solid sample of polystyrene longer than necessary at 80°C. He noticed that this resulted in the creation of a solid filament of foam. Further

Microplastic Pollutants. http://dx.doi.org/10.1016/B978-0-12-809406-8.00002-5

experimentation with pentane and air resulted in the creation of foam composed of many pre-expanded polystyrene beads. The resultant foam product was named Styropor by BASF AG and production began in Germany in 1951.

Note: Styropor (Fig. 2.1) is often referred to as Styrofoam (Fig. 2.2), albeit incorrectly. As it happens, Styrofoam only technically refers to the extruded closed-cell polystyrene foam created during 1941 by the Dow Chemical Company in the Unites States.

Figure 2.1 Styropor.

Figure 2.2 Styrofoam.

In 1951, the American oil company Phillips Petroleum was investigating ways in which it could utilise two of its refining products, ethylene and propylene gas, in the production of gasoline. At the time, American chemists John Paul Hogan and Robert Banks had been assigned the task and were studying various catalysts. While investigating the conversion of propylene to other liquid hydrocarbons, they had utilised a nickel oxide catalyst. Parenthetically, their work was based on a previous breakthrough eight years earlier by two other chemists at Phillips Petroleum, Grant C. Bailey and James A. Reid, who discovered the conversion of propylene to liquid hydrocarbons using a nickel oxide catalyst. Developing on from this work, Hogan and Banks noticed that the nickel oxide catalyst did not last very long and wished to find an alternative. On one occasion they decided to modify the catalyst by adding chromium oxide. Using propane gas the chemists then forced propylene through a 25 mm pipe at high pressure, which was packed with this new catalyst combination. However, the pipe quickly became blocked and the experiment had to be stopped. To their surprise, along with the usual liquid hydrocarbon products, a white-coloured solid substance was also produced. This proved to be the first synthesis of crystalline polypropylene. Recognising the magnitude of their discovery the substance was quickly patented by Hogan and Banks. Further investigation of their catalyst combination established that the chromium catalyst was solely responsible for the polymerisation of propylene. Thereafter the chemists decided to attempt to utilise this new chromium catalyst in polyethylene production. Importantly, at that time the manufacture of polyethylene (developed by ICI in the United Kingdom) required pressures in excess of 20,000 atm to produce. Consequently, Hogan and Banks set about modifying the existing process using their new chromium catalyst and ultimately this resulted in the synthesis of the first high-density polyethylene (HDPE), a straight chain polymer with very little branching and strong intermolecular forces. Subsequent testing revealed that HDPE possessed unprecedented tensile strength and resistance to heat. In light of this, Phillips Petroleum took a considerable risk and entered the plastics business, with no prior experience in the industry. Hoping that the unparallelled properties of their new polymer would be a success, they commercialised HDPE under the trade name Marlex. Unfortunately, sales of Marlex at that time proved to be fairly insignificant.

Coincidentally, around the same time, crystalline polypropylene was also independently discovered by two other researchers. In the first case, chemist Alex Zletz, while working at American oil company Standard Oil, was exploring the alkylation reactions of ethylene and the catalysis of these reactions. During his investigations, Zletz used a molybdenum catalyst and noticed that a solid substance was produced. This would later be identified as being partially crystalline polypropylene. Further experimentation by Zletz with a variety of other metal salts resulted in the development of a low-pressure technique for the production of polyethylene. However, Standard Oil decided to licence the process, as opposed to undertaking manufacturing and consequently a licenced chemical plant began production in Japan with moderate commercial success. In the second case, German chemist Karl Waldemar Ziegler, while working at the German research organisation the Max-Planck Institute, was attempting to synthesise a high molecular weight polyethylene. However, Ziegler was having difficulty with an unwanted elimination reaction that consistently produced 1-butene from ethylene. Eventually he established that the reason was a result of contamination with nickel salts. Ultimately, this led to the development of a catalyst for polymerising ethylene at low pressures using chromium, titanium or zirconium salts. In 1952 Ziegler collaborated with the Italian chemical company Montecatini and provided information regarding the catalysts he had developed to Italian chemist Giulio Natta. Interested in the potential of these catalysts, Natta investigated their properties further, and this eventually lead to the formation of crystalline polypropylene and the creation of stereospecific transition metal Ziegler–Natta catalysts. The pair were jointly awarded the Nobel Prize in Chemistry for their work on polymers. Ultimately, these simultaneous independent discoveries of crystalline polypropylene would eventually result in a complex and arduous legal battle over patents in the coming years. Around the same time, chemist Robert Neal Macdonald, while working at DuPont, first synthesised a high molecular weight hydroxyl-terminated (hemiacetal) polyoxymethylene (POM). However, the material was determined to be prone to molecular deterioration from overheating (thermal degradation), which begins at the hydroxyl groups on the ends of the polymer chain. Consequently the material was deemed as not being commercially viable.

In 1953, the first synthesis of ultra-high molecular weight polyethylene (UHMWPE) was achieved by Ziegler, while working at the Max-Planck Institute. Using his newly developed catalysts, Ziegler polymerised ethylene at a low temperature of 66–80°C and a low pressure of 6–8 atm. This was considerably lower than the dangerous pressures in excess of 20,000 atm used to produce regular polyethylene at that time. As a result of these mild reaction conditions, the polyethylene formed exceptionally long linear chains with minimal branching that efficiently packed into a crystalline structure. In fact, the linear chains were so long that the material had a molecular weight 10–100 times higher than standard high-density polyethylene (HDPE). Following testing, the material was found to be inert, extremely robust and resistant to chemicals while possessing a coefficient of friction value rivalling that of polytetrafluoroethylene (PTFE). Ziegler then passed the material to German company Ruhrchemie AG, who were interested in commercialising it. Ruhrchemie AG then introduced UHMWPE under the trade name RCH-1000 at K Trade Fair, a renowned international fair for plastics and rubbers in Dusseldorf, Germany. Shortly after, Ruhrchemie AG began large-scale production and marketing of RCH-1000. Ultimately, UHMWPE went on to be used in biomedical implants, and as a result of the materials inert frictionless properties and unsurpassed impact strength, it eventually replaced polytetrafluoroethylene (PTFE) as the bearing material of choice in artificial hips. In due course, high-strength UHMWPE spun fibrous products became available which possessed a strength-to-weight ratio 8–15 times higher than carbon steel, while exhibiting 10 times higher resistance to abrasion and possessing the highest impact resistance of any thermoplastic known.

Also in 1953, German chemist Hermann Schnell, while working at German company Bayer, synthesised the first linear polycarbonate. Concomitantly and independently, American chemist Daniel Fox, while working at American company General Electric Plastics, synthesised the first branched polycarbonate one week later. However, in 1955 both companies filed their patents in the United States at the same time and upon learning of each other's discoveries, a dispute erupted as to who was the rightful patent holder. The polycarbonate substances produced were demonstrated to be exceptionally robust and highly resistant against impact, while being transparent in appearance. Consequently, both Bayer and General Electric Plastics recognised the commercial potential of their discoveries. After some deliberation, both organisations decided to form an agreement whereby the company that was awarded the patent would licence manufacture of the substance to the other. In the end the patent was ruled in favour of Bayer and they began production of the polycarbonate material under the trade name Makrolon. Meanwhile, General Electric Plastics manufactured the substance under licence from Bayer and adopted the trade name Lexan. Ultimately, and many decades later, polycarbonate would go on to be used in a multitude of applications, including the production of high-density optical discs, such as Blu-ray Discs. Also in 1955 researcher Jim White, while working at DuPont, observed what appeared to be a white fluffy substance exiting a pipe in the research laboratory. Upon examination of this peculiar substance, it was ascertained that it was in fact polyethylene fibres. Consequently, DuPont began exploring the commercial possibilities of the substance and it was given the trade name Tyvek.

In 1956, DuPont patented the synthesis of high molecular weight polyoxymethylene (POM). Around the same time, Canadian chemist Allan Stuart Hay, while working at General Electric Plastics, first synthesised polyphenylene ether (PPE). The material was a high-temperature resistant thermoplastic which had a major advantage in that it was cheap to produce. Unfortunately, during development it was discovered that the plastic was difficult to process and the heat-resistant properties degraded over time. However, they soon discovered that blending the material with polystyrene in any ratio successfully counteracted this degradation by increasing the glass transition temperature (T_g) to more than 100°C. The relationship between the amount of polystyrene added and the T_g of the material was smooth and linear. Thus, as more polystyrene was added, the T_g increased in direct proportion. Furthermore, it was found that the addition of polystyrene formed a rare completely amorphous polymer blend. Incidentally, when any two polymers are blended together, they are typically incompatible with one another and tend to form separate phases. However, the presence of monomers with a benzene ring in both PPE and polystyrene enables their compatibility. Over time the discovery of this amorphous blend prompted numerous research studies in polymer miscibility. Consequently, General Electric Plastics began commercial development of the material.

By this point in time, plastics had gained wide acceptance as the wonder material of the future. Indeed, Disneyland California, in the United States, opened an attraction in 1957 named the Monsanto House of the Future. The entire building was constructed from polyester and was portrayed as a typical home in the then-distant future of 1987. Practically everything was synthetic, and the house featured many items that are commonplace today, such as a microwave oven, electric toothbrushes and plastic furniture and utensils.

Around the same time in 1957, a new worldwide fad emerged in the form of a circular plastic tube, termed the Hula hoop. Although similar purposed hoops had been around for thousands of years, made from a variety of items such as bamboo or grapevines, none had ever been constructed from a plastic material. In response to the popularity of bamboo hoops at the time, American inventor Richard Knerr and his associate Arthur Melin created and brought to market, a Hula hoop fabricated from Marlex. Incidentally, Phillips Petroleum, who produced Marlex, had experienced only limited commercial success with their product thus far. However, the Hula hoop proved to be incredibly popular, and production output quickly rose to 50,000 hoops per day with 25 million hoops being sold in a matter of months. Owing to this commercial success, Phillips Petroleum witnessed a dramatic increase in sales and eventually could not keep up with the demand for Marlex. Ultimately, this early use of HDPE in a circular shaped tubular toy would eventually lead to the development of HDPE plastic pipes. At the same time, production of isotactic polypropylene first began at Montecatini using the recently developed Ziegler–Natta catalysts.

In 1958, the United States Patent Office announced that there was an interference proceeding regarding multiple patent applications with respect to the discovery of polypropylene. In total, there were five opposing parties: Hogan and Banks, Ziegler, Zletz and two other companies. Incidentally, in the United States at that time, the Patent Office recognised that the first person to invent something was to be deemed as the rightful patent holder; as opposed to the first person to file for the patent. Consequently, the time at which the patent was submitted had little consequence. The resulting court battle spanned over three decades with a multitude of testimonies and scientific research cited. In the end, the court ruled that Hogan and Banks were the inventors of polypropylene.

Around the same time in 1958, American chemist Joseph Shivers was working at DuPont attempting to create a synthetic elastomer that could be used as a rubber alternative. After attempting to modify the thermoplastic polyester PET, he created a thermoplastic elastomer fibre which he observed to be exceedingly elastic. So much so, that it could be elongated to five times its original length without breaking. Further research on the fibre eventually resulted in DuPont terming the substance as Lycra in 1959. At the same time, DuPont patented the synthesis of a thermally stable melt-processable polyoxymethylene (POM) homopolymer. Chemists Stephen Dal Nogare and John Oliver Punderson, while working at DuPont, had discovered that by reacting acetic anhydride with the hydroxyl-terminated (hemiacetal) ends of the thermally unstable POM homopolymer, they could remove the hydroxyl groups by acetylation and produce a thermally stable plastic. Consequently, Dupont patented the POM homopolymer and began commercial development of the new material.

Around the same time in 1959 and following their discovery of the fluoroplastic polytetrafluoroethylene (PTFE) in the late 1930s, and its subsequent commercialisation as a non-stick coating under the trade name Teflon, DuPont became interested in creating a fluoroplastic that retained the advantageous properties of PTFE but was able to be processed like other thermoplastics. While PTFE was regarded as non-stick slippery substance that was highly resistant to chemical attack, the substance was not particularly melt-processable. This is a process in which the plastic material is melted with heat to form a liquid. This liquid can then be subsequently processed to create moulded artefacts, such as with injection-moulding. Thus, DuPont wished to produce fluoroplastics with outstanding melt-processability while still possessing properties rivalling those of PTFE. In 1960 the first such plastic was introduced, a co-polymer of tetrafluoroethylene and hexa-fluoropropylene, under the trade name Teflon FEP.

Also in 1960, DuPont began producing their polyoxymethylene (POM) homopolymer, which was end-capped with ester groups to prevent thermal degradation. As a result, a new engineering class of mouldable thermoplastic acetal resins was introduced, under the trade name Delrin. These highly-crystalline materials were announced as having a melting point of $175°C$ and exhibiting outstanding rigidity and resistance to wear while possessing exceptional resistance to fatigue and deformation. Around the same time, the American manufacturer Goodyear Tyre and Rubber Company began introducing blends of PVC with chlorosulphonated polyethylene (CSPE), a synthetic rubber which was produced by DuPont under the trade name Hypalon. Meanwhile, the German electrical engineering company Siemens and Halske AG announced that they had created the first blends of polyethylene terephthalate (PET) with polybutylene terephthalate (PBT).

In 1961, the American company Pennsalt Chemicals Corporation introduced the fluoropolymer polyvinylidene fluoride, under the trade name Kynar. This melt-processable fluoropolymer was particularly easy to process, in comparison to other fluoropolymers, mainly as a result of its fairly low melting point of $171°C$. Furthermore, the material had good impact, weathering resistance, tensile and mechanical properties, while being comparatively inexpensive. At the same time the American automotive company BorgWarner created the first blends of acrylonitrile butadiene styrene (ABS) with polyamide (PA). A year later in 1962, BorgWarner created a blend of ABS with poly(α-methylstyrene-co-acrylonitrile) which was announced as a heat-resistant

ABS material. This was a significant breakthrough as although ABS had high impact strength and good mouldable properties; ABS was susceptible to heat.

Meanwhile, ongoing research and development of Lycra at DuPont resulted in the introduction of the material to the mass market in 1962. Ultimately, Lycra went on to revolutionise the clothing industry and became known under several household names, such as Spandex and Elastane. Around the same time in 1962, DuPont released the first polyimide films to the mass-market under the trade name Kapton. As a result of their superior insulating properties and temperature characteristics, polyimide films went on to be used in the thermal micrometeoroid garment layer of astronaut's spacesuits and as thermal insulating blankets on satellites, telescopes and spacecraft. Furthermore, the light-weight characteristics of polyimide films have facilitated their use in civilian and military aircraft as wiring insulators. Around the same time, and concomitant to DuPont's research into polyoxymethylene (POM), the American company Celanese had been independently developing their own version of POM. However, unlike DuPont's version of POM which was thermally stabilised by end-capping with ester groups, their version had no such end-capping. Incidentally, DuPont's patent only covered the POM homopolymer and accordingly, research was carried out regarding the investigation of POM copolymers. Eventually, Celanese developed a thermally stable methylene oxide based copolymer that was easier to process but with similar properties to Dupont's homopolymer, albeit with a slightly lower melting point of 165°C, as opposed to 175°C. In 1962, and under a limited partnership, Celanese and the German company Hoechst AG began producing the material under the trade name Celcon.

1963–89

By this point the vast majority of popular commodity plastics, which we find littering the aquatic environment today, had now been synthesised and manufactured. Accordingly, the development of plastics had progressed rapidly. However, it was recognised that the inherent pliability of plastic meant it was simply no match for metals, especially in applications that required strength and durability. Consequently, there was a significant drive to increase the specific strength of plastics to equal, or even surpass, that of metal. At the same time the exponential growth in plastic materials, and their possible applications, resulted in dynamic research into improving the high-temperature characteristics and thermal-oxidative resistance of plastics. There was also significant progress achieved during this period with regard to creating blends or alloys of polymers, termed as copolymers. Blending was particularly useful as it enabled the manufacture of polymeric materials that exhibited better properties, such as heat resistance or durability, than the properties of each individual polymer alone (see Chapter 4 for more information on blended polymers).

In 1963 the Royal Aircraft Establishment in the United Kingdom made a significant breakthrough when they discovered that when plastic is mixed with carbon-fibre, the rigidity and strength of the plastic increases significantly. The strength of carbon-fibre-reinforced polymers was found to be on par with some metals but possessing a distinct advantage in that the substance was much lighter, thereby demonstrating a high strength-to-weight ratio. In 1965 a subsidiary of the Dow Chemical Company, named Union Carbide, introduced the first polysulphone (PSU), termed as Bakelite Polysulphone. The material was subsequently given the trade name Udel and was regarded as having properties akin to that of light metals. The polymer exhibited rigidity, high strength and solidity at temperatures of up to 180°C. Shortly after in 1965 American chemist Stephanie Kwolek, while working at DuPont, created a bullet-proof aromatic polyamide, termed poly-paraphenylene terephthaliamide, which was five times stronger than steel. DuPont marketed the substance under the name Kevlar and it was first used as a reinforcing agent in racing tyres. Eventually Kevlar went on to revolutionise the design of personal armour with its development into bullet-proof garments and military equipment.

Around the same time William Gorham, while working at Union Carbide, developed a poly-p-xylene plastic coating technique. The substance was capable of being deposited onto a surface as a uniform film. This was achieved via the thermal decomposition of paracyclophane at temperatures in excess of 550°C. The film coating was highly corrosion-resistant and Union Carbide commercialised the coating process in 1965. Eventually, Parylene films would go on to be used in a multitude of applications, such as printed circuit board coatings, military and space electronics, metal protection and medical applications, such as moisture barriers for biomedical tubes and friction reducers in microelectromechanical systems. In 1966 the Japanese companies, Mitsubishi Monsanto Kasei Co and Japan Synthetic Rubber Co, began the production of acrylonitrile butadiene styrene (ABS) resin. Around the same time, General Electric Plastics introduced a high-temperature modified polyphenylene oxide (PPO) thermoplastic, which consisted of a miscible amorphous blend of polyphenylene ether (PPE) and polystyrene, under the trade name Noryl. The material had good processability and was cheap to make. Furthermore, the material exhibited excellent electrical insulation, heat-resistance, hydrolytic stability and had a low density. In addition, the material exhibited outstanding retention of flexural and tensile strength, even at high temperatures. One particular application of Noryl was its use as the housing for the first Apple computer over a decade later.

In 1967, and after 12 years of research and development, DuPont introduced a flashspun high-density polyethylene fibrous sheet structure, under the trade name Tyvek. The sheet material proved to be incredibly resistant to puncture, ripping or tearing and was porous, yet exhibited unprecedented microbial penetration resistance. Ultimately the porosity and strength of Tyvek revolutionised the medical packaging industry, especially on account of its ability to withstand ethylene oxide gas sterilisation. The material was then subsequently utilised as a breathable membrane to wrap newly-built houses, in which moisture vapour was allowed to escape but water could not pass through.

In 1968 the Monsanto Company in the United States introduced a new class of polyphenylene resins that were promoted as being suitable for blending with carbon-fibre or asbestos to create a cross-linked laminate material that exhibited considerable heat-resistant properties. At the same time, Japanese company Toray Industries announced that it had successfully blended butyl rubber with polyethylene terephthalate (PET) and polybutylene terephthalate (PBT) to create a blend that had good mechanical properties and high impact strength. Meanwhile, Union Carbide announced the creation of the first blends of acrylonitrile butadiene styrene (ABS) with either polybutylene terephthalate (PBT) or polyethylene terephthalate (PET). In 1969, BorgWarner introduced a high-impact resistant ABS and PVC blend with flame-retardant properties, under the trade name Cycovin. Meanwhile, the German company Dynamit Nobel AG introduced the first amorphous aromatic polyamide under the trade name Trogamid. The material exhibited a high mechanical toughness and excellent resistance to chemicals while being highly transparent and resistant to ultraviolet fluxes. As a result of these properties the material went on to be used in many optical applications, such as in lenses for sunglasses and inspection glasses for flow metres.

Around the same time, a waterproof, breathable fabric membrane comprised of expanded polytetrafluoroethylene (ePTFE) was co-invented by Wilbert Lee Gore and his son Robert W. Gore while experimenting with using PTFE as a wiring insulator. The discovery of ePTFE occurred by happenchance when Robert W. Gore, while heating and gradually stretching a rod of PTFE, grew frustrated and pulled the rod apart very quickly. After further examination the Gores realised that a strong lightweight waterproof, yet air permeable, porous substance had been formed. Realising the market potential of these unique properties, they patented and marketed the material under the trade name Gore-Tex. At that time, many companies, as well as the military, had been searching for a lightweight breathable waterproof fabric and as a consequence, this new material was a great success. Ultimately, Gore-Tex went on to become a widely used breathable, waterproof and windproof fabric which was used in applications ranging from astronaut's spacesuits to outdoor garments and footwear.

Note: There was considerable dispute regarding who was the true inventor of ePTFE. This arose a result of the fact that in 1966, New Zealand engineer John W. Cropper, after being previously approached by DuPont, developed and constructed a machine to create Teflon (PTFE) tape. The machine Cropper built stretched polytetrafluoroethylene (PTFE) into a breathable flat thin material which possessed approximately 3 billion holes per square centimetre. Each hole was 20,000 times smaller than a water droplet and thus, liquid water could not pass through but water vapour could. However, Cropper decided not to patent the invention and instead, kept it a trade secret. Following the Gores' invention of Gore-Tex in 1969 and the subsequent patenting of the manufacturing process, it came to light that the American company Garlock Inc. was producing a similar ePTFE material using Croppers machine. Consequently, this sparked a long bitterly contested court battle over infringement of patent rights that would involve over 300 exhibits and 35 testimonies. In the end the court ruled that the Gores' patents were invalid, citing Cropper as the true inventor of ePTFE. However, following appeal, the decision was reversed as the court decided that since no patent was applied for by Cropper and because he kept the process a secret, he forfeited any superior claim to the machine and thus, could not legally be considered the inventor of the material. Consequently the appeal court ruled that the Gores were indeed the legal inventor of ePTFE.

In 1970, British company ICI created the first blends of polyamide (PA) and polysulfone (PSU). At the same time, British Gas in the United Kingdom introduced yellow HDPE piping for use in natural gas distribution pipelines. Meanwhile, and following on from DuPont's research and development of fluoroplastics and the subsequent introduction of Teflon EPA, DuPont introduced a new fluoropolymeric substance under the trade name TEFZEL. The substance was assessed as being particularly tough and exhibiting a high tensile strength. Furthermore, TEFZEL was rated as being capable of withstanding continuous temperatures of 155°C for a period of more than 2 years. In 1971 the American manufacturing corporation 3M company introduced an amorphous high-temperature and high-performance thermoplastic termed polyarylethersulphone (PAES), under the trade name Astrel. As a result of the materials exceptional thermal oxidative resistance, it proved to be especially useful in service environments that involved exposure to high temperatures for long periods of time. Meanwhile, further research and development of fluoropolymers by DuPont resulted in the introduction of Teflon PFA in 1972. This material was proclaimed to have outstanding melt-processability while still retaining many characteristics similar to that of PTFE. In addition, the material was regarded as being resistant to flexing and fracture and maintaining rigidity and strength at high temperatures.

In 1973 there was a global oil crisis and the price of crude oil increased by 300%, while ethylene increased by 200%. Accordingly, plastics produced from petroleum distillates increased by 50–100% in price. This fuelled rising speculation that it may be too expensive to produce oil-based plastics in the future. Consequently, the British company ICI to begin investigating methods in which the efficient bacterial production of the biodegradable polymer polyhydroxybutyrate (PHB) could be accomplished. Around the same time, the Japanese company Mitsubishi Gas Chemical Company created the first polycarbonate (PC) and polyamide (PA) blends.

In 1974, Japanese company Unitika Ltd introduced the first commercial aromatic amorphous polyester as a transparent engineering plastic, under the trade name U-polymer. The material was regarded as possessing the highest heat-resistance of any transparent resin at 175°C while being transparent as poly(methyl methacrylate) (PMA) or polycarbonate (PC) with a visible light transmission value of approximately 90%. Furthermore, the material exhibited excellent weathering resistance and an exceptional elastic recovery and allowable strain ratio. Consequently the material went on to be used in a multitude of applications, including automobiles, medical products, machines and precision devices. Meanwhile, American chemical manufacturer Hercules Incorporated introduced a thermosetting polyphenylene resinous material which they marketed as H-resin. The company announced that after suitable cross-linking and curing, the material was capable of being used in a service environment with air temperatures of up to 300°C. Furthermore, if the environment was deprived of oxygen, then the upper temperature limit was increased to 400°C. In 1975 Japanese researchers Yoshio Imai and Motokazu, while working at Yamagata University, reported a novel two step synthesis whereby the polymerisation of 4,4′-oxydianiline (ODA) and bis(phenylsulphonyl) pyromellitimide (BPSP) produced polyimide.

By 1976 the increasing use of plastic materials resulted in plastics being deemed as the most used material in the world. Incidentally, this still holds true to this day. Also in 1976 the high performance amorphous plastic polyphenylethersulphone (PPSF) was first introduced by Union Carbide under the trade name Radel. The impact-resistant material exhibited a tensile strength of up to 8000 psi and was chemical and heat resistant up to 207°C. Furthermore, the exceptional long-term hydrolytic stability of the material resulted in its development as hot water plumbing fittings and manifolds, as well as medical devices and dental or surgical instruments that required repeated sterilisation by steam, even upwards of 1000 cycles. Furthermore, it was developed as a low smoke generating, low toxic gas emitting, low heat releasing material for use in aircraft interiors and airline catering trays and trolleys.

In 1977, American company Bayer Corporation, under licence from BorgWarner, introduced an amorphous thermoplastic blend of acrylonitrile butadiene styrene (ABS) and PC, under the trade name Bayblend. The material exhibited heat resistance to 142°C, high impact strength and rigidity while having a low tendency to warp. The product went on to be used in applications such as automotive door handles, binocular housings, household items and in electronic and electrical equipment. Concomitantly, BorgWarner introduced their own high impact amorphous thermoplastic blend of ABS and PC, under the trade name Cycoloy. At the same time, the Japanese company Mitsubishi Rayon introduced a high-impact blend of polyethylene terephthalate (PET) and polybutylene terephthalate (PBT), while the American company Philips Petroleum announced that they had created the first blend of two dissimilar forms of linear low-density polyethylene (LLDPE), in which one form was an ethylene butene copolymer, while the other form was an ethylene hexene copolymer. The resultant blended copolymer was deemed to be suitable for both dynamic and screw extrusion. Meanwhile, researchers at the University of Pennsylvania developed the conductivity of polyacetylene, thereby discovering the first conductive polymer. Furthermore, they discovered that iodine vapour was capable of increasing the conductivity of polyacetylene by eight orders of magnitude (100 million times).

Also in 1977, the British company ICI first synthesised a highly robust thermoplastic polymer termed polyether ether ketone (PEEK) with a glass transition temperature (Tg) of 143°C and a melting point of 334°C. The material was assessed as being suitable for use in demanding environments. Consequently, ICI subsequently marketed the substance in 1978 under the trade name Victrex and in due course, the material went on to become regarded as one of the highest performing engineering plastics in the world. In 1979, DuPont's research and development of fluoroplastics had been particularly fruitful and that year, the company introduced a new material under the trade name Teflon EPE. This new fluoroplastic exhibited a melting point of 295°C. Around the same time, Japanese company Asahi Kasei Corporation introduced Xyron, a range of halogen-free flame-retardant thermoplastics which were blends composed of modified polyphenylene ether (mPPE) with varying amounts of polyamide (PA), polyphenylene sulphide (PPS), polypropylene (PP) and polystyrene (PS). The materials exhibited high resistance to acids and alkalis, outstanding heat resistance of 80–220°C and a low specific gravity of 1.03. Today, the material is gradually phasing out PVC in electrical wire insulation, with many computer and automobile manufacturers planning to switch to mPPE insulation in the near future. This is a result of the superior dielectric properties of mPPE $(3.94^{16}\,\Omega$-cm at 100v), in comparison to PVC $(2.7^{15}\,\Omega$-cm at 100v), which allows for up to approximately 25% reduced wall thickness of wire insulators. Furthermore, mPPE is more environmentally friendly since it lacks halogenation, while being lighter, tougher and exhibiting more abrasion and pinch-resistance than PVC.

In 1980, Union Carbide introduced a blend of acrylonitrile butadiene styrene (ABS) and polysulfone (PSU) with high impact strength, low cost and good processability properties, under the trade name Mindel. Also in 1980, the use of carbon-fibre composites as a direct replacement for the metal bodies in formula 1 race cars was first introduced. The carbon-fibre composites had two distinct advantages in formula 1; they were much lighter and stronger than metal and in a collision, the carbon-fibre composite shattered into small fragments thereby dissipating energy away from the driver. Also in 1980, significant manufacturing advancements led to the first full-scale production of low density polyethylene; the world's most popular plastic and the main ingredient of the infamous 'plastic bag'. At the same time, high-pressure HDPE pipes, and blue medium-density polyethylene (MDPE) pipes, were introduced in the United Kingdom for the supply of drinking water.

In 1981, French company Atochem introduced a thermoplastic elastomer product composed of block copolymers, which were in turn comprised of linear chains of soft polyether and rigid polyamide segments. The product was given the trade name Pebax and had a very low material density, in comparison to other thermoplastic elastomers, of 1.00–1.01 g/cm^3 which is practically equal to the density of water (1.00 g/cm^3). In 1982, Hoescht AG introduced various blends of polyoxymethylene (POM) with polyurethane. The resultant material exhibited many of the advantageous properties of POM but had enhanced toughness. At the same time, General Electric Plastics introduced a high-temperature polyetherimide engineering thermoplastic with exceptionally high tensile strength and flammability performance and with a glass transition temperature (Tg) of 215°C, under the trade name Ultem. The plastic possessed the advantageous properties of polyimides, such as toughness and thermal stability. However, in contrast to most polyimides, the plastic exhibited excellent resistance against chemical decomposition in the presence of water and had ether linkages which allowed the material to be melt-processable.

Also in 1982 and named after its American inventor Robert Jarvic, the Jarvic-7 artificial heart was first implanted in American dentist Barney Clark. The device was constructed from polyurethane and aluminium. This remarkable medical accomplishment represented the dawn of using plastics to replace biological structures within the human body.

In 1983, polyether ether ketone (PEEK) was marketed by ICI and went on to be regularly utilised in high temperature and demanding applications, such as valves, pumps and engine parts. Additionally, the material was subsequently used in medical implants. At the same time, Bayer launched polyphenylene sulphide (PPS), a completely solvent resistant high-performance engineering thermoplastic, and polyethersulphone (PES), an engineering plastic that can withstand long-term exposure to high temperatures in water and air while exhibiting minimal dimensional change. In 1984, BorgWarner introduced a blend of acrylonitrile butadiene styrene (ABS) and polyamide (Nylon 6) under the trade name Elemid. The material was reported to possess an outstanding balance of properties, including dimensional stability at high temperatures, as well as excellent chemical and impact resistance.

In 1985, very low-density polyethylene (VLDPE) was introduced by Union Carbide. This unique type of polyethylene is considerably linear and possesses a significant amount of short chain branches. The material subsequently went on to be used for food packaging and tubing, as well as being used as a highly stretchable film, termed stretch wrap, which was used to wrap around items, such as boxes on pallets to secure the cargo. Around the same time, Japanese company Toray Industries introduced blends of glycidyl methacrylate (GMA) with polycarbonate (PC), polyphenyl ether (PPE) and polyethylene terephthalate (PET). The material was destined to be used in demanding service environments within the automotive industry that required a high performance plastic. In 1986, a blend of acrylonitrile butadiene styrene (ABS) and polycarbonate (PC) was introduced by Dow Chemical Company, under the trade name Pulse. The material was intended for use within the interior of automobiles.

In 1987, DuPont introduced Kevlar HT which possessed 20% higher strength than normal Kevlar. Around the same time, DuPont also introduced Kevlar HM, which exhibited 40% higher stiffness than previous grades of Kevlar. In 1988, German company BASF AG introduced polypyrrole (PPy), a solvent resistant and infusible electrically conductive film, under the trade name Lutamer P160. The material was capable of being produced as powder, which upon blending with a thermoplastic, acted as a conductive component. Ultimately, PPy went on to revolutionise chemical sensors and electronic devices. In 1989, DuPont introduced a family of amorphous fluoroplastics, under the trade name Teflon AF. With a commercial price of one or two orders of magnitude higher than other amorphous fluoroplastics, it was considered as one of the highest-priced plastic substances at that time. The high cost of Teflon AF was attributed to its unique and unsurpassed properties. For example, in addition to retaining the highly advantageous properties of polytetrafluoroethylene (PTFE), the material exhibited exceptional optical clarity of more than 95%. At the time of writing, Teflon AF has the lowest refractive index of any known plastic material. Similarly, the material possesses outstanding electrical properties and has the lowest dielectric constant of any known solid plastic substance.

1990–present

By this point, a large assortment of new polymer blends and engineering plastics had been introduced. Thus, in terms of commodity thermoplastics, the general consensus at the time was that creating entirely new commercially viable polymers was unlikely to be a fruitful endeavour. Consequently, the emphasis had almost entirely shifted to compounding and creating composites with enhanced properties. Furthermore, owing to their unique ability to improve upon the unpredictability of plastic properties, the commercialisation of metallocene catalysts represented a significant leap forward in terms of polymer manufacture. Additionally, at that time there was growing awareness of plastic waste accumulating in the environment and consequently, particular emphasis was placed on the development and introduction of biodegradable plastics.

As such, and following 15 years of development, it was in 1990 that the first commercially available entirely biodegradable plastic, a copolymer of polyhydroxybutyrate (PHB) and polyhydroxyvalerate (PHV) which was insoluble in water, was first introduced to the world by British company ICI under the trade name Biopol. The synthesis of PHB relied upon growing the bacteria *Alcaligenes eutrophus*, which produced PHB as globules utilised for energy storage. These globules were then harvested to obtain the PHB. Incidentally, over subsequent years, the gene responsible for PHB production was identified and consequently, the use of faster growing bacterial strains to produce PHB became possible via genetic modification. One of the first uses of PHB was in the fabrication of hair shampoo bottles and initially, these bottles were first sold in Germany. Additionally, since PHB is produced by bacteria in the human gut, the substance is biocompatible.

Consequently, PHB went on to be utilised as a carrier in targeted drug delivery systems in which following introduction to the desired area of the body, the PHB carrier degrades and releases the drug. This allowed specific areas of the body to be targeted with a drug, thereby reducing the risk of the drug affecting unwanted areas of the body or undergoing undesirable metabolisation before it can have a therapeutic effect. Furthermore, the biocompatibility of PHB plastic also saw the substance go on to be utilised for internal biodegradable sutures and screws, as well as scaffolding for nerve tissue. Parenthetically, oligomers (chains consisting of only a few monomers) of PHB have been found in the cells of almost all biological organisms and consequently, it has been hypothesised that the biological ubiquity of PHB suggests that it must be a fundamental constituent of biological cells where it serves some sort of vital purpose. Nonetheless, no biological function of PHB has been unequivocally established thus far, and its presence in cells remains an enigmatic scientific curiosity.

Shortly after in 1990, American company Walter-Lambert introduced a starch-based biodegradable plastic under the trade name Novon. The first commercial domestic application of the plastic was the manufacture of golf tees. This was hoped to combat the problem of tees left behind in golf courses since exposure to water and carbon dioxide degraded the plastic at around the same rate as leaves and wood. Around the same time, Italian company Novamont introduced their own starch based biodegradable plastic under the trade name Mater-Bi. Also in 1990, American company Air Products and Chemicals first introduced a series of biodegradable thermoplastic copolymers, which were a blend of polyvinyl alcohol (PVOH) and poly(alkyleneoxy) acrylate, under the trade name Vinex. The material rapidly degraded in water and it was reported that a piece of film 1.5 millimetres in thickness would completely dissolve within approximately 30 seconds of being submerged in distilled water at room temperature.

In 1991, American company Amoco Corporation initiated the first commercialisation of polythalamide (PPA), a high performance polyamide, under the trade name Amodel. The material was capable of replacing metal in high temperature automotive applications of 120–185°C. Furthermore, the material provided improved properties over standard polyamides, such as increased strength and stiffness at high temperatures. Around the same time, American company Exxon introduced the first commercial metallocene-made polyolefin, polyethylene. Shortly after in 1992, Dow Chemical Company enhanced polyethylene under the trade name Elite. Additionally, they introduced a polyolefin plastomer under the trade name Affinity and a polyolefin elastomer under the trade name Engage.

In 1993, a significant breakthrough in producing plastic from plants occurred with the development of PHB producing cress plants (*Lepidium sativum*). This was accomplished by infecting the plants with the bacterium *Agrobacterium tumefaciens*, in which it acted as a 'Trojan horse' to introduce two new PHB producing genes to the plant. Consequently, each plant was capable of producing approximately 14% PHB, thereby possibly eliminating the need to produce PHB with bacteria instead. However, the possible commercial exploitation of using these genetically modified plants to create PHB was a controversial issue at that time due to worries about containment of the plants and the risk of the modified genes escaping into the wider environment. Indeed, many scientists reasoned that the risk to the environment was unknown since the biological function of PHB in cells was yet to be determined.

In 1994 a new fluoropolymer, described as having a significantly improved melt-viscosity, was introduced by Hoechst AG under the trade name Hostaflon PFA-N.

In 1995, and following their previous acquirement of the amorphous transparent polyamide Trogamid, the German company Evonik Industries introduced a crystallisable, permanently transparent polyamide under the trade name Trogamid CX. The material was developed by initially substituting the aromatic components of Trogamid with aliphatic monomers to give a material with increased resistance to ultraviolet fluxes. Thereafter, systematic selection of the 1,12-dodecanoic acid and cycloaliphatic diamine monomers resulted in the creation of a high clarity crystallisable form with 92% visible light transmission. The material exhibited a property known as microcrystallinity, whereby the crystallites (small crystalline regions) of the material were of such a small size, that visible light was not scattered and the material appeared highly transparent.

Note: When the crystallites of a plastic material are larger than the wavelength of visible light, the transparency of that material is significantly reduced due to the scattering of visible light.

Further development of Trogamid resulted in various grades of the material being produced to be used in a multitude of speciality transparent optical applications, such as lenses for sunglasses and sports glasses, as well as food and drinking water contact applications. Around the same time, the Japanese company Japan Synthetic Rubber Corporation announced new blends of acrylonitrile butadiene styrene (ABS) and polybutylene terephthalate (PBT) with improved resistance to chemicals and reduced susceptibility to fatigue fracture, under the trade name Macalloy B.

In 1996, Monsanto purchased all the patents from ICI for producing the biodegradable plastic Biopol, and began producing PHB from bacteria. In 1997, Italian company Novamont acquired Warner-Lamberts worldwide patents, ultimately becoming the global leader in manufacturing and supplying starch-based biodegradable plastics. In due course, their product range found use in a large array of applications, such as bags, plant pots, office stationary and packaging for food and hygiene articles. In 1998 the large-scale production of biodegradable polyester began by the American Company Eastman, under the trade name Eastar Bio. Also in 1998, the first commercial development and production of the biodegradable polymer polylactic acid (PLA), produced from natural resources such as corn or sugar beets, began by American company Cargill Dow Polymers LLC, under the trade name EcoPLA.

In 1999 a considerable global debate ignited over the potential toxicity of bis(2-ethylhexl) phthalate (DEHP), a potentially toxic plasticiser used in the manufacture of polyvinyl chloride (PVC). This resulted in many toy manufacturers switching to alternative plastics, while the medical industry raised concerns about the presence of DEHP in PVC medical products, such as intravenous bags. Incidentally, today DEHP is being phased out and many bans have since come into effect.

In 2001, American company Metabolix purchased the patents for the production of Biopol from Monsanto and began investigating ways of producing the biodegradable plastic PHB from genetically modified plants, in which natural sugars present within the plants could be converted to the substance. At the same time, the first self-healing plastics with imbedded adhesives were reported by the Sottos Research Group in America. In 2003, Belgian chemical company Solvay S.A. introduced a rigid highly transparent polysulfone (PSU) product with a high optical clarity rivalling that of polycarbonate, under the trade name Udel P-3799 HC.

In 2004, Monsanto shut down their bacterial fermentation reactors for producing PHB, instead shifting the focus to investigating the production of PHB from genetically modified plants. Around the same time, Novamont acquired Eastman's Eastar biodegradable polyester business. In 2005, Metabolix was awarded the American Presidential Green Chemistry Challenge Award for the development, efficient production and successful commercialisation of biodegradable polyhydroxyalkonoate (PHA) plastics, including PHB.

In 2005, Japanese company Fujitsu became the first company in the world to create computer cases made entirely from biobased polymers. Around the same time, the American space agency National Aeronautics and Space Administration (NASA) created a new polyethylene-based material termed RXF1 which possessed three times more tensile strength than aluminium, while being 2.6 times lighter. The material was intended as a radiation shield incorporated within the body of future spacecraft to protect against high-energy particles in space, as well as incoming micrometeorites.

In 2006, researchers at Case Western Reserve University in America first synthesised an aliphatic amine-based polybenzoxazine (PBZ). The thermoset material was reported to be much more flexible than the bisphenol-type polybenzoxazines. In 2007, Brazilian petrochemical company Braskem pronounced that they had developed a biobased polymer. Their innovative process comprised obtaining ethylene from Brazilian sugarcane, which could then be utilised to manufacture high-density polyethylene (HDPE), which was informally described at the time as 'Green Polyethylene' or 'Green Plastic'. The company partnered with Japanese company Toyota Tsusho Corporation to produce and market the biobased plastic in Asia.

Note: Biodegradable plastics are polymers which are capable of decomposing by way of the actions of living organisms, such as microbes that convert carbon into energy (See Chapter 4). These should not be confused with biobased polymers, which are plastics derived from renewable biomass resources, such as microorganisms or vegetable fats. While many biobased polymers may be biodegradable, not all are. Furthermore, with respect to degradation in the environment, it should not be assumed that a biobased polymer exhibits any superiority over a common plastic, unless this has been demonstrated to be the case. At the time of writing, biobased polymers still remain a niche industry and only represent 1% of the global market.

In 2008, the race to build lighter, stronger and more fuel efficient aircraft from composite materials began when Airbus announced that their new aircraft, the A380, was composed of 22% carbon-fibre reinforced polymer (CFRP). Shortly after in 2009, Boeing announced that their 787 Dreamliner had an airframe composed of CFRP materials representing 50% of the aircraft. Around the same time, the Airbus announced its A400M military transport aircraft had wings and four counter-rotating propellers composed of almost entirely CFRP materials. Also in 2009, DSM Engineering Plastics introduced a bio-based high-performance polyamide mainly produced from tropical castor beans, under the trade name EcoPaXX. The material exhibited excellent chemical resistance, excellent long and short term heat resistance and low moisture absorption. It was reported to have a melting point of up to 250°C, the highest melting point of any bioplastic. Shortly after in 2010, the company introduced a high-performance thermoplastic copolyester in which 20–50% of its content was derived from renewable resources.

In 2011, a tough and transparent composite material produced from the polysaccharide chitosan, which was derived from shrimp shells, and the silk protein fibroin was created and termed as Shrilk. The material was a flexible compostable and biocompatible composite material which was formed in films that were as strong as aluminium but only half the weight.

In 2012, the permanent effects of plastic pollution were highlighted when a new type of rock, termed 'plastiglomerate', was proposed following the discovery of peculiar plastic-based rocks on Kamilo Beach, Big Island, Hawaii 6 years earlier. Scientific analysis of the specimens determined that the rocks were composed of organic debris, sedimentary grains and melted plastic. It was then hypothesised that these materials had become agglutinated by camp fires or possible even lava flows to form the new type of rock. It has since been suggested that the rocks may even survive in the geological record indefinitely, thereby representing a marker horizon for anthropogenic plastic pollution. Around the same time, the London Olympic Games in the United Kingdom used 142,538 square metres of polyvinyl chloride (PVC) during the construction of the event venues, after which a large proportion was recycled.

In 2013, scientists at the Centre for Electrochemical Technologies in Spain reported the first self-healing polymer that required no intervention to induce repair of the material. It was demonstrated that when the material was cut in half and both parts were placed back together, the material fused back together with 97% healing efficiency after a period of 2 h. Following repair, significant stretching of the material was possible without any breakage where the join had occurred. In 2014, IBM announced that they had created two entirely new biodegradable thermoset plastics. The first, termed Hydro, was a flexible gel-like material that was capable of self-repair if the material was cut in half. Furthermore, the material completely degrades in water to revert back to its constituent components. The second was a different form of the material, termed Titan, which was produced at a higher temperature and had approximately one-third of the tensile strength of steel. However, blending Titan with 2–5% carbon nanotubes resulted in a material three times stronger than polyamide. To degrade the compound, a weak acid was required. Both compounds are unusual in that they are considered as reversible thermoset plastics. Incidentally, thermosets are usually irreversible reactions such that once they have cured to a shape, they form a solid material that cannot be recycled or reformed. However, both Hydro and Titan can be recycled and are considered to be a new class of thermosetting plastics with many possible applications, such as 3D-printing, adhesives and drug delivery.

In 2015, Airbus announced that their aircraft, the A350 XWB, had a fuselage and wing structures composed of carbon-fibre reinforced polymer in which this composite material represented 52% of the aircraft. Around the same time, Brazillian company Braskem and American company Genomatica announced that they had successfully produced butadiene, a precursor of the synthetic rubber polybutadiene, using renewable feedstocks. Following computational analysis of 60 possible biological pathways in which a microorganism could potentially bio-synthesise the substance, the company picked the best candidates and developed them. Ultimately, their commercial process relied upon a microorganism which converts sugar to butadiene.

By 2016 we had a large assortment of plastics with a wide variety of properties, some even exotic. We had plastics that were bullet-proof or impervious to solvents and gases, and we had created plastics that were capable of self-repair. Damage to these polymers allowed liquids contained within tiny microchannels (vascular) or microcapsules (capsule-based), throughout the polymer structure, to come into contact with one another thereby polymerising and self-repairing holes up to 3 centimetres in diameter. Another version of these self-healing plastics involved exposure to intense ultraviolet fluxes to initiate repair. The high intensity light weakened the bonds of intrinsic metal ions within the material, thereby liquefying the exposed area. Once the light was switched off, the bonds reformed and the affected area solidified, thereby repairing the damage. Additionally, an alternative self-healing version relied upon many molecular links between the polymer chains. If the plastic surface was scratched, the links broke and changed shape, which induced a colour change at the site of damage due to a change in the absorbance and reflection of light by the broken links. This change of colour could then be used to identify the damaged areas. If the areas were then exposed to light, pH or temperature changes, the broken links reformed and repaired the plastic.

Furthermore, we had shape-shifting polymers that, upon exposure to external influences such as light or temperature, underwent changes in the density of cross-linking within their polymer matrix. This transformed the polymer from a rigid substance to a viscoelastic substance, thereby allowing the polymer to revert back to its original predetermined shape. We also had created plastics which could conduct electricity in the form of organic light emitting diodes (OLED) and consequently, this revolutionised the image quality of television and computer monitor screens. We had also witnessed the rapid increase in the use of polymers within the human body, such as in artificial hip joints, tissue scaffolds, sutures and cosmetic surgery.

As such, plastics are here to stay and as we continue to develop these revolutionary polymeric substances, we must seriously consider their fate at the end of their service life. Plastics may well have revolutionised our lives, but at the same time, they have trapped us in an ever-increasing cycle of producing, consuming and disposing. At the time of writing, around 17% of all plastics worldwide are recycled, while aluminium and paper represent approximately 69% and 60%, respectively. The reason as to why there is such a low rate of recycling for plastics is mainly due to cost. Typically, there is about a 20% increase in manufacturing costs associated with using recycled plastics, as opposed to utilising virgin-plastic feedstock. Furthermore, a typical recyclable commodity plastic can only be recycled about 3 times. This is because repeated melting and remoulding of the plastic results in diminished mechanical properties. The plastic loses its flexibility and becomes brittle, discolours or becomes translucent, as opposed to transparent. Once the mechanical properties have diminished, the plastic can no longer be utilised in its original application and must either be reused elsewhere or discarded. Thus, as a result of this rising demand for commodity plastics and the low global rate of recycling, it is expected that increasing contamination of the aquatic environment with non-biodegradable plastics will be observed for the foreseeable future.

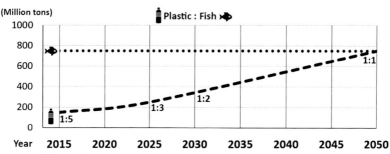

Figure 2.3 The predicted ratio of plastics to fish in the ocean (by weight).

At the time of writing, there is about 150 million tonnes of plastics already in the ocean and approximately 8 million tonnes of plastics enters the ocean each year. This is the equivalent of dumping around 15 tonnes of plastic into the ocean every minute. If no action is taken, the amount of plastic entering the ocean each year is predicted[312] to increase to around 16 million tonnes by 2030, and approximately 32 million tonnes by 2050. Indeed, it is estimated[312] that in terms of weight, there will be more plastics in the ocean than fish by 2050 (see Fig. 2.3). For this reason, it is essentially that concomitant to the development of plastics, novel approaches to dealing with plastic waste are studied, created and ultimately embraced by citizens, policy makers, engineers and scientists alike.

Plastic production, waste and legislation

3

Global production of plastics

Today, plastics are considered as the most used and versatile materials of the modern age, and inevitably, global production has increased prodigiously to meet the rapidly growing demand for these materials.[16,139] In 1950 the annual global production of plastics amounted to the diminutive weight of 1.5 million tonnes.[460] However, despite a dip in production during the 1973 oil crisis, as well as the 2007 financial crisis, by 2009 global production of plastic materials had increased substantially to 250 million tonnes.[319] Historically the global production of plastic materials has increased by around 9% each year.[191] Indeed, by 2014 the rate of global production had reached 311 million tonnes per year.[337] This represented an increase in annual global production of around 25% in just 5 years; while over the period of 65 years, annual global production increased exponentially on the order of an astounding 20,000%.

By 2014 the top three global producers of plastics were China, Europe and North America at 26%, 20% and 19%, respectively.[337] Five countries accounted for 63.9% of the total European demand for plastics: Germany (24.9%), Italy (14.3%), France (9.6%), the United Kingdom (7.7%) and Spain (7.4%).[337] The plastics in most demand worldwide were polyethylene and polypropylene, and the packaging industry was by far the biggest consumer of these materials. By 2015 the worldwide consumption of plastic materials was almost 300 million tonnes[306] (Fig. 3.1).

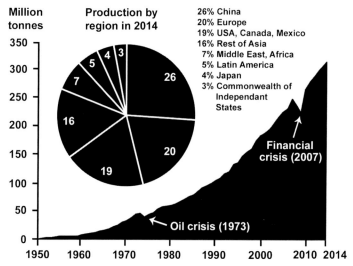

Figure 3.1 The global production of plastics from 1950 onwards.

Microplastic Pollutants. http://dx.doi.org/10.1016/B978-0-12-809406-8.00003-7

Long-term forecasts

The current long-term forecasts suggest that the production of plastic materials shows no sign of levelling off and is expected to surge exponentially.[312] The global population is projected to soar to around 10.85 billion by 2100.[24] This represents an increase of more than 50% over current figures. Thus, by 2050 it is anticipated that around an extra 33 billion tonnes of plastic will have been produced[356] and annual global production shall be between 850 million tonnes[383] and 1124 million tonnes.[312] Nevertheless, it is extremely difficult to predict this far into the future with any degree of certainty. Indeed, such a high level of production may sound unrealistic in the present day. Though, consider the possible reaction if you stated to someone in 1950 that the production of plastics would rise from the mere 1.5 million tonnes produced at that time to 311 million tonnes in just 65 years. They may well have reacted with bemusement or disbelief. Thus, although these estimates may sound improbable at present, if we look at the historic output and the rates of production of these materials today, it is entirely plausible that these distant forecasts may turn out to be accurate or possibly even too conservative.

Plastic waste

Of all the plastic ever produced, it has been estimated that approximately 10% has been released into the ocean.[234] Furthermore, of all the plastic produced each year, approximately 33% is designed to be non-reusable and is typically discarded within a 12-month period following manufacture.[228] One possibility to deal with plastic waste is to burn it. However, burning of plastic materials releases highly toxic chemicals, such as furans and dioxins which can have serious human health implications, as well as being highly detrimental to the environment. Furthermore, many countries do not impose legislation that requires recycling of plastic waste. Consequently, it is often cheaper and easier to choose landfill instead.

Of all the municipal waste produced worldwide, 16% is composed of plastic.[306] Inevitably, much of this land-based plastic waste ends up in the aquatic environment and current estimates are that 15–40% of all discarded plastic is released into the ocean. This typically occurs when landfilled or discarded plastic materials are carried on the wind or enter rivers and urban watercourses and are transported to the ocean (Fig. 3.2). A large amount of undeveloped and developing countries utilise open dumps due to the low cost associated with the practice.[306] One preventative measure to reduce the amount of plastic being washed into the ocean is to close landfill sites that are located in regions susceptible to flooding.[46]

An assessment by The United Nations Joint Group of Experts on the Scientific Aspects of Marine Pollution (GESAMP) concluded that 80% of waste in the marine environment originates from land (Fig. 3.3), while only 20% was a result of activities at sea. Furthermore, it has been estimated that 4.8–12.7 million tonnes of plastic waste entered the ocean in 2010 alone,[204] while in 2015 around 8 million tonnes was estimated to have been released.[312] By 2050 this amount is expected to increase to around 32 million tonnes per year.[312]

Figure 3.2 Plastic waste often enters urban watercourses and is transported to the ocean.

Figure 3.3 Plastic waste can substantially accumulate in the ocean.

Plastic waste largely enters the ocean by dumping or littering or unintentional spillage. The provision of satisfactory facilities in which plastics can be disposed in ports would help to alleviate littering directly into the seas and oceans.[46] However, one source of plastic pollution in the marine environment is the accidental loss of shipping containers from vessels during storms and groundings (Fig. 3.4). Between 2008 and 2013, the World Shipping Council reported that approximately 1679 containers were lost worldwide every year, although the grounding of MV Rena in 2011 resulted in the spilling of 900 containers while the sinking of MOL Comfort in 2013 resulted in the loss of 4239 containers. Incidentally, many shipping containers contain plastic materials, which can subsequently escape if the containers burst or erode. For example, during a stormy night in 1992, a single transport vessel in the eastern Pacific Ocean accidentally released 29,000 plastic turtles, frogs and ducks when a shipping container was washed overboard. For over 20 years,

Figure 3.4 Containers containing plastic materials can be lost at sea.

these plastic items were tracked by various researchers, based on reported sightings, and were confirmed to have travelled more than 17,000 miles around the world's oceans. Incidentally, at the time of writing they are still reported to occasionally appear on beaches worldwide and have provided valuable insight into the movement of plastic pollution by ocean currents and the capability of macroplastics to travel great distances.

What are macroplastics?

In this book a piece of plastic equal to or larger than 25 mm in size is termed as a macroplastic. Additionally, pieces of plastic smaller than 25 mm, depending on their size, are categorised as shown in Table 3.1.

Table 3.1 Size categories for pieces of plastic

Category	Abbreviation	Size	Size definition
Macroplastic	MAP	≥25 mm	Any piece of plastic equal to or larger than 25 mm in size along its longest dimension
Mesoplastic	MEP	<25 mm–5 mm	Any piece of plastic less than 25 mm to 5 mm in size along its longest dimension
Plasticle	PLT	<5 mm	All pieces of plastic less than 5 mm in size along their longest dimension

Table 3.1 **Size categories for pieces of plastic—cont'd**

Category	Abbreviation	Size	Size definition
Microplastic	MP	<5 mm–1 mm	Any piece of plastic less than 5 mm to 1 mm in size along its longest dimension
Mini-microplastic	MMP	<1 mm–1 μm	Any piece of plastic less than 1 mm to 1 μm in size along its longest dimension
Nanoplastic	NP	<1 μm	Any piece of plastic less than 1 μm in size along its longest dimension

We shall discuss these smaller plastics in greater detail within Chapter 5 but first, we shall examine floating, beached and submerged macroplastics.

Floating macroplastics

It has been estimated that 90% of all buoyant debris floating on the surface of the aquatic environment is composed of plastic[164] and there are more than 5 trillion pieces weighing almost 269,000 tonnes. Furthermore, it has been estimated that of all the waste present within the aquatic environment, 60–80% is plastic.[22,286] However, the United Nations Joint Group of Experts on the Scientific Aspects of Marine Pollution (GESAMP) estimate this could be as high as 95%. In the waters of the United States, leisure boats are responsible for around half of all the discarded plastic debris present there. Furthermore, large accumulations of plastic waste can also build up in freshwater lakes as a result of littering directly into the water from leisure boats, recreational activities or as a result of transport by urban water courses, drainage systems, rivers, flooding and storms (Fig. 3.5).

Figure 3.5 Plastic waste can substantially accumulate in large bodies of freshwater.

In the marine environment, floating plastic debris may be distributed by winds, tides, currents and even tsunamis.[284] Using computation simulations, it has been calculated that as a result of rotating surface currents, the Tyrrhenian sub-basin and the Gulf of Sirte appear to be areas where the retention of floating plastics could occur. Furthermore, the north-western Mediterranean may retain floating plastics due to weak currents. Additionally, the coastal regions of Tunisia and Libya were predicted to accumulate the highest amounts of floating plastic debris since weak currents coupled with the geometry of the coastline and the presence of a region of high accumulation favour the influx and deposition of plastic debris.[270]

Nevertheless, most plastic debris tends to congregate in vast areas of the ocean termed 'gyres'.[350,367] These gyres are vast swirling vortices created and driven by powerful ocean currents, which gather significant accumulations of plastic waste.[298] It is recognised that there are five main gyres in the world's oceans: the North Atlantic Gyre, the South Atlantic Gyre, the North Pacific Gyre, the South Pacific Gyre and the Indian Ocean Gyre.[245] The amount of plastics in these gyres is immense. For example, in the North Pacific Gyre, there is a colossal manifestation of marine plastic litter, termed the North Pacific Garbage Patch, which has steadily accumulated within this vast swirling vortex over many years and represents a profound environmental challenge.[294] In the North Pacific Gyre, there is approximately 330,000 pieces of plastic per km^2, whereas in the South Pacific Gyre and the North Atlantic Gyre, there are around 25,000 pieces per km^2 and 20,000 pieces per km^2, respectively.[117,243,298] In the South Pacific Gyre, high accumulations of floating plastics have been observed in the Polynesian regions.[293] Similarly, in a region of the South Pacific Ocean, located between Easter Island and the continental coastline of Chile, it was found that plastic was particularly abundant on oceanic islands, as opposed to the coastline, and was especially abundant at the centre of the South Pacific oceanic gyre where 62.3% were fragments larger than 2 cm in size.

The locations of these oceanic gyres and their proximity to shipping lanes are shown in Fig. 3.6. Looking at the map, we can see that both Northern gyres are

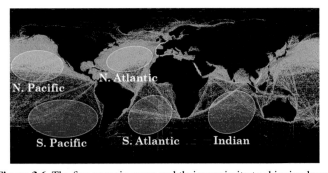

Figure 3.6 The five oceanic gyres and their proximity to shipping lanes.

situated in regions of extremely high shipping activity. This is a very important factor to consider since increased shipping activity has been correlated with increased marine contamination with chemical pollutants.[2,364,442] Furthermore, floating plastic debris can undergo adsorption or adherence of microbial communities, organic debris, clay particles, metals and persistent organic pollutants (POPs).[284] Chapters 6 and 7 discuss in detail: the growing realisation that dangerous interactions occur between plastics and aquatic chemical pollutants and the potential risk to life forms that this represents.[212,279,354] Shipping activities are also associated with increased marine litter. In the Mediterranean Sea, which is considered to be one of the most polluted regions in the world, plastic materials are considered to be one of the most abundant types of litter, with a large proportion of debris being derived from shipping activities in the region.[81]

The levels of plastic accumulation on the surface waters of some regions of the North American Great Lakes are on par with those witnessed in the oceanic gyres.[104] Studies have measured the average abundance of microplastics to be as high as 43,000 particles/km^2, while at some locations, the volume of plastic was an astounding 466,000 particles/km^2. Furthermore, a large proportion were very small spherical microbeads, of the type used in consumer products, such as face scrubs[116] (see Chapter 5).

Beached macroplastics

Floating plastic debris can be removed from the surface of the ocean by shore deposition, nanofragmentation, ingestion and sedimentation.[284] In a study which sampled four 100 metre sections of a North Sea Beach, four times annually, over a period of four years, it was determined that tourism and fishing were the greatest causes of plastic pollution in the region and that the vast majority of litter was a result of floating debris composed of polystyrene, other plastics and foam.[470] In an investigation of shore debris on Saint Brandon's Rock,[33] an isolated coral atoll within the Indian Ocean, it was found that 79% of all debris was composed of plastic, such as flip-flops and plastic bottles. It was hypothesised that many of the items originated from Southeast Asia, the Indian sub-continent and countries of the Arabian Sea. Similarly the most abundant type of marine debris present on the Argentinian coastline was plastic materials.[46] On the shorelines of the North American Great Lakes, it was found that over 80% of the debris deposited there was composed of plastic.[104] In shoreline cleanup operations of Lake Ontario, Canada between 2012 and 2014, almost half of all the debris observable on the shoreline was composed of plastic.[75] Furthermore, in 2012, between 77% and 90% of all observable debris present on the entire Canadian shoreline was comprised of plastic materials. Overall the relative amount of plastics, in comparison to other debris on shorelines, tends to fluctuate between approximately 62% and 92%[75,104,293] (Table 3.2).

Table 3.2 **The percentage of plastic in shoreline debris**

Location	Plastic content
Gulf of Oman	61.8%
Chile	65%
Brazil	69.8%
Australia	70.4%
Japan	72.9%
Canada	77–90%
North American Great Lakes	>80%
Salas & Gomez Island, South Pacific Ocean	87.6%
Baynes Sound, Canada	90%
Midway Atoll, North Pacific Ocean	91.9%

In terms of weight of plastic materials, busy areas of Georgian beaches and marches were found to contain 300–1000 kg of plastic debris, while more remote regions of the Georgian coast had an abundance of 180–500 kg. Furthermore, it was established that at a barrier beach, the additional accumulation of plastic at an 8 km^2 site was 0.18–1.16 kg every 30 days. Similarly, at an 8 km^2 marsh, the additional accumulation of plastic every 30 days was 0.6–1.61 kg.

Beach cleanups

Cleaning beaches of plastic is an effective way in reducing large amounts of accumulation, as well as acting as a preventive measure in avoiding plastic being washed back out to sea. For example, during annual beach cleanups in Baynes Sound (a channel between Denman Island and Vancouver Island, British Columbia, Canada) over a period of 10 years, it was reported that 3–4 tonnes of debris were collected by the local community, of which 90% was plastic (such as expanded polystyrene, bags, ropes and nets) and had been mainly attributed to the shellfish industry.[24] Community-based approaches like this which involves the removal of debris from shorelines by volunteers has been particularly effective in improving the aquatic environment. Other schemes, such as the Ocean Conservancy's International Coastal Cleanup and Beach Sweeps in the United States, are proving quite popular with local communities.

In a study[219] focussing on effective beach cleanup plans, (which on average, are usually conducted every 2 years) it was suggested that beach cleanups should be conducted on beaches where the average residence time of plastic litter on the beach is greater than period of time over which the plastic litter is deposited on the beach. Thus, the researchers suggest that regularly cleaning a beach where the residence time of the plastic on the beach is shorter than the average period of time over which the beach receives plastic influx would result in only a small effect on the beach environment. To establish residence times of plastic on beaches the researchers suggest utilising mark-recapture experiments.[220] While establishing the period of plastic influx is recommended to be ascertained by monitoring the beach using remote video cameras.

The influx of plastic litter typically exhibits variation as a result of seasonal weather patterns. Consequently, cleanup operations are generally more effective at the peak time of plastic influx. Ultimately, effective beach cleaning reduces the duration of leaching of toxic metal additives from plastic materials, such as lead and chromium, and reduces the potential for macroplastics to degrade into plasticles.

Submerged macroplastics

Somewhat counterintuitively, not all plastics will float on surface water and in reality, many will sink. Indeed, over 50% of all thermoplastics will sink in seawater.[296] Furthermore, in the case of small pieces of plastic, the rate at which they sink to deeper layers of the ocean, and thus their vertical distribution in the water column is significantly affected by their incorporation into marine aggregates[265] and the formation of biofilms. In a study[392] of litter at the bottom of the northern and central Adriatic Sea, plastic materials comprised the biggest percentage of materials recovered at 34%, while 36% of this amount was attributed to fishing gear. Furthermore, higher abundances of litter were found near coastlines. However, a study of the sea floor of the Wadden Sea in Germany observed a decrease in plastic pollution over the period of the study (1998–2007). Nevertheless, the researchers surmised that the action of waves and tidal currents resuspended the litter and transported it out of the region at a greater rate than the input of marine debris, which had decreased due to a decline in fishing activities.[470]

Sinking plastics can also descend into the depths of underwater canyons, which tend to act as pathways which channel submerged plastic litter into the open ocean.[81] Certainly, explorations of deep sea canyons[412] have unearthed vast accumulations of plastic litter there. For example, in an underwater study of La Fonera, Cap de Creus and Blanes canyons in the Mediterranean Sea, the canyons La Fonera and Cap de Creus were found to possess the greatest level of litter ever witnessed in the deep sea with 15,057 items/km² and 8090 items/km², respectively. The researchers even reported that on one occasion, up to 167,540 items/km² was observed and 72% of the litter was composed of plastic, whereas 17% comprised of lost fishing equipment. Furthermore, it was hypothesised that powerful currents resulting from severe storms in the coastal regions, as well as the seasonal movement of dense water down the walls of the canyons, and their close proximity to the shoreline amounts to an effectual transport mechanism for moving marine litter down to the deep sea.

Another possible region that plastic marine debris could accumulate, although less likely, is underwater sinkholes (blue holes). There are many well-known examples of blue holes, such as the Great Blue Hole of Belize, the Blue Hole of Dahab in the Red Sea, the Blue Hole of Gozo in the Mediterranean Sea, as well as both Watling's Blue Hole and the second deepest blue hole in the world at 202 metres (633 ft) deep, Dean's Blue Hole in the Bahamas. While there are currently no scientific studies of plastic litter at the bottom of blue holes, there are a few media reports of plastic litter at some of these sites, such as in the Blue Hole of Gozo, which is located in one of the most polluted seas in the world. Consequently, this could be an important area of research. For example, the world's deepest blue hole is Dragon Hole in Yongle, a coral reef located in the Xisha Islands, China at 300.89 metres (987.2 ft) deep, and with a 130-metre-wide entrance, it suffers from the restricted water exchange typically experienced in many blue holes. Consequently, a survey of Dragon Hole in 2016 by Chinese researchers found the bottom to be practically devoid of oxygen and thus, marine life. If dense plastic debris was to sink to the bottom of these regions, it could potentially lay there for thousands of years, if not indefinitely, trapped in an anoxic (oxygen-free) environment. Similarly, in some regions, such as Baltic Sea, anoxic basins occur in which bacteria are able to oxidise organic material faster than oxygen is able to be resupplied. Consequently, one would expect that in these anoxic underwater regions, plastic materials would remain relatively intact due to the lack of significant degradative or transport processes, such as sunlight, oxidation, mechanical damage, tidal currents and ingestion by organisms. Ultimately, considering that a large proportion of plastics sink, and that vast accumulations of plastic have been discovered hidden away in the deep sea, it may well be the case that submerged plastic debris exists in far greater abundance than floating debris.

Ultimately, submerged plastics can sink to deep sea and become incorporated into bottom sediments. Alternatively, they may exhibit near-neutral buoyancy as a result of their physiochemical properties, acquired biofilms or other accumulations of organic and inorganic material, thereby slowing their descent and facilitating movement around the water column. The cycling of oceanic currents and the ebb and flow of the tide can induce this movement and as a consequence, submerged plastics may be misidentified as food by aquatic species and subsequently ingested. For example, sea turtles may mistake polyethylene bags as jellyfish, a primary food source (Fig. 3.7).

Figure 3.7 Submerged plastic may be misidentified as food by aquatic species.

Macroplastic litter legislation

There is growing evidence that macroplastic litter, and more specifically marine litter, has a wide range of adverse environmental, economic, social and public health and safety impacts.[313] In polluted regions, such as the Mediterranean Sea and the Black Sea, many aquatic species have been noted to ingest plastic litter and suffer entanglement with debris.[81] For these reasons, it has become increasingly necessary to legislate against this issue. Specifically, marine litter is defined as 'any persistent, manufactured or processed solid material discarded, disposed of or abandoned in the marine and coastal environment'.[416,417] Comprehensive reviews of the policies and legislation relating to plastic litter, and specifically on marine litter, have been undertaken[56,254] and outlined in detail in the specific legislation currently in place at an international, regional and nation level. Additionally, there have been calls to standardise the way in which plastic debris is quantified, sampled and reported.[249]

International

There are a number of instruments used to tackle marine litter including conventions, agreements, regulations, strategies, action plans, programmes and guidelines.[56] At an international level the United Nations Convention on the Law of the Sea (UNCLOS) covers all aspects (environmental, scientific, economic, commercial and settlement) relating to ocean matters. More specifically the United Nations Environment Programme (UNEP) Global Programme of Action (GPA) for the protection of the marine environment from land based activities, and the UNEP Regional Seas Programme established a global initiative on marine litter in 2003. The UNEP and US National Oceanic and Atmospheric Administration (NOAA) formulated the Honolulu strategy as a global framework on possible actions to combat marine litter.[419] In 2012 the UNEP established the Global Partnership for Marine Litter (GPML) to act as a coordinating forum to build upon the Honolulu strategy. The UNESCO Intergovernmental Oceanic Commission and more recently the Joint Group of Experts on the Scientific Aspects of Marine Environmental Protection (GESAMP) have been active in the area of marine litter.[156]

Other international agreements include the International Convention for the Prevention of Pollution from Ships or MARPOL (73/78) Annex V under the International Maritime Organisation banning the discharge of rubbish from ships and the London Protocol (or Convention) on the prevention of marine pollution by the dumping of wastes and other matter at sea, which came into force in 2006. Guidelines on the surveying and monitoring of marine litter [Intergovernmental Oceanographic Commission (UNEP/IOC)] and on abandoned, lost or otherwise discarded fishing gear [Food Agriculture Organisation (UNEP/FAO)] have also been established to help reduce marine litter.

Regional

Within the European Union, there are numerous initiatives and directives relating to marine pollution and debris. The most important and relevant of these is European Directive 2008/56/EC or the Marine Strategy Framework Directive (MSFD).[119] The MSFD is an integrated policy for the protection of the European marine environment, with the aim of reaching good environmental status by 2020.

Other relevant EU directives include the Port Reception Facility (PRF) for the reduction of ship generated waste and cargo residues to sea (2000/59/EC) and the Water Framework Directive (WFD) (2000/60/EC) which aims for all aquatic ecosystems and wetlands in the EU to achieve good chemical and ecological status. The WFD covers transitional waters (estuaries) and coastal water up to a distance of one nautical mile and consequently, will impact upon marine litter within the European coastal marine environment. Less specific EU directives are the waste framework directive (2008/98/EC), packaging and packaging waste (2004/12/EC), landfill of waste (1993/31/EC), urban waste water directive (91/271/EC), bathing water directive (2006/7/EC), the EU fisheries policy and the integrated Maritime Policy.

Other regional actions include legislation from the Convention for the Protection of the Marine Environment of the North-East Atlantic (OSPAR) who have developed a system of Ecological Quality Objectives (EcoQOs) for the North Sea. Following a monitoring programme,[427] one EcoQO in particular concerns the presence of plastics in the stomach content of northern fulmars and was undertaken by the Institute for Marine resources and Ecosystem studies (IMARES). OSPAR have also developed guidelines for monitoring marine litter on beaches[324] and for the implementation of the Fishing For Litter (FFL) project.[325] In the Mediterranean, the UNEP Barcelona Convention for the protection of the Mediterranean Sea against pollution developed a Marine Action Plan (MAP) legal framework to address aspects of conservation, including the elimination of pollution from land based sources. In the Baltic, the Helsinki Convention on the Protection of the Marine Environment in the Baltic Sea Area (HELCOM) is applicable, specifically the Regional Action Plan for Marine Litter in the Baltic Sea, which was adopted in 2015,[184] and the Baltic strategy on Port Reception Facilities for ship-generated waste (Annex IV, regulations 4 and 6 of MARPOL 73/78). In Antarctica, the Commission for the Conservation of Antarctic Marine Resources (CCAMLR) has initiated the Marine Debris Program to reduce the amount of debris entering the marine system and mitigate its impacts.

National

There are numerous US national instruments[56] for the regulation of marine litter, including the Marine Plastic Pollution Research and Control ACT (MPPRCA), the Marine Debris Program (MDP), the National Marine Debris Monitoring Program (NMDMP), the Shore Protection Act and the Beaches Environmental Assessment and Coastal Health Act. Examples of other more general National legislation that impact on marine litter include the UK Environment Act and the Merchant Shipping Regulations, the Environmental Code in Sweden, the Canadian Environmental Protection Act and the Ocean Policy in Australia.

Plastics recycling

Plastics are generally classified into two main categories; thermosetting plastics, such as polyimides and Bakelite,[180] and thermoplastics, such as polypropylene and polyethylene.[175] The vast majority of thermosetting plastics are permanently set and cannot be melted and reformed.[101] This irreversible chemical change prevents thermosetting plastics from being recycled. On the contrary, thermoplastics, which are far more abundant, can be heated until melted and then cooled and formed into new shapes, thereby allowing many of them to be recycled.[461]

The thermoplastic polyethylene terephthalate (PET) is regarded as the most widely recycled plastic. Certainly, by 2014 over 200 recycling processes had been registered with the European Food Safety Authority and the United States Food and Drug Administration and 75–80% of them are related to the recycling of PET. This bias towards PET recycling occurred as a result of the materials widespread use in food and drink packaging. Consequently, rising concerns regarding the disposal of these packaging items, and their impact on the environment, spurred the development of many PET recycling processes. Incidentally, the recovery rate and subsequent recycling of plastic waste produced by plastic manufacturers is practically 100%.

The recycling of other plastic materials is more challenging. In the case of multilayer and composite materials, recycling of the plastic components can be prohibitively expensive. Furthermore, composite materials may contain thermoset plastics, long fibres and mixtures of many other materials. Additionally, polyolefins, such as polypropylene, have a tendency to succumb to oxidation. Nevertheless, the addition of additives can be utilised to combat this.[462] Ultimately, there are limits on the number of times plastic materials can be recycled before they inadvertently lose their properties and require to be repurposed, discarded or incinerated. Indeed, many recycling plants struggle financially. The main reason is that the collection and separation of plastics is particularly time consuming and expensive. Furthermore, cleaning and efficient sorting of waste plastic is not perfect and inevitably unwanted contaminants can be introduced during the manufacture of recycled plastics, such as from the processing of electrical and electronic equipment, degradation products, residues of chemicals and food components adsorbed to the plastic.

The possibility of the presence of unwanted contaminants may discourage the use of recycled products by food and drink packaging manufacturers. For example, plastic containers that hold carbonated beverages are considered to be pressurised vessels. If a contaminant was introduced during recycling, this could result in failure of the vessel's structural integrity during use. This would be deemed as unacceptable by a packaging manufacturer. Thus, if the safety or durability of the recycled material used in a packaging product is in question, the manufacturer would likely reject the material. As a consequence of these concerns the majority of recycled plastic is utilised in low-grade applications, such as in the production of fleeces, carpets, fillings for pillows and outdoor furniture. To strengthen recycled plastics, techniques have been developed which involve adding fibres or nanomaterials. However, the extent as to how these additives affect further recycling of the materials at the end of their service life has not been fully evaluated thus far. Another issue with unwanted contaminants is that they may be toxic. There are a number of studies that have identified the presence of toxic brominated flame retardants, phthalates and heavy metals within some recycled plastics.[57,248,370,452] Nevertheless, recycling processes are rapidly evolving and the introduction of more rigorous testing is beginning to counteract these issues.

Ultimately, to reduce the vast amounts of plastics entering the aquatic environment, recycling of plastic waste is a crucial step. However, the recycling of plastic materials is often regarded as uneconomical since the recovered material tends to exhibit low intrinsic value. Thus, without the intervention of legislative bodies, it is unlikely that the commercial attitudes towards the recycling of many plastic materials will substantially change in the near future. At present, many plastics are shipped to highly populated developing countries for recycling as this is seen to be a more cost effective route. Furthermore, with the growing production and development of composite materials, the identification of high value waste materials of good quality may be the first step in developing a cost effective recycling route for these materials. Indeed, glass and carbon fibre have considerable scrap value and may be of high value in certain niche markets. There is however promise on the horizon as conventional recycling processes evolve to become more efficient and cost effective. For example, there have been industry proposals to introduce bar codes imprinted onto the distinct types of plastic comprising multi-component artefacts to aid in identification and separation of individual plastic materials.

A conventional plastic recycling process typically involves pre-sorting, which removes metal, glass, paper and other non-plastic items. This may be done manually or by automation. The next stage involves a spectroscopic and optical analysis of each pre-selected item to determine its chemical composition and colour. The selected plastics are then crushed into flakes and washed with chemical cleaning agents. The flakes are then separated by density and dried. A second optical analysis is performed to separate the flakes by colour. An extruder then processes the flakes into granules which are then melted down and reformed into new artefacts.

Many governments have introduced recycling programmes for the collection of waste from consumers. Consequently, to aid consumers in sorting the materials for collection, colour coding system are typically utilised. Each colour indicates the type of material that should be deposited in the recycling bin. Although a universal international colour code system does not yet exist, many such programmes follow the colours indicted in Fig. 3.8. In this case a yellow recycling bin is used for plastic materials.

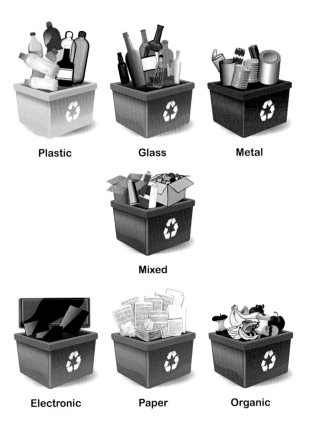

Figure 3.8 The most widely used recycling colour code system.

To identify specific types of plastics from one another, most plastic items, especially plastics used in food, drink and product packaging, display an internationally recognised code which identifies the type of plastic that the item is made from. The current coding system is administered by The American Society for Testing and Materials (ASTM).

The ASTM International Resin Identification Coding System

In 1998 the Society of Plastics Industry (SPI) issued the most common plastics (commodity plastics) with a designator code to aid reprocessing efforts, thereby enabling plastics to be easily identified and separated.[176] However, in 2008, ASTM International took over administration of the codes. In 2013, ASTM International decided to revise the symbols, moving away from the familiar three mutually chasing arrows, and replacing them with solid equilateral triangles as part of the newly revised ASTM D7611 standard. The reasoning behind the decision was that the original symbols closely resembled the universal recycling symbol. This was a source of confusion since many facilities and programmes at that time only accepted plastics for recycling with certain codes and refused other types of plastic, despite the mutually chasing arrows (Fig. 3.9). Consequently, consumers were confused as to why the plastics were refused despite having a recycling-like symbol. As such, ASTM International wished to ensure that the symbols only identified the type of plastic, as opposed to their ability to be recycled. Thus, to maintain effective and reliable use of the resin identification coding system across the entire stake holder community, the solid equilateral triangle system was created.

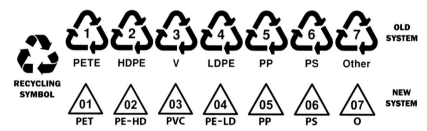

Figure 3.9 Comparison of the original Society of Plastics Industry system and new American Society for Testing and Materials system for plastic identification.

Applications of plastics

As a result of their versatile nature, plastics are used in many industries and applications. However, the main sectors that plastics are most commonly used in are packaging, electrical goods and electronics, building and construction, terrestrial, air and marine transportation, agriculture, medical and health applications, and sport and leisure (Table 3.3).

Table 3.3 **Applications of plastics**

American Society for Testing and Materials designator code	Plastic	Abbreviation	Typical application
01 PET	Polyethylene terephthalate	PET	Beverage bottles Food containers Film and sheeting Strapping Fibres Fleece Filling material
02 PE-HD	High-density polyethylene	HDPE	Chemical containers Beverage bottles Food containers Crates Buckets Tubing Pipes
03 PVC	Polyvinyl chloride (rigid)	PVC	Containers Electrical conduit Pipes Gutters Cladding Window frames
03 PVC	Polyvinyl chloride (plasticised)	PVC	Cable insulation Garden hose Footwear Sheeting Flooring Mats
04 PE-LD	Low-density polyethylene	LDPE	Bags Squeezable bottles Food packaging Carton linings Shrink wrap Pallet sheets Mulch film Waste bins Outdoor furniture

Table 3.3 **Applications of plastics—cont'd**

American Society for Testing and Materials designator code	Plastic	Abbreviation	Typical application
▲ 05 PP	Polypropylene	PP	Bottle caps and container lids Packaging tape Pipes Rope Automotive Applications Outdoor furniture
▲ 06 PS	Polystyrene (rigid)	PS	Disposable cutlery and tableware Rigid disposable food containers Other low-cost applications requiring rigidity
▲ 06 PS	Polystyrene foams (expanded or extruded)	EPS, XPS	Disposable food containers and cups Packaging foam Foam boxes Thermal insulation
▲ 07 O	Polycarbonate	PC	Information storage discs Traffic light lenses Riot shields Security windows Eye protection
▲ 07 O	Poly(methyl methacrylate)	PMA	Glazing Aircraft canopies and windows Fluorescent light diffusers Vehicle lighting covers and lenses Contact lenses

Continued

Table 3.3 **Applications of plastics—cont'd**

American Society for Testing and Materials designator code	Plastic	Abbreviation	Typical application
△ 07 O	Polytetrafluoroethylene	PTFE	Non-stick coatings for cookware and bakeware Breathable fabrics Plumbing tape Waterslides Low-friction coatings Automotive applications
△ 07 O	Acrylonitrile butadiene styrene	ABS	Electrical and electronic equipment Automotive applications Pipes
△ 07 O	Polyamide (nylon)	PA	Fibres Bristles Monofilament Bearings Automotive applications
△ 07 O	Polychloroprene (neoprene)	CR	Gloves Footwear Diving garments Medical bandages and supports Automotive gaskets Electrical insulation Weather balloons

The service environment in which a plastic material is suited, is to a large extent, determined by the physiochemical properties of the plastic. Other considerations, such as aesthetics and tactility, are of a secondary nature in the selection of a suitable material. For example, the brittle nature of polystyrene makes it unsuitable for applications requiring high tensile strength, while with polypropylene, poor impact resistance is exhibited at low temperatures. Ultimately, each plastic has its own distinct characteristics and these unique properties will dictate the persistence of the material in the aquatic environment and the rate, and mechanisms, by which it may potentially degrade.

Physiochemical properties and degradation

4

The structure of plastics

In the contemporary era, plastics have replaced many traditional materials, such as wood, glass and metal, and become an irreplaceable part of our modern existence (see Chapters 1 and 2). The main reason behind the success of plastic materials can be attributed to their properties. For example, many plastics offer performance characteristics that are simply not attainable with more traditional materials while the cost associated with their production can be orders of magnitude less. At the same time, the properties of plastic materials allow them to be moulded into specific single piece parts thereby reducing manufacturing time and facilitating mass production of identical artefacts. However, the very properties that confer plastics as highly desirable manufacturing materials, such as strength and durability, are the very same properties which hamper their degradation in the aquatic environment. Ultimately, all plastics are complex chemical structures composed of atoms. The way in which these atoms are configured with respect to one another has a direct bearing on the individual properties of a plastic. Properties such as rigidity, toughness and the glass transition temperature are directly attributed to the arrangements of the atoms that comprise a plastic material. Furthermore, the degree of structural order of the atoms and molecules in the material (crystallinity) will have a significant influence on the materials properties, such as density, transparency and hardness. Additionally, the types of bonds between the atoms will have a bearing on the materials resistance to heat and ultraviolet light. Consequently, these different properties directly determine the persistence and degradation of plastic materials in the aquatic environment. Thus, to appreciate the ways in which plastics degrade in this medium, it is necessary to examine the properties of plastic materials in more detail.

Polymerisation

As discussed in Chapter 1, plastics are a composed of long chain-like molecules (macromolecules), which are in turn, composed of many repeating small molecules (monomers) linked together in a sequence. While the molecular topology is not always completely linear in fashion, it is generally the case that the process of polymerisation imparts a large molecule with chain-like segments and character. There are primarily two types of polymer: chain-growth polymers and step-growth polymers.

Chain-growth (addition polymers)

In the case of chain-growth, the individual monomers are added to one another in sequence to form a long chain. A typical example of chain-growth polymerisation is free-radical polymerisation. In many plastics, alkene monomers are used and their

Microplastic Pollutants. http://dx.doi.org/10.1016/B978-0-12-809406-8.00004-9

polymerisation is initiated with heat and a catalyst, such as organic peroxide. Upon heating, the weak O—O peroxide bond breaks to form two radicals which add to the C=C bonds of the monomer, thereby creating a reactive intermediate (carbon free radical). This intermediate then reacts with the C=C of another monomer, linking the two monomers together. This continues until a long chain is formed and the reaction terminates in some fashion, such as two radicals reacting with one another. Importantly, each atom of each individual monomer that forms the chain is retained. Two typical examples of chain-growth polymers are polyethylene and polypropylene.

Step growth (condensation polymers)

Unlike chain-growth, which retains every atom of the constituent monomers, step-growth results in the loss of atoms in the form of a small molecule, such as water. Step-growth typically involves a chemical reaction between two different groups of specific atoms (functional groups). Thus, a functional group on one monomer reacts with the functional group on another, thereby linking them together and eliminating a small molecule in the process. A typical example of a step-growth polymer is polyamide (Nylon) (Fig. 4.1).

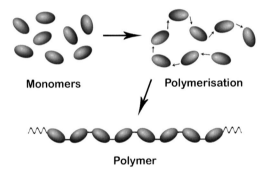

Figure 4.1 Small molecules (monomers) connect together in a repeating sequence (polymerise) to form a large chain-like molecule (polymer).

Stereochemistry

Stereochemistry (also known as 3D chemistry) concerns the three dimensional spatial arrangement of atoms in a molecular structure and the resulting effects of these different arrangements on the reactivity and properties of the molecule. For example, a common plastic monomer is ethenyl (also known as vinyl) (Fig. 4.2) and is the basic building block of polyvinyl chloride (PVC) and polypropylene (PP).

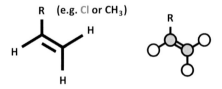

Figure 4.2 The ethynyl monomer (the R represents any functional group).

The R position in ethenyl can be substituted with different functional groups, such as chlorine (Cl) in the case of PVC. When ethyl is substituted with a single methyl functional group (CH_3) and then linked together with other ethenyl monomers to form a long chain (polymerised), polypropylene is formed. This long chain of carbon atoms is considered to be the polymer backbone and is the simplest representation of the molecule to which the functional groups are attached (pendant groups). When a carbon atom is attached to four different types of atoms, or functional groups, it is termed as an asymmetric carbon (also known as a chiral carbon). Each asymmetric carbon in polypropylene is considered to be a stereogenic centre (chiral centre) because the carbon atoms are characterised as having the methyl groups attached to them in such a manner that their mirror image is non-superimposable (see Fig. 4.3).

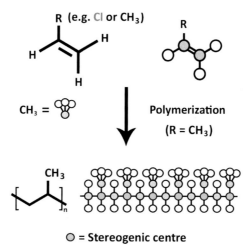

Figure 4.3 When ethenyl is substituted with a methyl group and then polymerised to polypropylene, every second carbon is a stereogenic centre.

These asymmetric carbons are capable of arranging the methyl groups in different configurations (tacticity), thereby giving rise to different forms of polypropylene. Thus, every second carbon is bound to a methyl functional group in such a way that its mirror image is not superimposable and thus lacks a plane of symmetry (asymmetric). Consider an observer looking along the length of the polypropylene chain, where A is the end closest to the observer and B is the end furthest away (see Fig. 4.4).

One conformational form is the mirror image of the other form and each form is termed as either left-handed (levorotary) or right-handed (dextrorotary). If the left form is superimposed over the right form, it becomes apparent that the mirror images are non-superimposable (Fig. 4.5).

If during polymerisation, these different conformations of the methyl groups have occurred in a regular arrangement along the polymer backbone, then the polymer chain is said to be stereoregular. However, if the methyl groups on the asymmetric carbons are in random orientations with respect to one another, and thus exhibit no

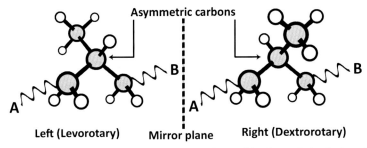

Figure 4.4 Since the two ends A and B are non-superimposable, the methyl-substituted carbon atom in polypropylene is asymmetric, having both a left-hand and a right-hand orientation.

Figure 4.5 The left-handed form is non-superimposable over the right-handed form because the asymmetric carbons are orientated in such a manner that the substituent methyl group gives rise to the asymmetry.

regularity, then the polymer chain is said to be stereorandom. In polypropylene, there are three distinct stereochemical arrangements[218];

1. Isotactic: All the asymmetric carbon atoms are configured identically in a regular arrangement (stereoregular) (Fig. 4.6).

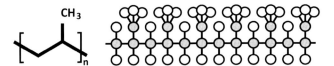

Figure 4.6 Isotactic polypropylene.

2. Syndiotactic: All the asymmetric carbon atoms regularly alternate in configuration (stereoregular) (Fig. 4.7).

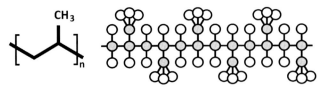

Figure 4.7 Syndiotactic polypropylene.

3. Atactic: None of the asymmetric carbon atoms exhibit any form of regularity and thus have random configurations (stereorandom) (Fig. 4.8).

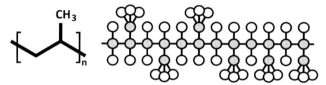

Figure 4.8 Atactic polypropylene.

In commercially produced polypropylene, it is typical that the polymer chain will exhibit some degree of different tactic forms at different sections of the chain (blocks). Depending upon the degree to which these different stereochemical arrangements are present, directly affects the physical properties of the polymer, despite the fact that it was polymerised from the same monomer. For example, the free radical polymerisation of propylene typically results in atactic polypropylene. However, the development of the Ziegler-Natta mixed organometallic catalysts in the 1950s revolutionised the manufacture of polypropylene (see Chapter 2). By means of coordination catalysis, it was possible for the first time to produce atactic polypropylene with blocks of isotactic and syndiotactic polypropylene in the chain, resulting in greatly improved properties over atactic polypropylene. Parenthetically, alterations to the conditions in which poly(methyl methacrylate) is polymerised can also impart a configuration which is either isotactic or syndiotactic. Ultimately, commercially available polypropylene is typically a mixture of tactic forms in the ratio 75% isotactic to 25% atactic. This mixture of tactic forms imparts a material with a high molecular weight and excellent tensile strength and stiffness, although impact strength is only moderate. However, manufacturers can alter the ratio to obtain the desired properties (Table 4.1).

Table 4.1 **The three tactic forms of polypropylene have differing properties**

Polypropylene tactic form	Properties
Isotactic	A highly crystalline rigid material with a high melting point of 160–166°C
Syndiotactic	A semi-crystalline substance with elastomeric properties. 30% crystallinity imparts a melting point of 130°C
Atactic	An amorphous and tacky rubber-like material with no specific melting point

Crystallinity

The crystallinity of a polymer refers to the degree as to which there are regions where the polymer chains are aligned with one another. However, in order for this to occur, some degree of stereoregularity is required. This is because the crystalline regions are formed from the stereoregular blocks in the polymer chain. When a polymer is in a melted state, the long polymer chains are entangled with one another in irregular coil-like structures. When the melted polymer is cooled to a solid, the chains of some polymers remain in this disordered tangled arrangement due to a large degree of irregularity in their polymer chains (stereorandom). Such polymers are termed as amorphous polymers.

Alternatively, in other polymers which contain chains with stereoregular portions, distinct ordered crystalline regions are formed (polymer crystallites) where, depending on the plastic, 10–80% of the chains fold and align themselves with one another (crystallisation) at intervals of approximately 10 nm to form structures termed lamellae. These are in turn, part of larger spherical structures termed spherulites. Since there are still some atactic (stereorandom) blocks in the chains, complete crystallisation does not occur. Furthermore, if there are any regions of the chain where branching has occurred, there will be a loss of stereoregularity and thus crystallisation is inhibited. As some fraction of the polymer remains un-crystallised, it is termed as a semi-crystalline polymer. Since all plastics are polymers, there are two main types of plastic materials: amorphous plastics, such as acrylonitrile butadiene styrene (ABS), polystyrene and polyvinyl chloride (PVC); and semi-crystalline plastics, such as the high temperature engineering plastic polyether ether ketone (PEEK), polyethylene terephthalate and polytetrafluoroethylene (Fig. 4.9).

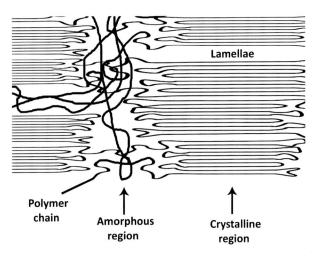

Figure 4.9 Amorphous regions in plastic materials have disordered entangled chains while semi-crystalline plastics are interspersed with ordered crystalline regions in which the chains have aligned.

The degree of crystallinity (amount of structurally ordered regions) in a plastic has a direct bearing on the materials mechanical, optical, chemical and thermal properties. For example, when an amorphous plastic is at room temperature (21°C), the polymer chains are in a highly disordered and tangled state. Consequently, the movement of the polymer is hindered and it exhibits rigidity and brittleness. However, when an amorphous plastic is heated, molecular motion is induced and the plastic gradually becomes softer and flexible as the temperature rises without exhibiting a definitive melting point. Conversely, a semi-crystalline polymer can withstand a given amount of heat without undergoing softening and flexing. However, once the melting point is reached, semi-crystalline plastics tend to rapidly undergo a phase transition in which they change state from a solid to a low viscosity liquid (Table 4.2).

Table 4.2 General properties of amorphous polymers versus semi-crystalline polymers

Amorphous	Semi-crystalline
High melt viscosity	Low melt viscosity
No distinct melting point	Distinct melting point
Low strength	High strength
Poor fatigue and wear resistance	High fatigue and wear resistance
Poor resistance to chemicals	High resistance to chemicals
Transparent	Translucent
Molecular orientation in molten phase: random	Molecular orientation in molten phase: random
Molecular orientation in solid phase: random	Molecular orientation in solid phase: presence of crystalline regions (crystallites)

Branching

Plastics are available in a multitude of varieties with properties suited to almost any service environment. Some properties of the material arise from what is known as the polymer architecture. This refers to the way in which branching of a polymer chain results in a deviation away from a completely linear structure. The physical properties of a plastic material, such as the glass transition temperature and melt viscosity, are affected by the extent of branching and the way in which this branching occurs (see Fig. 4.10).

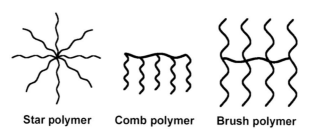

Star polymer   Comb polymer   Brush polymer

Figure 4.10 Examples of homopolymer branching.

For example, low-density polyethylene (LDPE) was often referred to as 'branched-polyethylene' due to the large number of chains stemming from the main polymer chain[153]. These branches reduce the compressibility of the polymer thereby providing larger gaps between the chains[329]. The gaps have the effect of reducing the density and mass of the material while increasing the flexibility of the polymer[329]. A slightly stronger and more rigid version of LDPE, identified as linear-LDPE, has many short branches connected to the main chain but no large branches (see Fig. 4.11). This variation permits tighter chain packing, and thus an increase in density and strength while still allowing a fair degree of flexibility[175].

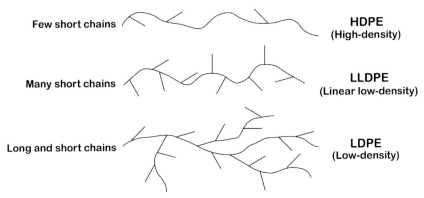

Figure 4.11 The effect of branching on density in polyethylene.

In contrast to LDPE, high-density polyethylene (HDPE) is composed of long linear chains with very few branches. This allows the chains to be packed very close together thereby significantly increasing the density, mass, strength and rigidity of the material[153,329].

Copolymers

In the case of homopolymers, there is just a single type of monomer which is polymerised to form the polymer chain. However, by utilising two or more different types of monomer, more complex structures can be formed with improved properties. There are four main types of copolymer; Alternating, Random, Block and Graft (see Fig. 4.12).

In the case of alternating, the chain is simply composed of two species of monomers in an alternating sequence. In the case of random, there is no regularity to the alternating sequence of different monomers in the chain. With block copolymers, a large portion of the chain will be a distinct type of monomer, followed by a large distinct portion of another monomer. A graft copolymer is when one or more blocks of a different monomer are connected onto the main chain as branches by taking advantage of the presence of functional sites on the main chain, such as double bonds. Block and graft polymers can be constructed in a variety of complex arrangements (see Fig. 4.13).

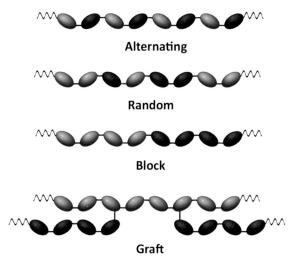

Figure 4.12 Four main types of copolymer.

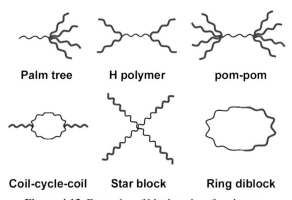

Figure 4.13 Examples of block and graft polymers.

Polymer blends

In the past, there was considerable emphasis on developing completely new types of plastic (see Chapter 1). However, over time the emphasis shifted towards producing novel combinations of different types of plastic to give new materials with the advantageous properties of both original materials (see Chapter 2). These types of plastics are known as polymer blends. For example, a polymer with soft elastic properties could be blended with a polymer which has brittle glass-like properties. The resultant blend would exhibit improved toughness and rigidity compared to the original brittle glassy material and the soft elastic material. One such is example of a common polymer blend is polypropylene and propylene-ethylene copolymers. At temperatures below 0°C, the polypropylene homopolymer tends to become particularly brittle and thus prone to fracture. However, by blending the material with propylene-ethylene copolymers, a material with low temperature resistance, as well as improved tensile strength and toughness, is produced (Table 4.3).

Table 4.3 Examples of plastic blends and their properties

Blend	Properties
PA (nylon)/elastomer	Good impact resistance
ABS/PC	Good heat and impact resistance
ABS/PVC	Good flame and impact resistance
PMA/PVC	Good flame and chemical resistance

Commodity plastics

In applications where mechanical properties and operating environment are not critical, plastics with wide-ranging applications, termed commodity plastics, are considered to be of the greatest commercial importance. Consequently, these common plastics have an exceptionally high rate of production and as a result of their worldwide abundance, commodity plastics are the most widespread plastics found littering the aquatic environment. In the preceding sections, the structures of individual commodity plastics are listed, as well as a description of their properties. As discussed in Chapter 3, individual commodity plastics have been allocated specific designator codes by the international standards organisation ASTM. These codes enable these plastic materials to be easily identified and separated.[176] While they can aid with recycling efforts, it is important to note that they are not an indication of materials recyclability.

Polyethylene terephthalate (PET)

Polyethylene terephthalate (PET) is a chemically stable polyester and its use has risen dramatically in the last few decades with a multitude of applications, ranging from food and drink containers to the manufacture of electronic components[176,196] and as fibres in clothes. Often, recycled PET bottles are used to make fleece garments, as well as plastic bottles. Indeed, one of the most common uses of PET is in the manufacture of drinking water bottles and these are regularly found littering the aquatic environment (see Chapter 3) (Fig. 4.14).

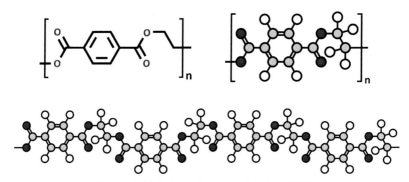

Figure 4.14 Polyethylene terephthalate (PET).

Chemical formula	$(C_{10}H_8O_4)_n$
Abbreviation	PET or PETE
Classification	Thermoplastic
Molecular orientation in solid phase	Semi-crystalline
Monomers	Terephthalic acid, Ethylene glycol
General properties	Transparency to visible light and microwaves. Very good resistance to ageing, wear and heat. Lightweight, impact and shatter resistant. Good gas and moisture barrier properties

Polyethylene (PE)

PE-HD

PE-LD

As the plastic with the greatest worldwide demand, polyethylene is inevitably the most abundant plastic found in the aquatic environment. With a density typically lower than that of water, as well as good water repellent characteristics, polyethylene is typically found floating on surface waters. The degree to which the material exhibits tensile and impact strength, as well as stiffness, is determined by the degree of crystallinity of the polymer matrix (Fig. 4.15).

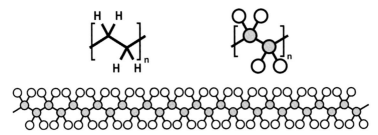

Figure 4.15 Polyethylene (PE).

Chemical formula	$(C_2H_4)_n$
Abbreviation	PE, LDPE, LLDPE, HDPE
Classification	Thermoplastic
Molecular orientation in solid phase	Semi-crystalline
Monomer	Ethylene
General properties	Depending on the degree of crystallinity, PE can be flexible (low-density) to rigid (high-density). When stabilised, PE exhibits good weathering resistance and good chemical resistance. Water repellent

Polyvinyl chloride (PVC)

Without the addition of additives, pure polyvinyl chloride (PVC) is a brittle white-coloured substance. During manufacture, PVC is produced in two varieties; rigid and flexible. The rigid type is a particularly durable hard material which exhibits fire retardant properties as well as demonstrating resistance to chemical degradation[175] and weathering, although tends to have low impact resistance at room temperature (21°C). Any weathering to the material tends to only occur on the materials surface.[85] The plasticised flexible form of PVC is created by the addition of phthalate plasticisers during manufacture to soften the polymer. When PVC is compounded with wood flour fillers, the resulting formulation may be susceptible to biological attack.[408] Consequently, antimicrobial additives may be added were necessary to combat biodegradative effects (Fig. 4.16).

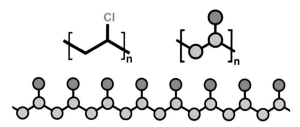

Figure 4.16 Polyvinyl chloride (PVC).

Chemical formula	$(C_2H_3Cl)_n$
Abbreviation	PVC
Classification	Thermoplastic
Molecular orientation in solid phase	Amorphous
Monomer	Vinyl chloride
General properties	The presence of chlorine provides fire retarding properties and imparts an ignition temperature of up to 455°C, as well as protecting against oxidation. Resistance to acids, alkalis and most inorganic chemicals. Considerable versatility and scope for adjustment of the materials properties using additives to modify flexibility and impact resistance. Easy to blend with other plastics

Polypropylene (PP)

PP

Polypropylene is regarded as one of the lightest and most versatile polymers. It is capable of undergoing a wide array of manufactured processes, such as injection moulding, general purpose extrusion, extrusion blow moulding and even expansion moulding. Although existing in three different tactic forms, commercial polypropylene is typically a mixture of 75% isotactic and 25% atactic. Demand for polypropylene is rapidly on the increase[19] and consequently, it is one of the most common types of microplastic found in the marine environment[355] (Fig. 4.17).

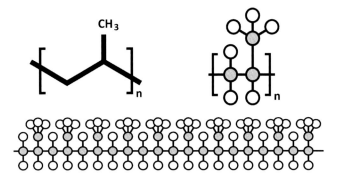

Figure 4.17 Polypropylene (PP).

Chemical formula	$(C_3H_6)_n$
Abbreviation	PP
Classification	Thermoplastic
Molecular orientation in solid phase	Semi-crystalline (isotactic, syndiotactic) Amorphous (atactic)
Monomer	Propylene
General properties	Resistant to many acids, alkalis and solvents. Low density, high stiffness, good balance of impact strength vs. rigidity, heat resistant and exhibits good transparency

Polystyrene (PS)

Representing a particular danger to marine biota, due to the possibility of the residual presence of the constituent oncogenic monomer styrene[260], polystyrene is typically found in the marine environment in either solid form or the foamed variety[356]. Upon heating, polystyrene tends to depolymerise to styrene. A more rigid form of polystyrene can be produced by the addition of p-divinylbenzene during manufacture. This results in crosslinking and produces the copolymer poly(styrene-co-divinylbenzene), which exhibits increased resistance to organic solvents. This crosslinked polystyrene is typically used to produce ion exchange resins, in the form of spherical beads 200–1200 μm in size. Polystyrene can also be mixed with a volatile solvent, typically 5% pentane by weight. When this mixture is heated, the pentane expands and bubbles to produce low-density polystyrene foam, termed Styrofoam. Alternatively, small spherical beads of polystyrene can be expanded with pentane to around 40 times their original size to produce a material which is composed of low-density spherical beads of polystyrene 0.5–1.0 mm in size, termed Styropor. Foamed, or expanded, varieties of polystyrene are often found floating on the surface of aquatic environments while solid polystyrene, being slightly denser than that of water, is generally found below the surface (Fig. 4.18).

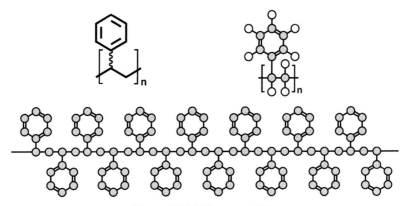

Figure 4.18 Polystyrene (PS).

Chemical formula	$(C_8H_8)_n$
Abbreviation	PS
Classification	Thermoplastic
Molecular orientation in solid phase	Amorphous
Monomer	Styrene
General properties	Rigid varieties are transparent, hard and brittle. Unfilled varieties have a sparkling crystal-like appearance. High impact polystyrene (HIPS) is produced by blending with a butadiene copolymer or rubber, thereby increasing impact resistance and toughness. Poor barrier to water and oxygen

Polycarbonate (PC)

Polycarbonate is a thermoplastic which contains organic functional groups connected together by carbonate groups. Due to the long molecular chain, the material can be easily thermoformed. Often, polycarbonate is blended with other plastics, such as ABS and rubber (Fig. 4.19).

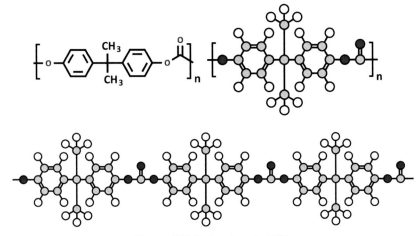

Figure 4.19 Polycarbonate (PC).

Chemical formula	$(C_{16}H_{18}O_5)_n$
Abbreviation	PC
Classification	Thermoplastic
Molecular orientation in solid phase	Amorphous
Monomer	Bisphenol A
General properties	Naturally transparent with light transmissibility similar to that of glass. Tough, strong, heat resistant and excellent dimensional stability

Poly(methyl methacrylate) (PMA) (Fig. 4.20)

Chemical formula	$C_5H_8O_2$
Abbreviation	PMA or PMMA
Classification	Thermoplastic
Molecular orientation in solid phase	Amorphous
Monomer	Methyl methacrylate
General properties	Extremely long service life, excellent light transmissibility and good resistance to ultraviolet light and weathering. Possesses the highest surface hardness of all thermoplastics

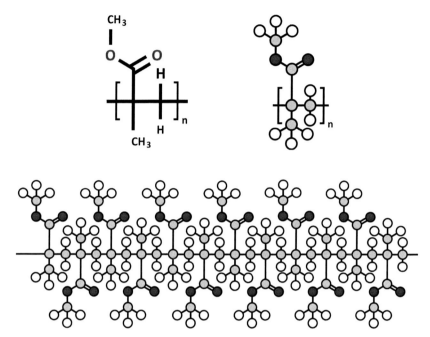

Figure 4.20 Poly(methyl methacrylate) (PMA).

Polytetrafluoroethylene (PTFE)

Polytetrafluoroethylene is a highly crystalline robust unique plastic with a high melting point. The material has a propensity to exhibit considerable elastic deformation under load. Consequently, in applications that require a frictionless surface and resistance to load, such as bearing surfaces, the deformation characteristics of PTFE are improved with the use of additives and fillers. PTFE has a high molecular weight and is generally an unreactive substance due to the highly stable nature of the fluorine and carbon bonds in its structure. Furthermore, the high electronegativity of the fluorine atoms gives the plastic an extremely water repellent surface and outstanding non-stick properties (Fig. 4.21).

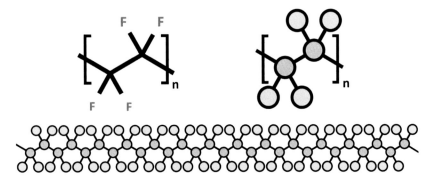

Figure 4.21 Polytetrafluoroethylene (PTFE).

Chemical formula	$(C_2F_4)_n$
Abbreviation	PTFE
Classification	Thermoplastic (fluoropolymer)
Molecular orientation in solid phase	Semi-crystalline
Monomer	Tetrafluoroethylene
General properties	White solid with high strength and toughness which possesses a waxy surface with a very low coefficient of friction. The material has outstanding chemical resistance and high resistance to weathering but exhibits high elastic deformation under load. Water absorption and flammability are low while thermal stability is high

Acrylonitrile butadiene styrene (ABS)

With a multitude of uses, acrylonitrile butadiene styrene (ABS) is the fifth most commonly used polymer after PE, PP, PS and PVC.[175] ABS is composed of a rigid matrix of polymerised acrylonitrile and styrene interspersed with particles of polybutadiene rubber. Different ratios of the three monomers produce different properties in the material.[175] Although ABS is regarded as an engineering plastic[85,463] and is a common plastic used in many electrical appliance cases,[175] the substance is prone to fire and smoking.[68] Consequently, ABS requires the addition of flame retardants to suppress this weakness[463] or alternatively, blending with PVC. In the environment, ABS is highly susceptible to weathering[175] and has a density greater than that of seawater,[84] tending to exist below the surface in aquatic environments (Fig. 4.22).

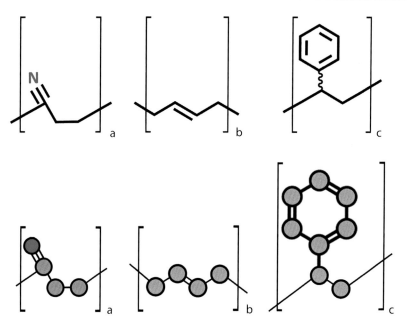

Figure 4.22 Acrylonitrile butadiene styrene (ABS).

Chemical formula	$(C_8H_8 \cdot C_4H_6 \cdot C_3H_3N)n$
Abbreviation	ABS
Classification	Thermoplastic
Molecular orientation in solid phase	Amorphous
Monomers	Acrylonitrile, butadiene, styrene
General properties	A high lustre opaque substance with a high surface quality. The high strength, hardness, impact resistance and rigidity of the material is a result of copolymerisation of styrene and acrylonitrile while the toughness is attributed to fine particles of polybutadiene rubber uniformly interspersed throughout the styrene-acrylonitrile copolymer matrix. The ratios of styrene, acrylonitrile and polybutadiene can be altered to produce different grades of ABS with different properties. Furthermore, ABS is often use in blends with other materials to impart toughness or with PVC to gain flame resistance

Polyamide (nylon)

Nylons are one of the most used widely used materials worldwide and are generally found in samples recovered from the aquatic environment as Nylon 6 or Nylon 6,6.[9] Synthesis of Nylons is achieved by way of a polymerising condensation reaction in which an amino group on one monomer reacts with the carboxyl group on another to create an amide bond and produce a polymeric translucent fibrous-like substance.[180] The structures of Nylon 6 and Nylon 6,6 are shown in Figs. 4.23 and 4.24, respectively. In the case of Nylon 6,6, the name derives from the fact that the two monomers of Nylon 6,6 (hexamethylenediamine and adipic acid) each contains six carbon atoms. Nylon fibres exhibit elasticity and have exceptional strength and toughness and are more robust than fibres of PET. Furthermore, Nylons have excellent wear resistance.

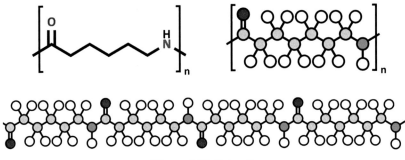

Figure 4.23 Nylon 6.

Figure 4.24 Nylon 6,6.

Chemical formula (nylon 6)	$(C_6H_{11}NO)_n$
Chemical formula (nylon 6,6)	$(C_{12}H_{22}N_2O_2)_n$
Abbreviation	PA
Classification	Thermoplastic
Molecular orientation in solid phase	Semi-crystalline
Monomers (Nylon 6)	Hexamethylenediamine, adipoyl chloride
Monomers (Nylon 6,6)	Hexamethylenediamine, adipic acid
General properties	Tough, high tensile strength and excellent abrasion resistance. However, prone to water absorption which decreases tensile strength

Polychloroprene (neoprene)

Although polychloroprene is regarded as a synthetic rubber and thus an elastomer as opposed to a plastic resin, the inclusion of the material here is important and shall be listed in the preceding sections. Owing to its low-density buoyant nature and widespread aquatic applications (wet suits/dry suits/flip flops), polychloroprene has the potential to contaminate aquatic environments. Furthermore, polychloroprene is often used as a material for weather balloons, which can subsequently descend into bodies of water and break apart over time.[323] When polychloroprene is cured, the tensile strength of the material is increased. However, both cured (vulcanized) and uncured polychloroprene exhibit a high degree of crystallinity due to the large extent of stereoregularity on the polymer backbone. Possessing a unique balance of properties, polychloroprene has good mechanical strength and ageing resistance as well a slow flammability as a result of the chlorine atoms attached to the polymer backbone (Fig. 4.25).

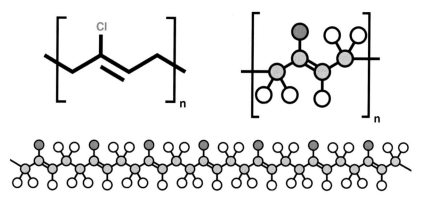

Figure 4.25 Polychloroprene (Neoprene).

Chemical formula	$(C_4H_5Cl)_n$
Abbreviation	CR
Classification	Elastomer
Molecular orientation in solid phase	Semi-crystalline
Monomer	Chloroprene
General properties	Used as a natural rubber substitute with good mechanical strength. Exhibits high temperature resistance. Paradoxically, thermal degradation results in hardening as opposed to melting. At temperatures below 0°C, the material starts to stiffen. Good weathering resistance and excellent water resistance. High elasticity and low flammability

Degradation

Once plastic waste enters the aquatic environment, various factors begin to have degradative effects on the material. The extent to which these have any effect, and the period of time over which they occur, is wholly dependent on the physiochemical properties of the material itself. Some plastics materials are inherently stronger than others and therefore are highly persistent in the aquatic environment. Consequently, the degradation of these materials may take many thousands of years. Other plastics are soft and brittle and readily break apart, while some plastics are designed to biodegrade. Since all plastic materials are comprised of large chains of molecules, the degradation of plastics can be defined as any process which results in the breakage of these large chains into short chains of lower molecular weight and consisting of only a few monomers (oligomers) or the constituent monomers. There are two main ways in which this can occur; biotic and abiotic.

Biotic degradation and biodegradable plastics

Biotic degradation of plastics refers to the deterioration of plastic materials by biological organisms. For example, degradation of biodegradable plastics relies upon biological processes that utilise the carbon present in the plastic as an energy source. However, in order for a plastic to be able to degrade, it must undergo a two-stage procedure;

Stage 1–Degradation
Oxygen, moisture, heat, ultraviolet light or microbial enzymes break the carbon–carbon bonds of the long polymer chains resulting in fragmentation of the plastic. The degree to which these various factors have any effect depends upon the molecular structure of the polymer. Importantly, the mere fact that a polymer has degraded does not mean that it has biodegraded.

Stage 2–Biodegradation
Once the polymer has fragmented sufficiently, the shorter carbon polymer chains are able to pass through microbial cell walls. The carbon in the chains is then utilised as a food and energy source by the microbes, before being converted to biomass, water, carbon dioxide or methane gases. Ultimately, this depends upon whether the conditions are aerobic or anaerobic. However, it is the conversion of this carbon by microbes that signifies that biodegradation has indeed occurred.

Importantly, to be considered as a suitable biodegradable plastic, this two-step process of degradation must take place at an acceptable rate and not have a negative impact on the surrounding environment in which the biodegradation is taking place. Thus, the inherent value of a biodegradable plastic depends upon its effects on the environment, such as the effect on biota, the stability of soil conditions, the emission of methane gas and the contamination of ground water. Today, considerable advancements have been achieved with biodegradable plastics (see Chapters 1 and 2 for a detailed history of plastics).

Polyhydroxybutyrate (PHB)

The most popular biodegradable plastic is polyhydroxybutyrate (PHB), a thermoplastic which belongs to the polyester class of compounds. PHB is insoluble in water and thus has better resistance against hydrolytic degradation than other biodegradable plastics, which are soluble in water or sensitive to moisture. Furthermore, the material exhibits good resistance to ultraviolet light. There are many items available which are made from PHB, such as shampoo bottles, cups and golf tees. While in Japan, there is a large range of PHB disposable razors on the market. PHB is also used within the human body as sutures which, owing to the non-toxic degradative properties of PHB, breakdown naturally and thus do not have to be removed (Fig. 4.26).

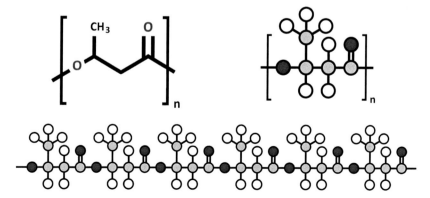

Figure 4.26 Polyhydroxybutyrate (PHB).

Chemical formula	$(C_4H_6O_2)n$
Abbreviation	PHB
Classification	Thermoplastic
Monomer	*beta*-Hydroxybutyric acid
General properties	Good oxygen permeability, melting point (175°C), glass transition temperature (2°C). Good resistance to ultraviolet light. Poor resistance to acids and alkalis. Insoluble in water. Biodegradable

Polycaprolactone (PCL)

Another common type of biodegradable plastic is polycaprolactone (PCL). The material has a low melting point of 60°C and is therefore unsuitable for high temperature applications. However, it is occasionally blended with other plastics to improve impact resistance or plasticise PVC. Owing to its biodegradative properties in the human body, which occurs at a slower rate than polylactic acid (PLA), there is considerable research underway to develop implantable devices or sutures that can remain in the body for long periods of time before breaking down when no longer required. Furthermore, the encapsulation of drugs with PCL for controlled and targeted drug delivery systems has been successfully accomplished (Fig. 4.27).

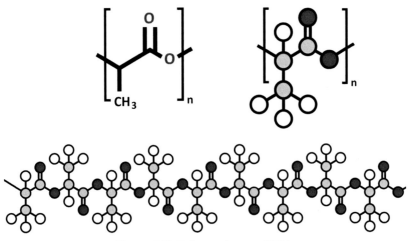

Figure 4.27 Polycaprolactone (PCL).

Chemical formula	$(C_6H_{10}O_2)n$
Abbreviation	PCL
Classification	Thermoplastic
Monomer	6-Hydroxycaproic acid or ε-caprolactone
General properties	Occasionally blended with other plastics to improve impact resistance or plasticise PVC. A tough material that has a glass transition temperature of −60°C and softens at 60°C thereby exhibiting poor heat resistance. Biodegradeable

Polylactic acid (PLA)

Polylactic acid (PLA) is the second most widely used biodegradable plastic and is derived from lactic acid, which can be produced by fermentation of renewable agricultural produce, such as sugar cane or corn starch. PLA has a large variety of applications, such as product packaging material, tableware and feedstock material for desktop 3D printers. Furthermore, owing to the substances ability to degrade to the innocuous lactic acid monomer, PLA is used within the human body for medical implants, such as pins, rods and screws. In the body, PLA completely degrades within 6–24 months, depending on the precise composition. The slow rate of degradation in the body is advantageous in terms of weight bearing structures like bone, since loading is gradually returned to the body as the polymer degrades and the body heals. Furthermore, lactic acid can be copolymerised with glycolic acid to create poly(lactic-co-glycolic acid) (PLGA), a biodegradable and biocompatible polymer. PLGA can be used for the targeted delivery of drugs within the body, such as the antibiotic amoxicillin, by creating microplastic-sized (~60 μm) solid PLGA microcapsules, which are loaded with the active drug, and are produced using micro-jetting technology[396]. Once inside the body, the solid PLGA microcapsules degrade when exposed to water via hydrolysis of their ester linkages, thereby releasing the active drug (Fig. 4.28).

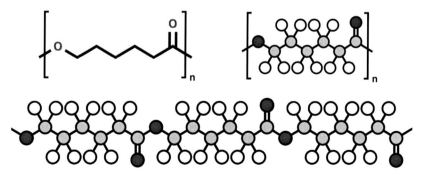

Figure 4.28 Polylactic acid (PLA).

Chemical formula	$(C_3H_6O_3)_n$
Abbreviation	PLA
Classification	Thermoplastic
Monomer	Lactic acid
General properties	A brittle substance with a glass transition temperature of 44–63°C. Although stable at room temperature (21°C), PLA is not suitable for high temperature environments and will start to soften within this temperature range. Biodegradable

When these biodegradable plastic materials are discarded into the environment, complete biodegradation takes about 2 weeks in the case of discard into a sewage treatment facility and approximately 2 months in the case of discard into soil or aquatic environments (Fig. 4.29).

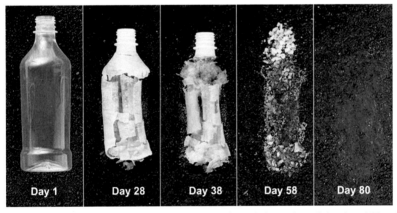

Figure 4.29 The degradation, and subsequent biodegradation, of a polylactic acid bottle.

Although biodegradable plastics may appear to be the perfect solution to plastic waste in the environment, in reality this is not the case. The main difficulty with biodegradable plastics is that they are unpopular with manufacturers. The compelling key properties of the highest-produced commodity thermoplastics are their low-cost, versatility and high strength. However, biodegradable plastics simply cannot compete with these commodity thermoplastics in terms of cost or mechanical properties. For example, polyhydroxybutyrate (PHB) is biodegradable, yet exhibits considerable brittleness, especially at low temperatures, owing to its glass transition temperature of 2°C. The point at which PHB breaks, following elongation, is only 6% of its original length. This is similar to polystyrene, which has an average value of 2.5%. In comparison, the average value for polypropylene is 450%. One solution to the brittleness of PHB is co-polymerisation with polyhydroxyvalerate (PHV), thereby imparting a material with reduced stiffness. As a larger proportion of hydroxyvalerate subunits are added to the PHB chain, the tougher and more flexible the material becomes.

Other issues arise with the poor resistance of PHB to solvents, acids and bases and its thermal stability, which may not be of a suitable specification for end products. To counteract these issues, there has been much research in trying to improve the properties of PHB, such as blending it with clay nanoparticles to increase its strength and rates of degradation. One particular advantage of PHB is that the material has a density of $1.25\,g/cm^3$ which is greater than that of seawater ($1.025\,g/cm^3$) thereby easily sinking and becoming incorporated into sediments where it can undergo biodegradation. Nevertheless, the bane of biodegradable plastics is that they are simply far more expensive to produce than non-biodegradable plastics, while the low yield associated with their production represents a major obstacle that manufacturers will have to overcome in the future.

Ultimately, the imperviousness of commodity thermoplastics to microbial attack in the environment is deemed as a highly advantageous property, especially with regards to structural integrity and durability. For this reason, biodegradable plastics may not be suitable for service environments in which there is likely to be exposure to significant microbial activity, such as in food storage and packaging or external environmental applications, such as automobiles and pipework. Indeed, an automobile manufacturer would not want to discover that their new plastic bumper has degraded within a year of use and needs replaced for safety reasons. Similarly, manufacturers of plastic pipework would not want to be the subject of claims for burst pipes and flooding as a result of rapid degradation. Consequently, until biodegradable plastics are developed with comparable properties to those of existing common thermoplastics, we shall continue to witness non-biodegradable plastics being produced in ever growing abundance and thus firmly remaining as the most popular plastics of choice.

Plastic degrading insects

At the present time, research is underway to find insects that are able to consume the common plastics littering our environment. Indeed, it has been demonstrated that some insects will consume plastics to reach a food source, such as cockroaches chewing through plastic bags containing bread. However, any ingested plastic is simply excreted and biodegradation does not occur. Similarly, woodworm have been found to have bored into PVC when the material has been in close contact with wood. While these insects are not consuming plastic materials for energy, an example of true ingestion and biodegradation of a plastic material by an insect have been recently documented. In 2014 it was reported that the larvae (waxworms) of the Indian mealmoth (*Plodia interpunctella*) are capable of consuming thin polyethylene films[466].

Note: These waxworms should not be confused with their relatives, the Greater and Lesser wax moth larvae, which are bred for animal food and also commonly referred to as waxworms.

Following ingestion of the plastic, the bacterial strains present in the gut of the waxworms *(Enterobacter asburiae* YT1 and *Bacillus* sp. YP1) were discovered to have degraded some of the polyethylene film. Upon inspection, by way of scanning electron microscope (SEM), it was observed that $0.3–0.4\,\mu m$ pitting had occurred on the films surface. The microbes were then isolated from the waxworms gut and cultivated. A 100 mg piece of polyethylene film was then introduced to a suspension culture of YT1 and YP1, containing 100 million bacterial cells per millilitre, and incubated for 60 days. Upon examination, it was determined that YT1 had degraded 6.1% of the polyethylene film, while YP1 had degraded 10.7%. It was subsequently determined that the bacteria had converted some of the polyethylene to biomass, carbon dioxide and biodegradable waste. Consequently, this is a promising area of research that may prove to be beneficial in dealing with plastic waste or developing new plastics that can be biodegraded more easily. Nevertheless, plastic eating insects should not be considered as a substitute for recycling, which is an indispensable practice in helping to reduce the accumulation of plastic waste in the environment (Fig. 4.30).

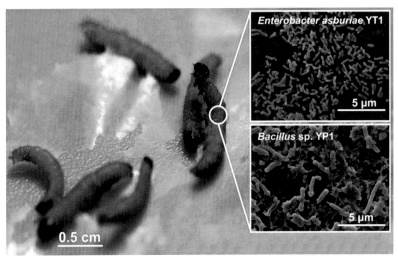

Figure 4.30 Waxworms consuming polyethylene film.
Reprinted (adapted) with permission from Yang J, Yang Y, Wu W, Zhao J, Jiang L. Evidence of polyethylene biodegradation by bacterial strains from the guts of plastic-eating waxworms. Environmental Science and Technology 2014;48(23):13776–84.

Plastic degrading microorganisms and fungi

Most plastic items recovered from the environment will have some degree of biological fouling (see Chapter 7). Nevertheless, many plastics are inherently resistant to biological attack or are protected with antimicrobial additives. However, microorganisms, such as sulphur bacteria, can form biofilms on plastic materials and secrete sulphuric acid. Similarly, acids can be secreted by some species of fungi. This can have a detrimental effect on a minority of plastics which are susceptible to acids, such as polyamides (Nylons). However, there are a few examples of specific plastic degrading microorganisms and fungi.

In 2011, it was reported that a new endophytic fungus (*Pestalotiopsis microspore*) was discovered in the Amazon rainforest in Ecuador[362]. The fungus was found to contain a serine hydrolase enzyme that allowed the fungus to feed solely on polyester polyurethane

(PUR) in an aerobic, as well as anaerobic (oxygen-free), environment. These characteristics may be especially beneficial as the biological degradation of PUR deep within landfill sites would require a microorganism that can operate when oxygen is absent.

In 2013, it was demonstrated that *Bacillus subtilus*, *Bacillus pumilus* and *Kocuria palustris* pelagic bacteria are capable of biodegrading low-density polyethylene (LDPE) in vitro.[179] In 2015, it was reported that a functional cutinase (a plant cuticle degrading enzyme), derived from the mesophilic fungus *Fusarium oxysporum*, was expressed in the bacterium *Escherichia coli* BL21. Upon exposure of the organism to polyethylene terephthalate (PET), it was found that the recombinant enzyme was capable of degrading the surface of PET via hydrolysis. The highest activity was found to occur at 40°C and pH 8.0.[100] In 2016, it was reported that a new bacterium (*Ideonella sakaiensis*) was discovered living on PET debris in soil and wastewater at a plastic bottle recycling facility in Japan.[469] The bacterium was capable of degrading PET via hydrolysis using two intrinsic enzymes to convert PET to its two basic monomers, ethylene glycol and terephthalic acid (Table 4.4).

Table 4.4 List of microorganisms and fungi reported to have a degradative effect on some common plastics

Plastic	Abbreviation	Degrading microorganism	Degrading fungi
Polyethylene terephthalate	PET	*Ideonella sakaiensis*[469]	*Rhizopus delemar* (copolymers with aliphatic dicarboxylic acids)[145,310,433]
Polyethylene	PE	*Pseudomonas chlororaphis*[377] *Brevibacillus borstelensis*[170] *Rhodococcus ruber*[159,386] *Bacillus subtilus*[179] *Bacillus pumilus*[179] *Kocuria palustris*[179] *Enterobacter asburiae* YT1[466] *Bacillus* sp. YP1[466]	*Penicillium simplicissimum* YK[464] *Curvularia Senegalensis*[194]
Polyester polyurethane	PUR	—	*Pestalotiopsis microspore*[362]
Polyvinyl chloride	PVC	*Ochrobactrum* TD[295] *Thermomonospora fusca*[224] *Pseudomonas fluorescens* B–22[224] *Pseudomonas putida* AJ[10]	—

Continued

Table 4.4 **List of microorganisms and fungi reported to have a degradative effect on some common plastics—cont'd**

Plastic	Abbreviation	Degrading microorganism	Degrading fungi
Polypropylene	PP	*Pseudomonas* sp.[43] *Vibrio* sp.[43]	*Aspergillus niger*[306]
Polystyrene	PS	—	—
Polycarbonate	PC	—	—
Poly(methyl methacrylate)	PMA	—	*Physarum Polycephalum*[230]
Polytetrafluoroethylene	PTFE	—	—
Acrylonitrile butadiene styrene	ABS	—	—
Polyamide (nylon 6)	PA	—	—
Polyamide (nylon 6,6)	PA	—	White rot fungi IZU-154[95]
Polychloroprene (neoprene)	CR	—	—

Often, fungi will readily grow on the surfaces of plastic materials in the environment. As the fungal filaments penetrate into the polymer matrix, they cause the plastic to swell up and burst, thereby inducing fracture and exposure of new surfaces. These freshly exposed surfaces are then vulnerable to further biotic and abiotic degradation processes (Fig. 4.31).

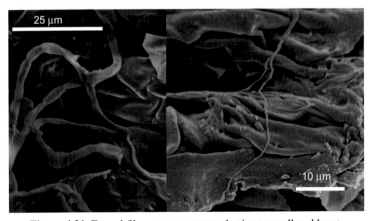

Figure 4.31 Fungal filaments can cause plastics to swell and burst.

Furthermore, fungi and bacteria often secrete organic chemicals, such as acids, as part of their decomposing activities. This can then create a micro-environment which is conducive to increased degradation of certain plastic materials, such as polyamides (Nylon), which are vulnerable to organic acids. However, the vast majority of plastic materials are impermeable to biotic degradation as fungi and bacteria have yet to evolve suitably effective enzymes which are capable of degrading these synthetic materials. Consequently, degradation of plastics tends to occur primarily via abiotic processes in which plastics are depolymerised to their constituent monomers. As a secondary process, microorganisms and fungi can then attack and utilise these constituent monomers as a source of energy.

Abiotic degradation

Abiotic degradation of plastics refers to the weathering of plastic materials from environmental factors, such as mechanical force, temperature, light, gases and water. In the early years of development of plastic materials (see Chapters 1 and 2), the degradation of plastics was regarded as a major problem because advantageous properties, such as tensile strength, impact resistance and rigidity were lost over time and the plastic material became inadequate for the desired service environment. Consequently, there was a large and highly competitive commercial drive in creating increasingly durable polymers that exhibited superior resistance to degradation and weathering.[85] Unfortunately, in modern times this pronounced durability is now the key factor in the persistence and ubiquity of plastics in the marine environment.[353]

Chemical additives

Often, plastics will contain a variety of chemicals which are added during the manufacturing process to change the properties of the polymer or to increase their stability and durability, and to protect against environmental influences and biological attack. Additives may also be added to facilitate manufacturing processes. For example, during processing, polypropylene is particularly susceptible to oxidative degradation at high temperatures since the tertiary carbons, on which the pendant methyl groups are attached, provide a site for oxidation. At this site, a free radical is formed, which further reacts with oxygen, resulting in breaking of the polymer chains (chain scission) and the production of carboxylic acids and aldehydes. Consequently, all commercial polypropylene homopolymer and copolymers will contain antioxidant compounds. A list of common additives and their purpose is provided in Table 4.5.

Table 4.5 **Common plastic additives and their purpose**

Additive	Purpose
Plasticisers (e.g., bis(2-ethylhexyl) phthalate	Gives flexibility and softens the polymer
Reinforcement (e.g., carbon fibres, wollastonite)	Used to increase tensile/flexural strength
Impact modifiers (toughners) (e.g., glass fibres)	Used to improve impact strength
Colourant (e.g., titanium dioxide, iron oxide)	Pigments used to impart colour/effects
Filler (e.g., calcium carbonate)	Used as a bulking agent to increase mass
Antistatics (e.g., ethoxylated esters)	Dissipation of built up static charge
Heat stabiliser (e.g., metal powders)	Used to dissipate any build-up of heat
Oxidative stabiliser (e.g., benzofuranones)	Protection from exposure to air
Surface modifier (e.g., metal stearates)	Can impart a rough or lubricated surface
Flame retardant (e.g., antimony trioxide)	Used as a flame and smoke suppressant
Light stabiliser (e.g., benzophenone)	Absorbs damaging UV radiation
Biological protection (e.g., silver ions)	Protects from microorganisms and fungi
Chemical blowing agent (e.g., azodicarbonamide)	Used to generate gas to expand plastics and rubber into foam
Chemical blowing agent activator (e.g., urea and zinc compounds)	Used to adjust the decomposition temperature of the blowing agent
Lubricants and slip additives (e.g., erucamide)	Used to improve processability by increasing flowability and lowering heat dissipation and viscosity

Characteristic weaknesses of common plastics

Despite the addition of chemical agents to improve their properties, different types of plastic materials have inherent weaknesses which are directly related to their chemical composition. Thus, in the aquatic environment, some plastics will demonstrate poorer resilience to environmental stressors and break apart more readily than others[84,369] (Table 4.6).

Table 4.6 **The inherent weakness of common plastics**

Plastic	Abbreviation	Environmental weakness
Polyethylene terephthalate	PET	Can hydrolyse in water, wet or humid conditions at high temperatures (>73–78°C)
Low-density polyethylene	LDPE	Prone to cracking under stress. Photo-oxidises
High-density polyethylene	HDPE	Prone to cracking under stress. Crazes in sunlight

Table 4.6 **The inherent weakness of common plastics—cont'd**

Plastic	Abbreviation	Environmental weakness
Polyvinyl chloride	PVC	Most rigid types have low impact strength at room temperature (21°C)
Polypropylene	PP	Poor resistance to bending, breaking, crushing (tenacity) at cold temperatures. Develops brittleness around 0°C. Oxidation from heat and sunlight results in crazing and fine cracks that deepen and become more severe over time
Polystyrene	PS	Brittle substance. Readily degrades in sunlight and turns yellow
Polycarbonate	PC	Prone to abrasion and sensitive to strong alkalis
Poly(methyl methacrylate)	PMA	Brittle and prone to chipping
Polytetrafluoroethylene	PTFE	Low resistance to pressure and high tendency to permanently deform under stress
Acrylonitrile butadiene styrene	ABS	Develops brittleness and yellows in sunlight
Polyamide (nylon)	PA	Decrease in tensile strength from water absorption. Sensitive to acids and alkalis
Polychloroprene (neoprene)	CR	Susceptible to ozone in the atmosphere

Thermal degradation

Thermal degradation refers to the breakdown of plastic materials by changes in temperature. Thus, the thermal properties of a plastic material can give an indication as to the materials susceptibility to thermal degradation in the environment. In the case of high temperatures, degradation occurs by way of a thermo-oxidative reaction. However, in order for a plastic to undergo thermal oxidative degradation, there has to be a sufficient input of energy, in the form of heat in order to break chemical bonds. When sufficient heat is applied to plastic materials, this initiates the polymer free radical chain reaction, a self-propagating oxidative reaction (see Fig. 4.32). This reaction has three distinct phases, initiation, propagation and termination.

Initiation

The input of energy, in the form of heat, overcomes an energy barrier and results in the breakdown of the long polymer chains to form small reactive molecules, termed radicals. Importantly, as well as heat, the radical chain reaction can also be initiated by the sufficient input of energy in the form of ultraviolet light (see Photo-oxidative degradation).

Propagation

These radicals produced in the initiation phase react with oxygen in the atmosphere to create peroxide radicals, which subsequently decompose further to form highly reactive hydroxyl free radicals and alkoxy radicals. These radicals then react with oxygen in the atmosphere and create more free radicals, and so on, thereby propagating the chain reaction.

Termination

The reaction will continue to self-propagate until the initial input of energy or the supply of oxygen is discontinued, thereby terminating the reaction. It is also possible for the radicals to react with one another and terminate the reaction by forming stable non-radical adducts. However, this is less likely due to the low probability of two radicals colliding with one another.[380]

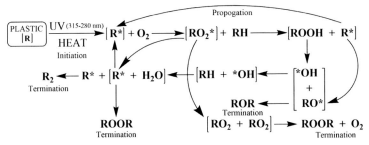

Figure 4.32 Polymer free radical chain reaction.

The temperature at which this oxidative reaction occurs depends upon the individual thermal properties of the plastic material and the availability of oxygen. For example, the limiting oxygen indices provide an indication of flammability by expressing as a percentage, the minimum amount of oxygen required to allow sustained combustion for 3 min. Thus, low numbers indicate that a smaller amount of atmospheric oxygen is required and therefore flammability is higher for these materials. However, ratings are often increased by the addition of halogenated flame retardants, such as polybrominated diethylethers (PBDEs). This group of chemicals is considerably toxic and is a common pollutant in the aquatic environment (PBDEs will be discussed in more detail within Chapters 6 and 7). Furthermore, materials such as polyvinyl chloride (PVC) and polychloroprene (Neoprene) are inherently flame resistant because they are halogenated with chlorine atoms on the polymer backbone. These halogen atoms serve as flame retardants by acting in the vapour phase against the flame, thereby reacting with free radicals and terminating the free radical chain reaction. If the limiting oxygen index values for a material is higher than 21, then this indicates that the material will not burn at room temperature as the Earth's atmosphere typically contains around 21% oxygen at sea-level. If the indicated values exceed 25–27, then combustion will only occur under the sustained application of extreme temperature.

One of the most important thermal properties, and one that is unique to polymers, is the glass transition temperature (T_g). This indicates the temperature region at which the plastic material transforms from a hard glassy state to a soft flexible state. Some plastics are used above their T_g, such as polyethylene and polypropylene, and are therefore soft and flexible at room temperature (21°C), while others are used below their T_g and are therefore hard and brittle at 21°C, such as ABS, polystyrene and rigid PVC. However, by adding certain chemicals to PVC, the material can be made more flexible and resilient by lowering the T_g and allowing the material to be workable. Such chemical compounds are known as plasticisers. Although there are over 100 commercially used PVC plasticisers, including adipates which impart resistance at low temperature and trimellitates which increase resistance to heat, by far the most commonly used class are phthalates, which provide a balanced solution. Since phthalates are not bound to PVC, they have the potential leach from the material and induce biological effects. However, for many years' fierce debate has ensued as to exactly what risk phthalates represent to human health and the aquatic environment, and whether they are considered to be persistent environmental pollutants or not (see Chapters 6 and 7).

With regard to the maximum service temperature limit of a plastic, the value is dependent upon whether the material is amorphous or semi-crystalline. In the case of semi-crystalline plastics, the material is able to absorb a given amount of heat before it quickly changes state from a solid to a low viscosity liquid. Thus, the most important temperature that determines the maximum service temperature in this case is the melting point. Conversely, amorphous plastics tend to get softer as they are heated, as opposed to possessing a specific melting point. Therefore, the most important temperature that relates to the service working temperature is the glass transition temperature.

As a result of prolonged exposure to the radiative heat of the sun, softening can be induced in many plastic materials, thereby resulting in morphological changes and affecting their resistance to mechanical degradation. A test can be undertaken which determines the temperature at which a plastic material becomes soft. This test involves heating a specimen of plastic until a needle, with a cross-sectional area of 1 mm^2, is capable of penetrating 1 mm into the surface of the plastic. The temperature at which this is occurs is termed as the Vicat softening temperature (also known as the Vicat hardness). There are generally two tests, depending on the force applied to the needle. In the Vicat A test, 10 newtons (N) of force is applied, while on the Vicat B test, the force is 50 N. On the other hand, plastics in the environment can be exposed to extreme cold and if temperatures fall below their glass transition temperature, brittleness will occur, thereby facilitating fracture and fragmentation (Table 4.7).

Photo-oxidative degradation

Photo-oxidative degradation refers to the breakdown of plastics by light. Microplastic litter on land, such as beaches, or floating on the surface of the ocean will inexorably be exposed to large amounts of direct sunlight and thus suffer the effects of exposure to high intensity ultraviolet light (UV) for significant periods of time. Most plastics

Table 4.7 The thermal properties of common plastics

Plastic	Abbreviation	Glass transition temperature (T_g)	Minimum service temperature (°C)	Maximum service temperature (°C)	Limiting oxygen index range (%)
Low-density polyethylene	LDPE	−110	−70	80–100	17–18
Linear low-density polyethylene	LLDPE	−110	−70	90–110	17–18
High-density polyethylene	HDPE	−110	−70	100–120	17–18
Polyamide (nylon 6)	PA	−60	(−40)–(−20)	80–120	23–26
Polyamide (nylon 6,6)	PA	(−55)–(−58)	(−80)–(−65)	80–140	21–27
Polychloroprene (neoprene)	CR	(−49)–(−20)	(−34)–(−57)	95–120	26–40
Polypropylene	PP	(−20)–(−10)	(−40)–(−5)	90–130	17–18
Polyvinyl chloride (plasticised)	PVC	(−50)–(−5)	(−40)–(−5)	50–80	20–40
Polyethylene terephthalate	PET	73–78	−40	80–140	23–25
Polystyrene	PS	90	20	65–80	17–18
Polystyrene (high heat)		90	20	75–90	17–18
Polyvinyl chloride (rigid)	PVC	60–100	(−10)–1	50–80	40–45
Acrylonitrile butadiene styrene	ABS	90–102	(−40)–(−20)	85–95	18–28
Poly(methyl methacrylate)	PMA	90–110	−40	70–90	19–20
Acrylonitrile butadiene styrene (high heat)	ABS	105–115	(−40)–(−20)	75–110	18–19
Polytetrafluoroethylene	PTFE	115	−200	260–290	95–96
Polycarbonate	PC	150	−40	90–125	30–40
Poly(methyl methacrylate) (high heat)	PMA	100–168	−40	100–150	19–20
Polycarbonate (high heat)	PC	160–200	−40	100–140	24–35

tend to be susceptible to UV light because they contain photo-reactive groups, termed chromophores. These groups readily absorb high-energy UV radiation, which results in the breaking of chemical bonds. However, there are some exceptions. For example, PVC has good resistance to UV light simply because it does not possess the relevant UV chromophore. However, PVC still exhibits some degree of photo-sensitivity. This is suspected to occur as a result of abnormalities in the polymer matrix, such as the presence of C=O and O—O groups.[85]

Photo-oxidative degradation of a plastic material occurs when photons of ultraviolet (UV) light, particularly in the UVB region (315–280 nm) of the electromagnetic spectrum (see Chapter 10), initiate the process of decomposition of the plastic by way of the free radical polymer chain reaction (Fig. 4.19). While UV light tends to have a minimal effect on the propagating steps of the radical chain reactions, progression of the degradation can continue, even after UV exposure has been discontinued, as a result of induced thermal-oxidative decomposition. As a result, the mechanical properties of the plastic can be significantly impaired, even from as little as 1% oxidation. Common signs of photo-oxidative damage to plastics are yellowing, hazing, cracking and embrittlement.[380] The plastic may also start to exhibit a colour shift and take on a noticeably bleached chalky appearance on any areas which have been exposed to sunlight, in comparison to areas which have been shielded. For example, the small fragment shown in Fig. 4.20 was recovered from the marine shoreline. The fragment had a patch of fouling covering part of its surface. Once this fouling was removed, it became apparent that the dark region underneath the fouling was protected from sunlight while the regions exposed to the sun had oxidised and turned white. Thus, to protect against such oxidative damage, plastics are typically stabilised with the addition of chemicals, such as carbon black, hydroxybenzophenone or amines which have been sterically hindered to impart ultraviolet resistance and permit outdoor use[85] (Fig. 4.33) (Table 4.8).

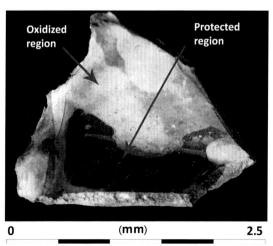

Figure 4.33 A fragment of plastic exhibiting both a photo-oxidised region and a region previously protected from ultraviolet by a layer of fouling.

Table 4.8 **The resistance of common plastics to ultraviolet light**

Plastic	Abbreviation	Resistance to UV light
Acrylonitrile butadiene styrene	ABS	Poor
Polystyrene	PS	Poor
Polyamide (nylon 6,6)	PA	Poor
Polyamide (nylon 6)	PA	Fair
Polyethylene terephthalate	PET	Fair
Polyethylene	PE	Fair
Polypropylene	PP	Fair
Polycarbonate	PC	Fair
Polyvinyl chloride	PVC	Good
Poly(methyl methacrylate)	PMA	Good
Polytetrafluoroethylene	PTFE	Good
Polychloroprene (neoprene)	CR	Good

Atmospheric oxidation and hydrolytic degradation

The presence of oxygen in the atmosphere can catalyse the breakdown of some plastic materials. For example, PVC degrades via the removal of hydrogen chloride (dehydrochlorination) to form double bonds (Fig. 4.34). However, this decomposition process tends to only occur down to a maximum depth of 1 mm from the surface exposed to the environment. This is because the oxygen in the atmosphere is unable to permeate to any further reaction sites beyond that depth.[85] As such, PVC is only partially vulnerable to atmospheric oxidation. However, if the environment is particularly dynamic, such as the interactions of waves and rocks, the surface of the PVC may become pitted and abraded, thereby revealing new reaction sites which are suitable for oxidation.

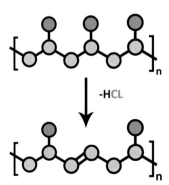

Figure 4.34 Dehydrochlorination of PVC.

Nevertheless, oxygen is not the only gas that can attack plastics. In urban environments, high levels of industrial and domestic activity result in greater levels of pollutants in the atmosphere, in comparison to rural and remote regions. Consequently, ozone and oxides are produced and several of these pollutants can have degradative effects on plastic. For example, when polyethylene and polypropylene are exposed to sunlight in the presence of atmospheric oxygen and the pollutant sulphur dioxide, crosslinking of the polymer chains occur. Similarly, Nylon 6,6 is susceptible to attack by the pollutant nitrogen dioxide.

However, the most significant damage tends to occur from the presence of highly reactive ozone in the atmosphere which can attack the double bonds of some plastics and elastomers (ozonolysis) resulting in degradation. The reaction occurs when a molecule of ozone reacts with the double bond to produce an unstable reactive polyatomic anion (ozonide). The rapid decomposition of the ozonide results in cleavage of the double bonds and thus breaking of the polymer chain. The long chains in polymers are what give them their high molecular weight and strength. Thus, the longer the chain, the greater the strength of the material. However, when ozonolysis results in breakage of the chains, the molecular weight decreases, thereby reducing the strength of the material until it becomes brittle and cracks. For example, in elastomers, ozone can induce cracking on surfaces exposed to the atmosphere and this ozone-cracking effect is often seen in old car tyres. However, some elastomers, such as polychloroprene (Neoprene), have good ozone resistance because the double bonds in the polymer chain are protected from attack by ozone. This arises from the presence of chlorine atoms on the polymer backbone, which decrease the electron density in the double bonds, thereby reducing the tendency of the double bonds to react with ozone.

When some plastic materials are submerged in water, diffusion of the water into the amorphous regions of the plastic occurs. In some plastics, such as polytetrafluoroethylene, there is only negligible absorbance while in others the rate of absorption is considerable, such as with Nylons. However, the diffusion of water into the polymer matrix can result in the addition of water molecules to the polymer by way of the cleavage of chemical bonds (hydrolysis). For example, the polyester polyethylene terephthalate (PET) is hydrolysed at temperatures above the glass transition temperature (73–78°C) in which a scission reaction, catalysed by oxonium ions or hydrogen ions produced by the carboxyl end groups, breaks the primary bonds of the polymer chain resulting in irreversible damage. Other plastics that suffer the effects of moisture are polyurethanes (Table 4.9).

Table 4.9 The percentage weight gains of plastics submerged in distilled water for 24 h and the effects of ozone

Plastic	Abbreviation	Water absorbance (% weight gain in 24 h at 21°C)	Effect of ozone (>1000 ppm)
Polytetrafluoroethylene	PTFE	0.005–0.01	No effect
Polyethylene	PE	0.005–0.02	Moderate effect (in air) Minor effect (in water)
Polypropylene	PP	0.01–0.03	Moderate effect
Polystyrene	PS	0.01–0.3	Moderate effect
Polyvinyl chloride	PVC	0.01–1.0	Minor effect

Continued

Table 4.9 **The percentage weight gains of plastics submerged in distilled water for 24 h and the effects of ozone—cont'd**

Plastic	Abbreviation	Water absorbance (% weight gain in 24 h at 21°C)	Effect of ozone (>1000 ppm)
Acrylonitrile butadiene styrene	ABS	0.05–1.8	Minor effect
Polyethylene terephthalate	PET	0.1–0.2	No effect
Polycarbonate	PC	0.1–0.4	No effect
Poly(methyl methacrylate)	PMA	0.1–0.8	No effect
Polychloroprene (neoprene)	CR	0.6	Minor effect
Polyamide (nylon 6,6)	PA	1.0–3.0	Moderate to severe effect
Polyamide (nylon 6)	PA	1.6–1.9	Moderate to severe effect

Mechanical degradation

Once in the aquatic environment, large pieces of macroplastic debris are subject to different forms of mechanical stress[73,285] (Table 4.10).

Table 4.10 **The forms of mechanical stress on plastic debris**

Stress	Definition	Example
Abrasion	A wearing action resulting in the creation of plastic fragments less than 5 mm in size (plasticles)	Scraping
Compressive	An inward directed force that acts to shorten or compress the plastic	Crushing
Flexural	Flexing which produces compression and tension on adjacent sides of the plastic	Bending
Impact	A high force or shock applied to the plastic over a short period of time that can result in fragmentation	Striking
Sheer	A force acting on the plastic which is perpendicular to the extension of the plastic	Tearing
Tensile	A force that acts to lengthen or expand the plastic	Stretching
Torsional	A type of stress or deformation that occurs when one end of the plastic is twisted while the other end is held motionless or twisted in the opposite direction	Twisting

These stresses occur on account of ocean currents, waves, collisions and abrasion from rocks and sand.[40,62] Over time, this culminates in the breakdown of macroplastics into progressively smaller pieces[8,73] (Fig. 4.35).

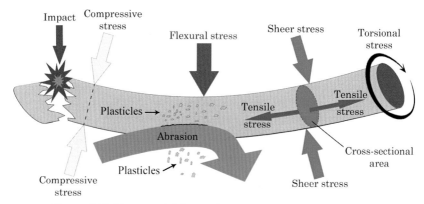

Figure 4.35 The potential forms of stress on a piece of macroplastic.

The degree to which these different forces have any effect depends upon the mechanical properties of the plastic. For example, elongation at break indicates the percentage of the initial length of a plastic specimen that it can be elongated by before it breaks and thus the degree to which it can resist changes to its shape without suffering fracture. For example, a brittle plastic like polystyrene has a maximum value of 4% thereby indicating that the material can only be marginally elongated before it fractures under the strain. However, polyethylene can be elongated by up to 900% of its initial length thereby indicating that it can resist great changes in shape without cracking. Thus, plastics with a high elongation at break value are much less likely to fragment from tensile forces. When a plastic material such as high-impact polystyrene suffers stress, for example tensile stress, craze lines will form where there has been an increase in entanglement of the polymer chains as a result of the applied stress (see Fig. 4.36). Continued stress results in the breakage of carbon–carbon bonds on the polymer backbone and thus fracture occurs (Table 4.11).

Figure 4.36 Fracture of a high impact polystyrene tube.

Table 4.11 **The elongation at break (%) ranges of common plastics**

Plastic	Abbreviation	Elongation at break (%)
Polystyrene (30% glass fibre)	PS	1–1.5
Polystyrene (crystal)	PS	1–4
Nylon 6,6 (30% glass fibre)	PA	2–2.2
Polypropylene (30–40% glass fibre)	PP	2–3
Polycarbonate (20–40% glass fibre)	PC	2–4
Polycarbonate (20–40% glass fibre and flame retardant)	PC	2–4
Polypropylene (10–20% glass fibre)	PP	3–4
Polyvinyl chloride (20% glass fibre)	PVC	2–5
Polyethylene terephthalate (30% glass fibre)	PET	2–7
Polyethylene terephthalate (30% glass fibre and impact modified)	PET	6–7
Poly(methyl methacrylate)	PMA	2–10
Nylon 6,6 (impact modified and 15–30% glass fibre)	PA	3–10
Polypropylene (10–40% talc)	PP	20–30
Nylon 6,6 (30% mineral filled)	PA	2–45
Acrylonitrile butadiene styrene	ABS	10–50
Polypropylene (10–40% mineral filled)	PP	30–50
Polystyrene (high impact)	PS	10–65
Poly(methyl methacrylate) (impact modified)	PMA	4–70
Polyethylene terephthalate	PET	30–70
Polyvinyl chloride (rigid)	PVC	25–80
Acrylonitrile butadiene styrene (flame retardant)	ABS	2–80
Acrylonitrile butadiene styrene (high impact)	ABS	2–100
Polycarbonate (high heat)	PC	50–120
Nylon 6,6 (impact modified)	PA	150–300
Nylon 6,6	PA	150–300
Polytetrafluoroethylene (25% glass fibre)	PTFE	100–300
Nylon 6	PA	200–300
Polyvinyl chloride (plasticised)	PVC	100–400
Polytetrafluoroethylene	PTFE	200–400
Polypropylene (copolymer)	PP	200–500
Polyvinyl chloride (plasticised and filled)	PVC	200–500
Polypropylene (homopolymer)	PP	150–600
Low-density polyethylene	LDPE	200–600
High-density polyethylene	HDPE	500–700
Polypropylene (impact modified)	PP	200–700
Polychloroprene (neoprene)	CR	100–800
Linear low-density polyethylene	LLDPE	300–900

Susceptibly of plastics in different marine zones

On account of these various degradative processes, the speed and effects of plastic degradation in different marine zones such as beaches, sediment and the deep ocean are very different. For example, beached microplastics are exposed to waves, wildlife and significant amounts of heat and cold, as well as considerable amounts of sunlight. As such, it would be expected that beached plastics would be susceptible to thermo-oxidative and photo-oxidative processes, as well as being exposed to acidic environments following ingestion by wildlife. Conversely, plastics in deep ocean sediments would be immune from sunlight and alternatively, would experience sustained cold and extreme pressure. However, it is possible for deep ocean plastics to experience extreme heat if they, however unlikely, enter the vicinity of hydrothermal vents. In these regions, water can exist as a supercritical fluid at temperatures of up to 464°C. It is likely that, owing to the extreme hydrostatic pressures and temperatures in these underwater regions, practically all plastics in the vicinity of these vents would undergo complete thermal degradation to their constituent monomers. Although some of these monomers are highly toxic, some can be subsequently degraded by microorganisms and fungi.

Thus, in some rudimentary form, nature has an inherent degradative mechanism in the depths of the ocean, but only for plastics that are able, and dense enough, to reach these regions. Interestingly, by plotting all the known hydrothermal vent fields in relation to the oceanic gyres (see Fig. 4.37), it appears that many hydrothermal vents are located directly under the oceanic gyres. Since these oceanic gyres are regions where vast amounts of plastic waste accumulate, it may be possible that biofouling of plastics (see Chapter 7) may allow some buoyant plastics to gain mass and sink down to these depths where these vents exist. However, the scarcity of these underwater volcanic regions and the powerful underwater currents that exist there may make such

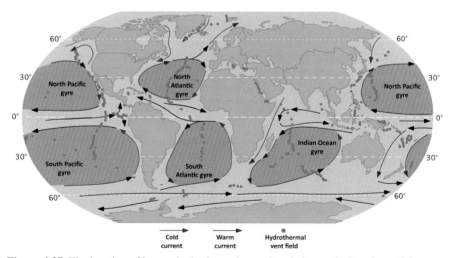

Figure 4.37 The location of known hydrothermal vents in relation to the locations of the oceanic gyres.

an encounter a rare occurrence. Nevertheless, this could be an area of possible study, especially with regard to thermal degradation of plastics to their substituent monomers and the subsequent release of toxic chemical pollutants at these sites. One could envisage a scientific study which involves testing the water and organisms around hydrothermal vents for known plastic degradation products and their effects on the unique ecosystem there. However, in terms of reducing the amount of plastic litter, it would seem that the impact of hydrothermal vents on plastic pollution in the oceans would likely be negligible at the present time. Nevertheless, since 32 million tons of plastic waste is expected to enter the ocean each year by 2050, this rapidly increasing volume of plastic waste, and its tendency to gather in regions above vents, may mean that the high temperature breakdown of plastics into more toxic persistent pollutants by hydrothermal vents may well be a problem in the future (Table 4.12).

Table 4.12 The susceptibility of plastics in the environment by location

Plastic location	Susceptibility
Beach/terrestrial	Sunlight, heat, cold, wind, waves, acids and enzymes (ingestion by wildlife), damage from anthropogenic activities (e.g., vehicles, construction, camp fires)
Surface water	Sunlight, heat, cold, wind, waves, rocks, acids and enzymes (ingestion by wildlife and aquatic organisms), aquatic contaminants, biofouling, damage from vessels (e.g., propellers, bilge pumps, ship strikes)
Sediment	Heat, cold, acids and enzymes (ingestion by aquatic organisms), salinity changes (estuaries), biofouling, damage from dredging
Deep ocean sediment	Extreme pressure, cold (~2°C), extreme heat from hydrothermal vents (60°C–464°C), aquatic contaminants, acids and enzymes (ingestion by aquatic organisms), low levels of oxygen

Ultimately, the rates and processes by which macroplastics degrade in the aquatic environment are not yet fully determined by the scientific community and consequently, continued research in the area is required. However, what is certain is that the degradation of plastic materials results in the breakdown of macroplastics and mesoplastics into increasingly smaller plastic fragments, termed microplastics.

Microplastics, standardisation and spatial distribution

What are microplastics?

The previous chapter discussed that large pieces of plastic litter, termed macroplastics, degrade in the environment into increasingly smaller pieces of plastic. Unfortunately, it appears that as the production of plastics has exponentially increased over the last half-century, these small pieces of plastic have become progressively more abundant in the aquatic environment. In fact to such an extent, that they are now becoming recognised worldwide, with rapidly growing concern, as one of the most ubiquitous and formidable threats ever to endanger the marine environment. Not only have they been discovered on practically every beach in the world,[7,96,257] they have been detected on the surface of every open ocean.[141] Furthermore, they have been discovered on the surface water,[116] shorelines and sediments of freshwater lakes,[75] intermixed with marine sediments[61] and even incorporated within the polar ice caps.[183] In fact, in the aquatic environment they are practically everywhere and what is more, the abundance of these pollutants can be exceptionally high. For example, 82% of the debris recovered from the surface water of the Tamar Estuary in the United Kingdom was reported to be composed of small pieces of plastic. These small plastic particles (plasticles) can be categorised, based on their size, as either microplastics, mini-microplastics or nanoplastics.

The term microplastic (MP) generally refers to any piece of plastic smaller than 5 mm to 1 μm in size along its longest dimension. However, if one wishes to expressly indicate that the microplastic in question is smaller than 1 mm in size along its longest dimension, then the term mini-microplastic (MMP) is used instead. Any piece of plastic smaller than 1 μm in size is considered to be a nanoplastic (NP). However, due to incredibly small size of nanoplastics and thus, the difficulties in detecting and recovering them, most studies of the aquatic environment tend to ignore nanoplastics and only focus on microplastics and mini-microplastics. As such the remainder of this book shall mainly focus on these.

What are plasticles?

The term plasticle (PLT) is a shortened version of the expression 'plastic particles' and was first introduced with this book as part of a newly developed standardised size and colour sorting (SCS) system for the effective categorising of microplastics, based on their size and appearance (see Fig. 5.8). Thus, the word plasticle is an all-inclusive term used to describe any piece of plastic smaller than 5 mm in size along its longest dimension and therefore includes microplastics, mini-microplastics and nanoplastics. These standardised size categories are listed in Table 5.1.

Microplastic Pollutants. http://dx.doi.org/10.1016/B978-0-12-809406-8.00005-0

Table 5.1 **The standardised size categories of pieces of plastic**

Category	Abbreviation	Size	Size definition
Macroplastic	MAP	≥25 mm	Any piece of plastic equal to or larger than 25 mm in size along its longest dimension
Mesoplastic	MEP	<25 mm–5 mm	Any piece of plastic less than 25 mm to 5 mm in size along its longest dimension
Plasticle	PLT	<5 mm	All pieces of plastic less than 5 mm in size along their longest dimension
Microplastic	MP	<5 mm–1 mm	Any piece of plastic less than 5 mm to 1 mm in size along its longest dimension
Mini-microplastic	MMP	<1 mm–1 μm	Any piece of plastic less than 1 mm to 1 μm in size along its longest dimension
Nanoplastic	NP	<1 μm	Any piece of plastic less than 1 μm in size along its longest dimension

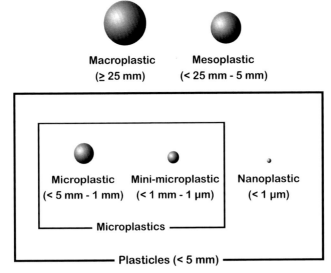

Figure 5.1 The standardised size categories of pieces of plastic.

The history of microplastics

The term 'microplastic' first appeared in 1968 in a publication by the US Air Force Materials Laboratory.[200] At that time the term was used to describe the deformation of a plastic material, on the order of microinches per inch, in response to an externally applied load. Incidentally, in the same publication, the term macroplastic also first appeared and was used to describe the deformation of a plastic material resulting from loading to significantly higher strains. Since then, these terms no longer apply and have been widely adopted by scientists to take on a new meaning in which they generally refer to the physical size of a piece of plastic. The impetus behind this change resulted from the discovery of small pieces of plastic in the aquatic environment in the 1970s.

It was in 1972 that the world first became aware of micro-sized pieces of plastic in the aquatic environment when it was reported that vast quantities of small plastic particles (plasticles) were found floating on the surface of the Sargasso Sea.[48] Incidentally, the Sargasso Sea is located in the middle of the North Atlantic Ocean Gyre and is the only sea on Earth which does not have a coastline. At that time, these micro-sized pieces of plastic were simply referred to as 'plastic particles', and it was not until a publication in 2004[406] that the modern use of the term microplastic was introduced. Within the publication the researchers used the term 'microplastics' to describe small pieces of plastic that had been collected from beaches and sediments in Plymouth, United Kingdom. This new use of the term eventually became adopted by the scientific community and thereafter, a microplastic was defined by the Steering Committee of the National Oceanic and Atmospheric Administration (NOAA) Marine Debris Program[12] as being a piece of plastic smaller than 5 mm in size along its longest dimension. However, no lower size limit was set. Furthermore, the precise definition of various sized pieces of plastic had, until now, been quite ambiguous and this was a considerable source of difficulty when comparing and reporting results.

For example, it was widely accepted that a nanoplastic was smaller than a microplastic. However, no lower size limit of a microplastic was previously set, nor any size boundary at which a nanoplastic was considered to be a microplastic. Furthermore, a few examples in the literature had referred to the lower size limit of a mesoplastic as 4.75 mm, while at the same time, there was no defined upper limit of a mesoplastic. Consequently, this overlapped with the definition of a microplastic and a macroplastic. To add further confusion to the issue, occasionally the upper size limit of a microplastic had been referred to as being 1 mm in size, depending upon the author. Again, this had presented difficulties in comparing results. Furthermore, defining a microplastics as being less than 1 mm in size had meant that there was no definition for a piece of plastic between 1 mm and 4.75 mm in size. Nevertheless, there had been a growing call from many researchers for an intuitive size cut-off of 1 mm for small pieces of plastics and a distinct definition. This was because such a definition would exclusively refer to microplastics which were only capable being ingested by very small aquatic organisms. As such, a standardised system for categorising pieces of plastic based on their size was desperately needed.

To address the situation and facilitate the effective comparisons of results from future research and environmental monitoring, an effective and logical standardised system for categorising microplastics based upon their size and appearance was introduced with this book. As part of that standardised system the term mini-microplastic was introduced to define the much needed size cut-off of less than 1 mm in size. Furthermore, the new standardised system provided an effective categorisation of pieces of plastic recovered from the environment based on their size, appearance and colour, as well as generating simple codes that greatly assist in data processing and analysis of samples. A basic outline of the size categories of the standard system is shown in Tables 5.1 and 5.4, as well as visual representations in Figs. 5.1 and 5.8. For the full procedure on how to use the standardised size and colour sorting (SCS) system to categorise pieces of plastic of any size, see Chapter 10.

Although a major source of microplastics in the aquatic environment is a result of the breakdown of larger pieces of plastic, microplastics are also purposefully manufactured. Indeed, we encounter one form of these manufactured microplastics every day, the synthetic fibres in our clothes. Thus, in order to delineate between degradative and industrial sources of microplastics, there are two main categories: primary and secondary.[251]

Primary microplastics

Primary microplastics are typically small spherical microbeads which are intentionally manufactured by the plastics industry for use in cosmetics, personal care products, dermal exfoliators, cleaning agents and sandblasting shot. Many primary microplastics are often carelessly released directly into the marine environment,[251,254] such as during sandblasting activities.[350] Instead of using the traditional sand based shot, modern sand-blasting approaches now involve the use of manufactured microplastics as a result of their greater durability in comparison to sand.[254] Consequently, these plastic shot particles can be swept up and carried on the wind and then deposited in bodies of water. Alternatively, they may be washed into urban watercourses where they can be carried to both freshwater and marine environments.[166]

Another form of primary microplastics is industrial feedstock (see Fig. 5.7). These small coloured pellets of plastic are manufactured globally by the plastics industry for the purposes of being melted down and moulded to form larger plastic artefacts.[254] A major source of plastic pollution in the aquatic environment stems from industrial plastic feedstock,[188] and the unintentional release of microplastics by industry sources is considered to contribute greatly to the abundance of microplastics observed in the marine environment.[435] In some cases, this may occur due to the direct release of these primary microplastics from industrial effluent pipes,[247] while in other cases their entry into the aquatic environment may be a result of industrial spillage (Fig. 5.2). For example, it has been estimated[63] that in the United Kingdom 105–1054 tonnes of primary microplastics of the pellet type are accidentally emitted into the environment every year, which equates to between 5 and 53 billion pellets annually.

Figure 5.2 Industrial spillage of microplastics.

The synthetic fibres used to produce garments are also considered to be primary microplastics, since they were purposely manufactured to be of a small size. In samples recovered from the environment, fibres are one of the most abundant categories of plastic pieces found. One particularly significant source of microplastic fibres is from the washing of fabrics, not only in the industrial settings but also by consumers. Indeed, a single item of synthetic clothing can release more than 1900 plastic fibres in just one wash cycle.[38] With the large drum capacity of modern washing machines, it is entirely plausible that up to 19,000 fibres could be released from a single wash of just 10 garments. More recently this figure was re-evaluated and a figure of 2000 fibres per garment per wash cycle was concluded.[9] Furthermore, it has been estimated that waste water contains up to 100 fibres/ litre.[110] Ultimately, microplastic fibres are able to be carried by water courses and sewage systems before being deposited in the seas and oceans via discharge outlets.[182,169] Certainly, synthetic microplastic fibres have been discovered in large abundance in soil that has been treated with collected sewage sediments, thereby adding credence to the growing evidence that synthetic clothing fibres represent a considerable environmental hazard.[415] Moreover, large amounts of microplastics have been observed to be able to pass straight through collection filters at municipal sewage treatment plants.[430]

Aside from fibres, waste water often contains large amounts of other plasticles, such as microbeads (small spherical particles of plastic less than 1 mm to 1 μm in size) and nanoplastics (see Fig. 5.8). Manufacturers of cosmetics often incorporate these tiny plasticles into many of their products. This may be for a variety of reasons, such as to facilitate control over the viscidness of the product and for the creation of films.[253] After use, the cosmetic products are typically wiped off the face with cleansing wipes. Often, consumers will then flush the used cleansing wipes down the toilet, resulting in their entry into sewer systems.[327] Since microbeads and nanoplastics are very small, they are easily washed off the cleansing wipes and subsequently escape capture by filtration at municipal sewage treatment plants.[307,410,430]

Furthermore, manufacturers of cleansing products used on human skin incorporate microbeads in their products to aid in deep cleansing and dermal abrasion of the skin[391] (see Fig. 5.3). Traditionally, such cleansers relied upon naturally abrasive materials such as oatmeal,[126] ground pumice[65] or walnut shells.[13] However, within the last 40 years, cleansing product manufacturers have increasingly switched to using plastics instead.[126,292] Typically, the microplastic of choice is polyethylene, which owing to the smooth nature of its surface, tends to have a gentler effect on the skin, thereby reducing damage in comparison to rough surfaced natural abrasives.[166]

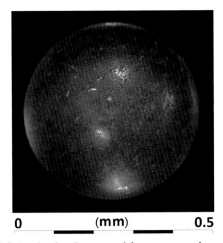

0   (mm)   0.5

Figure 5.3 A microbead recovered from a personal care product.

Furthermore, microbeads are often added to toothpastes for whitening purposes and even to deodorants to block pores and prevent perspiration.[13] During use of these products by consumers, microbeads are washed down drains and consequently travel through sewage systems to municipal treatment plants.[166] It has been estimated[165] that the average individual in the United States will use approximately 2.4 mg of polyethylene microplastics contained in personal care products each day. This equates to the discharge of around 263 tonnes of polyethylene plastic per year for the entire populace. Since treated wastewater effluents are typically discharged into rivers or the sea,[341,307] the suspended plasticles are emitted via the discharged water directly into the aquatic environment.[166,410] It has been estimated[382] that Europe emits between 80,042 and 218,662 tonnes of primary microplastics into the ocean every year and that between 2461 and 8627 tonnes (3.2–4.1%) of these primary microplastics are of the microbead type and originate from personal care products. Importantly, these estimates do not include nanoplastics, which are regularly incorporated in personal care products and cosmetics, and do not take into account other microbead containing products, such as abrasives. Thus, the actual amount of plastic entering the ocean from microbead and nanoplastic containing products is likely to be far greater. Indeed, a single industrial effluent pipe in Austria was alleged to potentially emit 95.5 tonnes of microplastics every year.[247]

During periods of high demand or heavy rainfall, wastewater from treatment plants may be emitted directly into the aquatic environment without any sort of filtration whatsoever. Importantly, in some cosmopolitan areas, such as London,[341] domestic tap water is composed of a large percentage of water previously discharged from wastewater treatment plants. Since plasticles have been found to escape filtration at wastewater treatment plants,[307,410,430] one could hypothesise that microplastics, and especially nanoplastics which are practically invisible, may be able to bypass further water treatments processes and find their way into municipal drinking water. Consequently, this may require rigorous scientific investigation to determine if any health risks are posed.

Table 5.2 **The purposes of different types of plastic microbeads in cosmetics and personal care products[253]**

Plastic	Use in cosmetics and personal care products
Polyethylene terephthalate	Hair fixative; adhesion; film formation; viscosity control; aesthetic agent (sparkles)
Polyethylene isoterephthalate	Bulking agent
Polybutylene terephthalate	Viscosity control; film formation
Polypropylene terephthalate	Emulsion stabiliser; skin conditioner
Polypentaerythrityl terephthalate	Film formation
Polyethylene	Film formation; viscosity control; abrasive; binding agent
Polypropylene	Viscosity control; bulking agent
Polystyrene	Film formation
Polyurethane	Film formation
Poly(methyl methacrylate)	Sorbent material for active ingredient delivery
Nylon 6	Viscosity control; film formation
Nylon 12	Viscosity control; bulking agent; opacifying agent
Polytetrafluoroethylene	Bulking agent; binding agent; slip modifier; skin conditioner
Ethylene/acrylate copolymer	Film formation; gelling agent
Ethylene/methacrylate copolymer	Film formation
Styrene/acrylate copolymer	Aesthetic agent (colour)
Styrene/ethylene/butylene copolymer	Viscosity control
Styrene/ethylene/propylene copolymer	Viscosity control

Following several decades of widespread use, microbeads (Table 5.2) have now become highly prevalent in marine surface waters and sediments,[126] and it is unsurprising that the some of the most common types of microplastics found in many aquatic environments are fibres and microbeads.[152] Importantly, areas in close proximity to effluent discharges from wastewater treatment plants are considered to be significant exposure sites of aquatic biota to contaminants that traditional water treatment processes have failed to remove.[343] Indeed, the source of microplastics present within the Solent estuarine complex in the United Kingdom were ascertained to be wastewater treatment plants and industry.[152]

Consequently, many of the world's major manufacturers have now promised to cease incorporation of microbeads in their product lines, partly in response to pressure from NGOs, consumers, legislative bodies and an increasing number of retailers who have taken the initiative and pledged to discontinue procurement of microplastic containing products. Furthermore, several governmental authorities are introducing outright bans on the production of microplastic containing products.[410] However, although these are promising steps in reducing the vast amounts of plastic discharged into aquatic environments every year, at the present time there is no effective method, of any kind, that is capable of removing the vast amounts of primary microplastics (Fig. 5.4) that are already there.

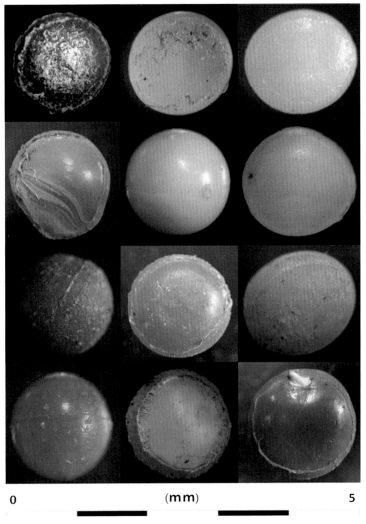

0   (mm)   5

Figure 5.4 Primary microplastics recovered from the aquatic environment.

Secondary microplastics

Secondary microplastics are irregular pieces of plastic that have been unintentionally produced as a result of the degradation of larger pieces of plastic, such as plastic bags, crates, bottles and especially ropes and nets.[350] Over a period of time, these large pieces of plastic litter will degrade as a result of exposure to ultraviolet light from the sun[62] and by mechanical means, such as tidal waves,[73] to form smaller and smaller pieces of plastic (see Chapter 4). For example, it has been very recently demonstrated that a 1 cm^2 piece of polystyrene coffee cup lid placed in demineralised water at 30°C for a period of 24 h, and exposed to ultraviolet light at 320–400 nm, is capable of producing 1.26×10^8 nanoparticles per millilitre after 56 days, and with an average size of 224 nm.[236] This is concerning as polystyrene nanoplastics are capable of distribution throughout the water column[29] and therefore are available for ingestion by a large variety of biota, which mistakes them as a food source (see Chapter 7). Furthermore, as a result of their very small size, as well as the difficulties in detecting and recovering them, very little research on nanoplastics within the aquatic environment has been carried out. Another source of secondary microplastics is discarded carpets[62] and anti-predator netting,[24] which owing to its fibrous nature, can release large volumes of fibres into the surrounding water. Consequently, it has been estimated that 18% of all microplastics originate from the degradation of plastic ropes (Fig. 5.5) and nets.[7] Additionally, it has been estimated that Europe contributes between 68,500 and 275,000 tonnes of secondary microplastics (Fig. 5.6) to the ocean annually.[382]

Figure 5.5 The degradation of plastic ropes is one of biggest sources of microplastics.

Since oil based microplastics are exceptionally resistant to biodegradation, they tend to be highly persistent and become distributed via ocean currents throughout the marine environment.[368] Nevertheless, at the current time there is a considerable knowledge gap concerning the spatial distribution, pathways and fate of microplastics in the aquatic environment.[81]

0   (mm)   5

Figure 5.6 Secondary microplastics recovered from the aquatic environment.

Microplastic litter legislation

Microplastics are essentially an element of plastic litter, thereby fitting under the broader umbrella of marine litter. Thus, they are covered under several global, international, regional and national legislative agreements. In terms of international legislation, microplastics are covered under numerous international laws, conventions, agreements, regulations, strategies, action plans, programmes and guidelines drawn up by international and regional agencies. These include the United Nations (UN), the International Maritime Organisation (IMO), the Convention for the Protection of the Marine Environment of the North-East Atlantic (OSPAR), the European Commission (EC) and the Baltic the Helsinki Convention on the Protection of the Marine Environment in the Baltic Sea Area (HELCOM).

In the European Union, microplastics are covered under descriptor 10 of the Marine Strategy Framework Directive (MSFD) for the establishment of their baseline quantities, properties and potential impacts.[147,305] This provides the legal framework that underpins microplastic research within Europe. Also in Europe, and in response to the OSPAR request 2015/1 relating to the 'development of a common monitoring protocol for plastic particles in fish stomachs and selected shellfish on the basis of existing monitoring surveys', the International Council for the Exploration of the Sea (ICES) have recently provided a preliminary protocol for the monitoring of plastics in the stomach of fish.[197]

The rise in concern regarding the impact of microplastics on the environment has resulted in increasing pressure on the cosmetic industry to refrain from the production and use of primary microplastics in personal care products (PCPs). To protect marine ecosystems from contamination, the Netherlands, Austria, Luxembourg, Belgium and Sweden have issued a joint call to the EU for the banning of microbeads in PCPs. This call followed a European Commission green paper published in 2013 on a European strategy on Plastic waste in the Environment, with microplastics being of particular concern.[108]

In the United States, Illinois became the first state to enact legislation banning the manufacture (by 2017) and sale (by 2018) of products containing microbeads. Similar legislation was subsequently passed in New Jersey, New York, Ohio and California with bans being proposed by numerous other states. Towards the end of 2015, President Barack Obama signed into law the Microbead-Free Waters Act of 2015 (H.R. 1321) which 'prohibits the manufacture and introduction into interstate commerce of rinse-off cosmetics containing intentionally-added plastic microbeads'. Ultimately, this federal law aims to ban the manufacturing of microbeads in 2017, with product specific manufacturing and sales bans in 2018 and 2019.

Other countries legislating specifically against microplastics in personal care products (PCPs) include the Canadian government who published a proposal in 2015 to add microbeads to a list of toxic substances, under the Canadian Environmental Protection Act, and intend to ban the use of microplastics in PCPs. Recently in India, the National Green Tribunal (NGT) has asked the ministries of environment and water resources to respond to the case for a ban on microplastics in PCPs, while in Australia, a Senate committee has recently called for the ban of microbeads in PCPs. Incidentally, Australia had previously favoured a voluntary phase-out of microbeads by the industry. This was also the approach favoured by the UK government, who have been coming under increasing pressure to legislate against the issue. However, one of the issues with the legislative route is that there is typically a considerable delay between the gathering of scientific evidence to build an argument that guides policy and the writing, passing and implementation of a law. Nevertheless, this process can be accelerated by public pressure on both industry and legislative bodies.

Recently, the United Nations Environment Programme (UNEP) published a report on plastic titled 'Cosmetics: are we polluting the environment through our personal care'[418] and have developed an app and website titled 'beat the microbead' to allow customers to check if products contain microbeads. Furthermore, the promotion of greater awareness of the impacts of microplastics on the marine environment was one of several policy related recommendations[156] made by UN Joint Group of Experts on the Scientific Aspects of Marine Environmental Protection for the reduction of microplastics in the environment. Highly important work is also undertaken by non-governmental organisations (NGOs) and can draw attention to crucial issues, as well as exerting great influence on both policy makers and industry. A small selection of influential NGO organisations and their mission statements, as of 2016, are listed in Table 5.3.

Table 5.3 **A selection of some influential NGOs and their mission statement**

NGO	Mission
Ceres	'Mobilizing investor and business leadership to build a thriving, sustainable global economy'
Conservation International	'Building upon a strong foundation of science, partnership and field demonstration, CI empowers societies to responsibly and sustainably care for nature, our global biodiversity, for the well-being of humanity'
Food & Water Watch	'Food & Water Watch champions healthy food and clean water for all. We stand up to corporations that put profits before people and advocate for a democracy that improves people's lives and protects our environment'
Greenpeace	'Greenpeace is an independent campaigning organisation, which uses non-violent, creative confrontation to expose global environmental problems and to force the solutions which are essential to a green and peaceful future'
Natural Resources Defence Council	'NRDC works to safeguard the earth – its people, plants and animals – and the natural systems on which all life depends'
The Nature Conservancy	'The mission of The Nature Conservancy is to conserve the lands and waters on which all life depends'
Ocean Conservancy	'Ocean Conservancy is working with you to protect the ocean from today's greatest global challenges. Together, we create science-based solutions for a healthy ocean and the wildlife and communities that depend on it'
Sierra Club	'To explore, enjoy and protect the wild places of the earth; to practice and promote the responsible use of the earth's ecosystems and resources; to educate and enlist humanity to protect and restore the quality of the natural and human environment; and to use all lawful means to carry out these objectives'
The World Wildlife Fund	'To stop the degradation of the planet's natural environment and to build a future in which humans live in harmony with nature, by: • Conserving the world's biological diversity • Ensuring that the use of renewable natural resources is sustainable • Promoting the reduction of pollution and wasteful consumption'

Table 5.3 **A selection of some influential NGOs and their mission statement—cont'd**

NGO	Mission
Marine Conservation Society	'To achieve measurable improvements in the state of our seas, marine biodiversity and fish stocks through changes in government policy, industry practice and individual behaviour'
Fauna & Flora International	'To act to conserve threatened species and ecosystems worldwide, choosing solutions that are sustainable, based on sound science and take into account human needs'

The crucial work undertaken by NGOs has played a key role in successfully raising public awareness of plastic pollution in the aquatic environment. Indeed, several multinational companies, such as Unilever, Johnson & Johnson and Proctor and Gamble, have taken the initiative and announced their intention to stop using microbeads in their products by 2016/17. Furthermore, community-based approaches involving removal of debris from shorelines have also been greatly beneficial, such as the Ocean Conservancy's International Coastal Cleanup and Beach Sweeps in the United States. Ultimately, it is clear that a successful approach to talking microplastic pollution in the aquatic environment should not solely rely upon policy makers and legislation. What is needed is a steadfast commitment from both the private sector (manufacturing, retail, tourism, fisheries) and society as a whole to reduce the manufacture, use and disposal of primary microplastics.

Microplastic characteristics

Microplastics collected from the aquatic environment exist in all manner of shapes, colours and sizes with some exhibiting a spherical appearance, while others appear to possess a fibrous or random shape.[41] Primary microplastics will tend to have a manufactured appearance, exhibiting either a spherical or fibrous shape, and have a consistent even surface. On the contrary, secondary microplastics will tend to have a more random appearance and are therefore more difficult to categorise.[65] One particularly difficulty is that weathering can dramatically change the appearance of both types of microplastic.[9] Furthermore, microplastics display a vast variety of different colours.[66] As such, these different colours are used as part of the standardised system for categorising microplastics recovered from the environment (see Chapter 10). It is worth noting here that the colour of microplastics often provides an indication as to the degree at which they are contaminated with chemical pollutants (see Chapter 6). Indeed, researchers found the highest levels of pollutants on yellow and black microplastics.[80,112,217,320] In a study of microplastics present in the sediments of the Humber River and Humber Bay, Canada,[75] the amount of white coloured plastics significantly outweighed other colours, with the next most abundant being grey and black pellets, followed by green, blue and very small amounts of pink and purple plastics. Moreover, the colour of microplastics has a significant bearing on the degree to which they are confused with natural food sources by marine biota and thus, consumed (see Chapter 7).

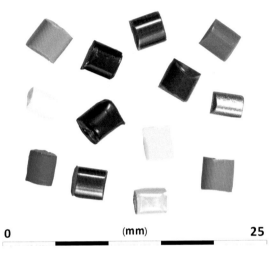

0   (mm)   25

Figure 5.7 Primary microplastics are produced in a large variety of colours, such as the pellets used as industrial feedstock to manufacture larger plastic artefacts.

As part of the standardised size and colour sorting system (SCS) system that was introduced with this book, microplastics are further subdivided into 10 types based upon their size and appearance during identification and categorisation in the laboratory. This ensures that reported results are standardised and importantly, comparable. These 10 types are listed in Table 5.4, as well as illustrated in Fig. 5.8. For the full procedure on how to use the standardised size and colour sorting system (SCS) system to categorise pieces of plastic of any size, see Chapter 10.

Table 5.4 Microplastics are further subdivided into 10 types based upon their size and appearance

Abbreviation	Type	Size	Definition
PT	Pellet	<5 mm–1 mm	A small spherical piece of plastic less than 5 mm to 1 mm in diameter
MBD	Microbead	<1 mm–1 μm	A small spherical piece of plastic less than 1 mm to 1 μm in diameter
FR	Fragment	<5 mm–1 mm	An irregular shaped piece of plastic less than 5 mm to 1 mm in size along its longest dimension
MFR	Microfragment	<1 mm–1 μm	An irregular shaped piece of plastic less than 1 mm to 1 μm in size along its longest dimension

Table 5.4 **Microplastics are further subdivided into 10 types based upon their size and appearance—cont'd**

Abbreviation	Type	Size	Definition
FB	Fibre	<5 mm–1 mm	A strand or filament of plastic less than 5 mm to 1 mm in size along its longest dimension
MFB	Microfibre	<1 mm–1 μm	A strand or filament of plastic less than 1 mm to 1 μm in size along its longest dimension
FI	Film	<5 mm–1 mm	A thin sheet or membrane-like piece of plastic less than 5 mm to 1 mm in size along its longest dimension
MFI	Microfilm	<1 mm–1 μm	A thin sheet or membrane-like piece of plastic less than 1 mm to 1 μm in size along its longest dimension
FM	Foam	<5 mm–1 mm	A piece of sponge, foam, or foam-like plastic material less than 5 mm to 1 mm in size along its longest dimension
MFM	Microfoam	<1 mm–1 μm	A piece of sponge, foam, or foam-like plastic material less than 1 mm to 1 μm in size along its longest dimension

Microplastic density

The density of a microplastic is a key factor that will affect its spatial distribution in the aquatic environment. In a study[72] involving the collection of microplastics by neuston nets from the North Atlantic, microplastics with a greater density than that of seawater were found floating on the surface waters. Similarly, in a study[303] of plastic litter on surface waters, 99% of the microplastic recovered by neuston nets in the western North Atlantic Ocean had an average density less than that of seawater, with the density of the microplastics ranging from 0.808 to 1.238 g/cm^3. Incidentally, the density of seawater is considered to be 1.025 g/cm^3.[353] These results may be surprising since intuitively, one would not typically expect to find material floating on the surface with a density greater than that of seawater. Thus, it is possible that microplastics with a density significantly greater than that of seawater can be found on surface waters, albeit in small quantities, and not exclusively in the bottom sediment. There are two main reasons as to why this can occur:

1. The occurrence of such high density microplastics in surface waters can result from powerful upward and downward movements of water, as result of temperature differences at different depths (vertical mixing).
2. Microplastics denser than seawater may contain pockets or bubbles of air within them, thereby increasing their buoyancy and allowing them to float on the surface.[356]

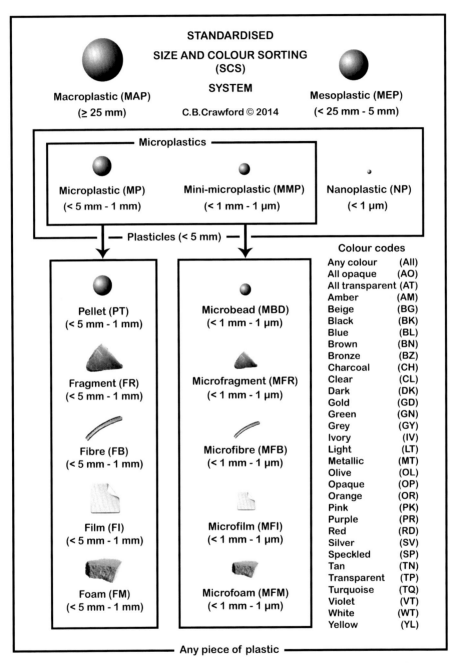

Figure 5.8 The standardised size and colour sorting (SCS) system is used to categorise any piece of plastic, based on size, colour and appearance.

Interestingly, although polyvinyl chloride (PVC) and polyamide (nylon) have high densities of (1.15–1.70 g/cm^3) and (1.12–1.38 g/cm^3), respectively, it has been deduced that wind and tidal currents are likely contributing factors in the transportation of these microplastics to various marine regions, as opposed to density alone.[368] This would suggest that at least some dense microplastics are being actively mobilised by dynamic atmospheric and oceanic currents, as opposed to simply sinking down to the benthos and becoming incorporated into sediments. As well as experiencing a decrease in density, microplastics can also increase in density. There are two main reasons for this:

1. Some microplastics can increase in density as a result of environmental factors. For example, polyethylene, which possesses a density less than that of seawater (0.92–0.97 g/cm^3), can undergo an increase in density as a result of weathering.
2. The accumulation of fouling material (see Fig. 5.9), such as biomass, can lead to an increase in density of the microplastic, thereby resulting in sinking.

Damage to plastics results in abrasion, pitting and cracking of their surfaces and material can collect in these openings and cavities, thereby increasing density. Furthermore, practically anything in the marine environment will start to accumulate a layer of fouling material, such as biomass.[7,309,350] Indeed, biofouling of microplastics less dense than water facilitates an increase in density which enables the microplastics to achieve neutral density within the water column, before slowly sinking to the benthos.[77] Importantly, microplastics can also be incorporated into marine aggregates. These are agglomerations of particulate organic detritus and other debris that form pellet-like precipitates which settle to the sea floor. However, due to their varying buoyancies, microplastics and other plasticles are having a direct impact on the settling velocity at which marine aggregates sink to the benthos. Indeed, diatom aggregates have been observed to experience decreased settling rates while the velocity at which cryptophyte aggregates sink has been observed to considerably increase by more than an order of magnitude (ten times).[265,304]

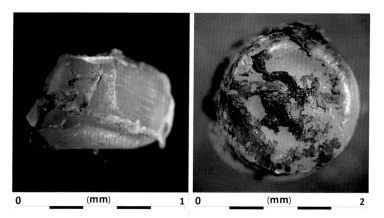

| 0 | (mm) | 1 | 0 | (mm) | 2 |

Figure 5.9 A biofouled secondary mini-microplastic (left) and a fouled primary microplastic (right).

Table 5.5 Density and buoyancy of common microplastics

Substance	Abbreviation	Density (g/cm^3)	Direction in column of seawater
Seawater	SW	1.025	–
Polystyrene (expanded foam)	EPS	0.01–0.05	Ascend ↑ (due to trapped gas)
Polystyrene (extruded foam)	XPS	0.03–0.05	Ascend ↑ (due to trapped gas)
Polychloroprene (neoprene) (foamed)	CR	0.11–0.56	Ascend ↑ (due to trapped gas)
Low-density polyethylene	LDPE	0.92–0.94	Ascend ↑
Linear low-density polyethylene	LLDPE	0.92–0.95	Ascend ↑
High-density polyethylene	HDPE	0.94–0.97	Ascend ↑
Polypropylene	PP	0.88–1.23	Ascend ↑ or descend ↓ depending on the composition – see Chapter 9 (density separation)
Acrylonitrile butadiene styrene	ABS	1.03–1.21	Descend ↓
Polyamide (nylon 6)	PA	1.12–1.14	Descend ↓
Poly(methyl methacrylate)	PMA	1.10–1.25	Descend ↓
Polychloroprene (neoprene) (solid)	CR	1.20–1.24	Descend ↓
Polyamide (nylon 6,6)	PA	1.13–1.38	Descend ↓
Polystyrene (solid)	PS	1.04–1.50	Descend ↓
Polycarbonate	PC	1.15–1.52	Descend ↓
Polyethylene terephthalate	PET	1.30–1.50	Descend ↓
Polyvinyl chloride	PVC	1.15–1.70	Descend ↓
Polytetrafluoroethylene	PTFE	2.10–2.30	Descend ↓

Note: The densities listed above take into account that many plastics are blends of glass fibre or other materials, as well as being plasticised and impact modified. Thus, the minimum and maximum density values relating to plastic blends with and without additional material content, plasticisation and impact modification, see Chapter 9.

Nevertheless, the density of microplastics is a pivotal factor in determining where they are likely to be found in the water column, and the general principle is that microplastics denser than seawater shall sink to the bottom, while those with a lower density than seawater shall float on the surface. Table 5.5 illustrates the densities of various plastics and whether they would ascend or descend if a single 1 mm spherical microplastic was released at the centre-point of a self-contained 1 metre column of non-turbulent seawater with a density of $1.025\,g/cm^3$ at $21\,^\circ C$:

Generally, a typical sample of microplastic will be composed of several different types of plastic. However, the most common types of plastic encountered in the aquatic environment are polyethylene, polypropylene, polystyrene, polyethylene terephthalate and polyvinyl chloride.[355] Nevertheless, some plastics, such as polystyrene,[62] tend to be far more common than other types of plastic and some will be more abundant in some regions, while less abundant in others.[257] Nonetheless, the three most common types of microplastic in the aquatic environment are polyethylene, polypropylene and polystyrene.[250]

Once microplastics enter the aquatic environment, their behaviour tends to falls into three categories[284]:

1. Physical behaviour, such as accumulation, sedimentation and migration.
2. Chemical behaviour, such as the adsorption and absorption of pollutants.
3. Biological behaviour, such as ingestion by biota, translocation and trophic transfer.

This chapter shall focus on the physical behaviour of microplastics in the aquatic environment, while Chapter 6 shall focus on their chemical behaviour and Chapter 7 shall focus on their biological behaviour.

Microplastics in the terrestrial environment and the atmosphere

At the present time, there is very little data on the extent of microplastic pollution in the terrestrial environment, as well as limited information thus far relating to their effects on terrestrial fauna.[107,195,235,352] However, soil applied with sewage sludge for 15 years had much greater concentrations of plastic fibres than soil where no sewage had been applied.[415] Furthermore, a study[195] of earthworms (*Lumbricus terrestris*) found that $<150\,\mu m$ polyethylene microplastics in the litter reduced growth rate and increased mortality of the worms at concentrations in the litter of 28%, 45% and 60% dry weight. In a study[254] of industrial sites in California and Hawaii, the samples of microplastics collected were found to be composed of polypropylene (80–90%) and polyethylene. In a study[105] which collected atmospheric fallout using stainless steel funnels in 20 L glass jars and subsequently identified the material with infrared spectroscopy, between 2 and 355 microplastics per m^2 per day, were reported to deposited. Furthermore, it was estimated that 29% of the microplastics were synthetic fibres and over the course of a year, between 3 and 10 tonnes of fibres would be deposited from the atmosphere over a $2500\,km^2$ area. Consequently, atmospheric fallout may well be another route in which microplastics can reach the aquatic environment in significant volumes.

Microplastics in lakes

Microplastics, such as microbeads from consumer products, are often transported into large bodies of freshwater by urban watercourses and tributaries[41,281,368] as well as being deposited directly via industrial effluents and treated outflows from water treatment plants.[410] Indeed, some lakes have been found to contain vast quantities of microplastics.[116,138] A study[116] of the surface waters of the Laurentian Great Lakes in North America (Lake Erie, Lake Superior and Lake Huron) found significant amounts of contamination from microbeads derived from consumer personal care products. Similarly, a study[336] of the Yangtze River, China found large amounts of microplastics in close proximity to the Three Gorges dam, where the concentrations of microplastics were three times higher than the rest of the river. Interestingly, the researchers concluded that based on the size and shape of the microplastics collected, it is likely that they were mainly composed of secondary microplastics produced from the degradation of larger pieces of plastic. Normally, secondary microplastics are typically associated with the marine environment where powerful tidal dynamics actively degrade macroplastics by mechanical means. However, the discovery of a large abundance of secondary microplastics near the Three Gorges Dam dissipates this notion and proves that secondary microplastics can be just as abundant in freshwater environments. Indeed, rivers can be fast flowing turbulent environments and many would be quite capable of mechanically degrading macroplastics into secondary microplastics.

In a study[75] of microplastics present in the sediments of Lake Ontario, Canada, it was established that in the centre of the lake, no microplastics were found in the sediment below 8 cm in depth. Accordingly, the researchers deduced that microplastics started to accumulate in the lake from around 1977. The most abundant types of microplastic found in the sediments were polyethylene, which represented 74% of the plastics present, followed by polypropylene (17%) and then nitrocellulose (9%) and were between 0.5 mm and 3.0 mm in size. Although polyethylene and polypropylene are buoyant materials with a density typically less than water, when they are blended with mineral materials, such as glass fibre, they have a density greater than water and shall sink in the aquatic environment (See Chapter 9).

Microplastics in rivers and estuaries

Plastic pollution originating in lakes may be transported out of these bodies of water and into the marine environment by naturally flowing watercourses, such as rivers and estuaries.[41,368] The estuarine environment is a unique and fruitful aquatic habitat, as a result of their variable salinity.[259] However, estuaries are also dynamic fluctuating transition regions where currents can flow both in and out as a result of tidal changes.[368] This turbulent environment can result in the congregation of colloidal suspensions of particles which materialise into a larger mass that subsequently sinks to the bottom and contaminates sediment.[456] Indeed, in the Goiana Estuary in Brazil, high amounts of rainfall were positively correlated with the flushing of microplastics into the lower parts of the estuary near the sea, where higher concentrations were detected during the rainy season than in the dry season.[258] Since higher concentrations of microplastics were found in the upper and lower parts of the estuary, as opposed to the middle section where the waterfronts meet, it is hypothesised that during low rainfall, the middle section of the estuary acts as a barrier preventing the progress of microplastics towards the sea. However, when rainfall is high and freshwater surges down the estuary, the microplastics are able to pass the barrier and be carried out to sea. Similarly, increases in 2 mm–5 mm polystyrene microplastics were observed at the Nakdong river estuary, located in the Southeastern Sea of Korea after the rainy season in July, as opposed to the levels of polystyrene microplastics before the rainy season in May.[215]

An investigation of primary microplastics present on the shoreline of Humber Bay, Canada[75] established that the rate of accumulation of primary microplastics, in the form of pellets from industrial sources, was dependent on the amount of rainfall in the region. Thus, high periods of rainfall resulted in a greater deposition of pellets along the shoreline of the bay. The researchers attributed this to increased flows of water through tributaries that feed into the lake and hence, transport the pellets there. In the St Lawrence River in Québec, Canada a study[52] of the river sediments found that the average density of 0.4 mm–2.16 mm polyethylene microbeads was $13,832 \pm 13,677$ per m^2, while one of the 10 sites sampled on the river had a concentration of microplastics at 140,000 per m^2, which is similar in amount to the most contaminated sediments in the world. Interestingly, the size of the microbeads at sites in the vicinity of industrial and municipal effluents was on average smaller (0.7 mm) than microbeads at sites free from effluents (0.98 mm).[52]

In the river Danube in Austria, Europe's second largest River (after the River Volga in Russia), it was alleged that under normal conditions, 200 grams of microplastics were emitted each day (during a monitoring period in 2010) into the river via industrial wastewater at a manufacturing site, and in line with local legislation, which specified the maximum allowable amount of plastic that can be released into running water to be 30 mg/L.[247] Consequently, under normal conditions, it was estimated that at that time, 95.5 tonnes (equivalent to 2.7 million non-returnable 1.5 L bottles) of plastic could potentially be released into the water annually at this site.[247] Incidentally, the River Danube drains into the Black Sea, a large body of water between Southeastern Europe and Western Asia, which is considered to be one of the most polluted regions in the world.[81] Asides from direct sampling of rivers and estuaries, computational modelling can be utilised to deduce the levels of plastics entering a given area. For example, a computational simulation of plastic debris in the Adriatic Basin over 2009–2015 concluded that the Po Delta, an expanse of marshland and lagoons where the River Po splits into several channels that drain into the sea, received a plastic flux of around 70 kg per kilometre per day.

On the shore of the Humber River, Canada, 95% of plastics were polyethylene while only 5% were polypropylene. However, on the beach of Humber Bay, Canada, 73.5% of plastics were polyethylene while 26.5% were polypropylene.[75] Similarly, in the Tamar Estuary in the United Kingdom, a study[368] found that the most ubiquitous types of microplastic recovered were polyethylene, polystyrene and polypropylene, while the most common size of microplastic was 1 mm–3 mm in size. However, nylon was only present in samples of microplastics that were 3 mm or smaller, while PVC was only found as fragments between 1 mm and 5 mm in size. Similarly, in a study[215] of the Nakdong river estuary, which involved the use of 330 μm Manta trawl and 50 μm hand nets, microplastics smaller than 2 mm were observed at all areas tested. The researchers noted that when using the smaller sized hand net, a two-fold increase in collection of these microplastics was observed. Furthermore, the most common types of microplastic (<2 mm) were polyethylene, expanded polystyrene and polyester fibres and paint particles. Conversely, in a study of the Solent estuarine complex in the United Kingdom, microbeads and fibres of around 0.5 mm in size were found to be the most common types of microplastic present.[152]

Microplastics in marine surface waters

Since polystyrene production began in 1930[393] (see Chapter 1), around four decades later in the early 1970s researchers began detecting polystyrene microplastics ranging from clear to white in colour, and with a diameter of 0.2 mm–2.5 mm, on the surface waters of the North Atlantic Ocean.[48,72] Since then, polystyrene has become one of the most frequently detected types of microplastic on the surface of the oceans.[7,202,358] Indeed, a study[298] which sampled the surface waters of the North Pacific Gyre found that the samples that were recovered were mainly composed of polystyrene, thin plastic films and most commonly polypropylene and monofilament fishing line. Incidentally, monofilament fishing line is generally composed of polyamide (nylon).[269] Since then, the amount of plastic entering the ocean each year has been increasing steadily.[312]

An analysis of data[115] collected between 2007 and 2013 from 1571 sites around the world concluded that there are now over 5.25 trillion pieces of plastic present on the surface of the world's oceans with a mass of at least 268,940 tonnes and of which 35,450 tonnes are in the form of microplastics. Furthermore, a study[298] in 2001 which involved sampling the surface water at 11 indiscriminate locations in the North Pacific Gyre, using 333 μm neuston nets, discovered the greatest profusion of microplastics ever documented in the Pacific Ocean at that time with 334,271 microplastics per km^2 and with a mass of over 5 kilograms per km^2. Eleven years later in 2013, a study[162] reported the median concentration of microplastics in the Northeast Pacific Ocean to be 0.488 microplastics per m^2, which potentially equates to 448,000 microplastics per km^2 (Table 5.6). Furthermore, a study[303] of plastic litter in the Northwest Atlantic Ocean identified that 69% of all pieces of plastic collected were in the size range 2–6 mm and 88% were smaller than 10 mm.

Table 5.6 **The reported abundance of debris in marine surface waters at various locations, in ascending order**

Region	Net size	Abundance of debris
California Current (Pacific Ocean)	0.505 mm	0.011–0.033 items/m^3 (median)[160]
United States (West coast)and Southeast Bering Sea	0.505 mm	0.004–0.19 items/m^3 [103]
California Current (Pacific Ocean)	0.333 mm	3.29 items/m^3 [240]
Southern California (Nearshore waters)	0.333 mm	7.25 items/m^3 [299]
Southeast Pacific Ocean (Chile) (40°S and 50°S)	No net–visual only	<1 item/km^2 [404]
Ligurian Sea (Arm of the Mediterranean Sea)	No net–visual only	1.5–3 items/km^2 [3]
Bay of Bengal (Northeast Indian Ocean)	No net–visual only	8.8 ± 1.4 items/km^2 [365]
Southeast Pacific Ocean (Chile) (Nearshore waters)	No net–visual only	>20 items/km^2 [404]
Ligurian Sea (Arm of the Mediterranean Sea)	No net–visual only	15–25 items/km^2 [3]
Northwest Pacific Ocean	0.5 mm	<37.6 items/km^2 [88]
Southern Chile (Fjords, gulfs and channels)	No net–visual only	1–250 items/km^2 [190]
Straits of Malacca (Northeast Indian Ocean)	No net–visual only	578 ± 219 items/km^2 [365]
Caribbean Sea	0.335 mm	1414 ± 112 items/km^2 [243]
Gulf of Maine (North Atlantic Ocean)	0.335 mm	1534 ± 200 items/km^2 [243]
Northeast Pacific Ocean	No net–visual only	0.0014–0.0032 items/m^2 [163]
North Atlantic Ocean (Nearshore waters)	0.333 mm	3537 items/km^2 [48]
Australia	0.333 mm	4256.4–8966.3 items/km^2 [350]
Northeast Pacific Ocean	No net–visual only	0–15,222 items/km^2 [407]
Northeast Pacific Ocean	0.333 mm (surface) and 0.202 mm (subsurface)	0.021 items/m^2 (median)[163]
North Atlantic Gyre	0.335 mm	20,328 ± 2324 items/km^2 [243]
South Pacific Gyre	0.333 mm	26,898 items/km^2 (mean)[117]
Bay of Calvi (Mediterranean Sea)	0.2 mm	6.2 items/100 m^2 [70]
North Pacific Gyre	0.333 mm	174,000 items/km^2 [465]
North Pacific Gyre	0.333 mm	334,271 items/km^2 [298]
Northeast Pacific Ocean	0.333 mm (surface) and 0.202 mm (subsurface)	0.448 items/m^2 (median)[163]

The mystery of the missing plastic

One of the biggest fundamental mysteries in the field of microplastic research is that based on the amount of plastic entering the ocean, 99% of the plastic that should be present is missing and cannot be accounted for.[77] It appears that the smaller a piece of plastic is, the less likely it is to be found in the surface waters of the aquatic environment. Indeed, a study carried out in 1999 by the Algalita Marine Research Foundation (AMRF) concluded that in the North Pacific Central Gyre, there was a correlation between the size of microplastics and their abundance. Thus, the AMRF discovered that the smaller a microplastic was, then the lower its abundance was.[296] Similarly, an analysis of sampling data[243] relating to the abundance of plastic in the surface waters of the Caribbean Sea and the North Atlantic Ocean over 12 years appears to suggest that the quantities of microplastics in regions in which one would expect high accumulations are actually remaining constant, despite more plastic entering the oceans every year. More recently, it was estimated that the amount of plastic in the world's oceans is two orders of magnitude (100 times) lower than one would expect, considering the amount of plastic regularly released into the marine environment.[77]

Since we know that microplastics are regularly released into the marine environment and that large pieces of plastic tend to degrade into smaller pieces of plastic, it appears that these very small pieces of plastic are leaving by some unknown mechanism. Furthermore, a study which involved the analysis of six years of global oceanic sampling data appears to show that microplastics smaller than 4.75 mm are leaving the surface waters of the ocean by some currently unidentified route or mechanism.[115]

Thus, one can postulate seven possibilities:

1. The vast majority of large plastics have not yet been in the ocean long enough to degrade to such a small size, hence the greater proportion of larger sized plastics.
2. Once in the ocean, the plastic rapidly degrades to such a small size that it can no longer be easily detected.
3. Once a buoyant microplastic degrades down to a suitably small size, it is able to be picked up by the wind from sea-spray and become subsequently deposited on land, with some being too small to detect.
4. The accumulation of fouling and biofouling leads to an increase in density of the microplastics, thereby resulting in sinking.
5. The microplastics are incorporated into marine aggregates, thereby forming pellet-like precipitates and sinking.
6. Once microplastics reach a suitably small size, they are consumed by marine life and translocate to their tissues. Alternatively, they are excreted and sink as a result of fouling.
7. The smaller pieces of plastic are being degraded/biodegraded by some unknown process.

Any, or a combination, of these seven possibilities would provide a mechanism that could explain the observed incidences in samples of the very smallest microplastics being in far lower abundance than the larger-sized pieces. It may well be the case that many microplastics are deposited on land by waves or picked up from within sea-spray by the wind and carried away from the ocean, perhaps even great distances if the plastics are small enough. Certainly, very small microplastics have been discovered on shorelines. For example, several samples collected from beaches and sediments in the Tamar Estuary, Southwest England revealed that the bulk of plastic particles found in their samples were predominantly fibres of approximately 20 μm in diameter,[406] while samples obtained from two Portuguese beaches contained fibres as small as 1 μm in diameter.[80]

The possibility that the unexplained loss of microplastics from surface waters may be attributed to microplastics sinking as a result of fouling, or becoming incorporated into marine aggregates, seems probable. Certainly, microplastics recovered from sediments will tend to be considerably denser than those floating on the surface,[40,41,319] and plastics with a higher density than seawater,[84] such as nylon and acrylonitrile butadiene styrene (ABS) have been found in marine sediments at St John Island, Singapore.[314] Furthermore, this possibility is lent credence by the alarming discovery that submarine canyons and deep sea sediments have recently been found to contain vast amounts of plastic.[332,458] Indeed, deep sea core samples taken from the Mediterranean Sea, the Atlantic Ocean and the Indian Ocean revealed that the abundance of microplastics was four orders of magnitude (10,000 times) higher in the deep sea core samples than in samples taken from the associated surface waters, and around 4 billion plastic fibres per km^2 are estimated to be present in the sediments of the Indian Ocean.[458] As such, further research is urgently needed to establish if this high abundance of plastic is having an impact on the deep sea ecosystem and whether microplastics are exhibiting detrimental effects on specific deep sea organisms. Thus, it may be the case that the deep ocean is where much of the plastic has gone and is therefore acting like a vast sink for plastic waste. Ultimately, until further research is carried out, the fate of the missing plastic still remains a mystery.

Microplastics in marine sediments and shorelines

Microplastics are now ubiquitous in marine sediments. However, in the vicinity of Plymouth in the United Kingdom, a study[406] concluded that microplastics were in much higher abundance in subtidal regions as opposed to estuarine and sandy regions and that 33% of samples were composed of nine distinct types of plastic. Furthermore, levels of microplastics at various sites within these different regions were reasonably constant. However, a study[61] found that microplastics were particularly abundant in the coastal sediments of Belgium at a concentration of up to 390 microplastics per kilogram of dry sediment, with an important source being freshwater rivers. The concentrations of microplastics, especially in harbour areas, were between 15 and 50 times higher than the concentrations of microplastics reported[406] in Plymouth, United Kingdom, as well as being considerably higher than the concentrations of microplastics reported[314] in the coastline sediments of Singapore. The researchers postulated that the enclosed nature of the Belgian harbours prevents adequate rates of flushing and thus, facilitates the deposition of microplastics in the sediment. Furthermore, based on the different abundances of microplastics at different depths in the sediments, the researchers state that the concentration of microplastics appears to be increasing.[61] More recently, much higher abundances of microplastics have now been reported. For example, a study[32] at two locations (Horseshoe Bay and Cates Park) in Burrard Inlet in British Columbia, Canada, intertidal sediments were found to contain microplastics at a concentration of 5560 microplastics per kilogram of wet sediment in Horseshoe Bay and 3120 microplastics per kilogram of wet sediment in Cates Park. Of all the microplastics that were recovered, 75% were identified as microbeads derived from consumer products. Similarly, a study of microplastics in sediments of the Venetian lagoon, a bay in the Adriatic Sea in Northern Italy in which Venice is located, it was found that the sediment contained between 2175 and 6725 microplastics per kilogram.[431] Importantly, the size of the grains in sediments or sands does not appear to have any relationship with the abundance of different sizes of microplastics in those sediments or sands.[41,345]

In a study[51] of core sediment samples obtained from Kamilo Beach, Hawaii, it was determined that the amount of microplastics present in the samples was up to 30.2% by weight. Furthermore, 95% of the microplastics were found in the top 15 cm of the core sample and 85% of them were identified as polyethylene. Furthermore, it was observed that beach sediment which contained microplastics tended to become warm more slowly with a maximum decrease in thermal diffusivity of 16%, while the maximum temperature that the beach sediment reached was lower. Thus, it was postulated that the presence of microplastics in sediment could have an adverse effect of marine species in which temperature plays a role. For example, the temperature at which buried sea-turtle eggs are exposed to during development determines the gender of the hatchlings,[51] with warmer temperatures producing females and colder temperatures producing males.

In samples collected from coastal locations in Portugal, it was discovered that 72% of plastic fragments where in the microplastic size range of less than 5 mm.[274] In another study[80] of Portuguese beaches, it was found that polyethylene, polypropylene and polystyrene were the most common microplastics found although there were around three times more polyethylene and polypropylene than polystyrene. Similarly, a study[217] which obtained samples of plastic pellets recovered from four different beaches in Greece found that the pellets were predominantly composed of polyethylene and to a lesser extent, polypropylene (Table 5.7).

Table 5.7 The reported abundance of benthic debris at various locations, in ascending order

Region	Depth (m)	Abundance of debris
United States (West coast)	55–1280	67.1 items/km^2 [222]
Eastern Mediterranean Sea (Greece)	–	72–437 items/km^2 [231]
Bay of Biscay (Northeast Atlantic Ocean)	0–100	0.263–4.94 items/ha [146]
Portugal (West coast)	850–7400	1100 items/km^2 [301]
Atlantic Ocean	200–2800	0.59–12.23 items/ha [457]
Indian Ocean	1320–1610	0.75–17.39 items/ha [457]
Northwest Mediterranean Sea	750	19.35 items/ha [148]
Antalya Bay (Eastern Mediterranean Sea)	200–800	115–2,762 items/km^2 [168]
Gulf of Mexico	359–3724	<28.4 items/ha [447]
Belgium continental shelf	–	3125 ± 2830 items/km^2 [424]
Eastern Fram Strait (The Arctic)	2500	3635–7710 items/km^2 [26]
French coast (Mediterranean Sea)	100–1600	0–78 items/ha [150]
United States (West coast, California)	20–365	1.7 items/100m [438]
European coast	< 2200	0–1010 items /ha [149]
Northeast Gulf of Aqaba (Red Sea)	–	2.8 items/m^2 [1]

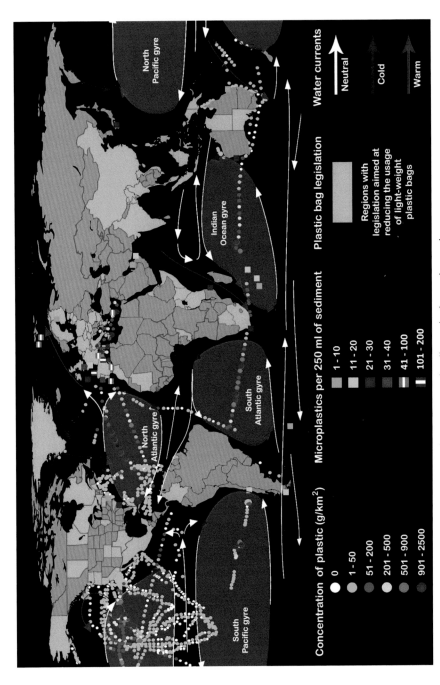

Figure 5.10 The global abundance of plastics in surface waters and sediment in the marine environment. Created using data from Refs.[38,71,77,79,91,103,116,161,242,243,280,350,458]

The global abundance of microplastics in marine surface waters and sediments

Ultimately, the concentration of microplastics in the aquatic environment is expected to increase dramatically in the coming years, mainly as a result of the increasing production of plastic materials, the mismanagement of plastic waste,[204] the influx of microbeads from industry and consumer products via effluent and the degradation of large plastic litter.[244] Nevertheless, many regions of the world have introduced legislation aimed at reducing the use of light-weight plastic bags (see Fig. 5.10). Consequently, this legislation significantly helps towards alleviating the number of plastic bags which end up in the aquatic environment and the potential for their breakdown into microplastics or their ingestion by marine biota.

Weathering of microplastics

Microplastics in the aquatic environment will tend to undergo weathering (see Fig. 5.11). This may be caused by mechanical means, such as the abrasive action of waves in the ocean, as well as from oxidative damage as a result of exposure to

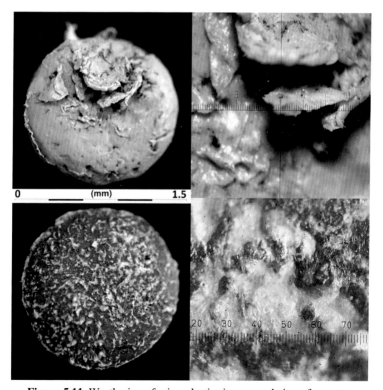

Figure 5.11 Weathering of microplastics increases their surface area.

heat and ultraviolet light, such as when sunlight degrades the plastic and makes it weak, brittle, discoloured and susceptible to damage (see Chapter 4). However, the accumulation of biomass and other material on the surface of microplastics may lead to an increase in density, thereby facilitating the sinking of buoyant plastic materials.[7,309,350] Certainly, analysis of polyethylene submerged in seawater for 3 weeks showed a significant increase in biofilm formation over time.[262] Once incorporated into sediments, microplastics are subjected to lower temperatures, decreased levels of oxygen and shielded from damaging ultraviolet light and mechanical weathering, thereby greatly facilitating their persistence in the aquatic environment.

Importantly, there are numerous toxic chemical pollutants which persist in the aquatic environment and these generally exist as a result of human activities, such as industrialisation and internal combustion engines. Microplastics have the potential to concentrate these pollutants onto their surface, or absorb them into the bulk of the material, with the levels of pollutants sorbed to the microplastics tending to reach much higher concentrations than the surrounding water (see Chapter 6). When these contaminated microplastics are mistaken for food and consumed by aquatic organisms, the pollutants are also ingested (see Chapter 7). Thus, the degree to which a microplastic is weathered is an important factor because the weathering of microplastics greatly increases their surface area. Consequently, the reaction sites at which the sorption of dangerous toxic pollutants can occur is potentially increased, thereby altering the chemical behaviour of the microplastics and increasing their contamination with dangerous pollutants that are present within the aquatic environment.

The interactions of microplastics and chemical pollutants

6

Persistent organic pollutants (POPs)

Of all the diseases affecting human health, it is estimated that more than 50% are purely the result of environmental factors,[171] while industrial, agricultural and other anthropological activities on land result in significant pollution of the aquatic environment with hazardous chemicals.[444] Consequently, the contamination of the world's aquatic environments with noxious chemicals is an area of serious and growing concern. For example, in Europe it is estimated that there are over 700 different chemical pollutants present within the aquatic environment, in 20 different chemical classes.[155] The number of contaminants of emerging concern, present in the global environment, amounts to more than 40,000 different chemicals. Alarmingly, this number increases by six additional chemicals each and every day.[171] It is estimated that every year, 700,000 barrels of petroleum pollutants are spilled into the waters surrounding North America, while less than 10% resulting from oil spills (Fig. 6.1).[311] Furthermore, a volume of oil, equating to approximately 625,000 barrels, is steadily discharged into coastal waters every year as a result of run-off from land.[18] Moreover, assessed plastic materials have been estimated to release between 35 and 917 tonnes of chemical additives into the global marine environment every year, with the vast majority being released from plasticised PVC[395].

Figure 6.1 Petroleum pollution on a beach.

Microplastic Pollutants. http://dx.doi.org/10.1016/B978-0-12-809406-8.00006-2

Many of these aquatic pollutants are considered to be persistent because they are highly resistant to environmental degradation. As a result, they remain in aquatic environments for considerable periods of time.[288,321,443] Indeed, a study[171] of emerging contaminants established that the time period in which a contaminant of concern decreases in magnitude from being regarded as being of the highest concern to a baseline level of the lowest concern is typically 14.5 ± 4.5 years. Furthermore, many of these pollutants possess the ability to bioaccumulate in many organisms, including humans, and are able to traverse the food chain, as well as being passed from mothers to their offspring within the womb.[210]

Consequently, hazardous chemicals of this nature are termed as persistent organic pollutants (POPs) and the considerable danger they pose resulted in the creation of the Stockholm Convention on Persistent Organic pollutants by the United Nations Environment Programme in Sweden, in 2001. The purpose of the convention was to create a list of pollutants of concern among participating countries and to attempt to reduce, restrict or eliminate the manufacture of these persistent pollutants.[177] Initially, the list contained 12 pollutants. However, this has been increased over time to include other pollutants of concern and at the time of writing, there were 26 chemicals on the conventions list (see Table 6.1).

Table 6.1 POP targeted by the Stockholm Convention

Annex A (Elimination)			
Aldrin	Chlordane	Chlordecone	Dieldrin
Endrin	Heptachlor	Hexabromobiphenyl	Hexabromocyclododecane (HBCD)
Hexabromodiphenyl ether	Heptabromodiphenyl ether	Hexachlorobenzene (HCB)	Alpha hexachlorocyclohexane
Beta hexachlorocyclohexane	Lindane	Mirex	Pentachlorobenzene
Polychlorinated biphenyls (PCB)	Endosulfan and isomers	Tetrabromodiphenyl ether	pentabromodiphenyl ether
Toxaphene			
Annex B (Restriction)			
DDT	Perfluorooctane sulfonic acid and salts	perfluorooctane sulfonyl fluoride	
Annex C (Unintentional Production)			
Hexachlorobenzene (HCB)	Pentachlorobenzene	Polychlorinated biphenyls (PCB)	Polychlorinated dibenzo-p-dioxins (PCDD)
Polychlorinated dibenzofurans (PCDF)			

For a pollutant to make the list, it has to fit specific criteria, namely; possessing a half-life of at least two months in the aquatic environment, remaining stable and showing evidence of widespread dispersal from its point of origin, as well as demonstrating bioaccumulative and toxicological effects.[177] Ultimately, the overriding purpose of the risk assessment is to safeguard human health and ecological communities from the dangers posed by a chemical contaminant.[155]

Classes of persistent organic pollutants (POPs)

Although there are many classes of persistent organic pollutants (POPs) which are of environmental concern and which contaminate the aquatic environment, this book shall focus on those classes of pollutants which have been demonstrated to interact with microplastics, and which represent the greatest danger to biological organisms.

Chlordane

Chlordane was a common pesticide which was used for several decades on crops, orchards and gardens until its use was restricted in the early 1980s to controlling termites, after which the substance was banned in the late 1980s. Despite this, the chemical is highly persistent and has a half-life of up to 30 years. Consequently, chlordane still contaminates aquatic environments several decades later and due to the hydrophobic (water repelling) nature of the compound, it readily adheres to hydrophobic surfaces, such as plastics.[264] Furthermore, chlordane tends to bioaccumulate in many organisms and is highly toxic to many species of fish and is classified as a probable human carcinogen (Fig. 6.2).

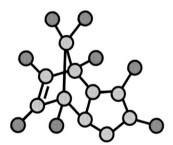

Figure 6.2 Chlordane.

Dichlorodiphenyltrichloroethane (DDT)

Agrarian activities in the 1960s frequently employed dichlorodiphenyltrichloroethane (DDT) as a highly effective pesticide until worldwide bans on manufacture and utilisation were instigated in the 1970s.[112] Furthermore, DDT has been widely utilised in the treatment of encephalitis and for the suppression of malaria.[422] Consequently, DDT and its related compounds dichlorodiphenyldichloroethylene (DDE) and dichlorodiphenyldichloroethane (DDD), are globally distributed and the substances are frequently detected as contaminants of the aquatic environment, even in isolated regions and areas considered to be pristine.[34] DDT tends to result in thinning of the egg shells of birds and long-term human exposure has been associated with chronic health effects (Fig. 6.3).

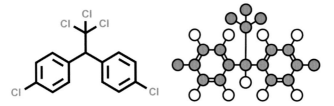

Figure 6.3 Dichlorodiphenyltrichloroethane (DDT).

Hexachlorocyclohexane (HCH)

Hexachlorocyclohexane is a cyclic saturated chlorinated compound that persists in the environment and has eight possible isomers. Despite being a banned toxic substance in the vast majority of countries, HCH is regularly detected in the aquatic environment.[455] Only four isomers were considered to be of the most commercial importance (α-HCH, β-HCH, δ-HCH and γ-HCH). The latter isomer, gamma-hexachlorocyclohexane (γ-HCH) was a purified form which was widely utilised as a pesticide, due to its potent insecticidal properties. Consequently, when an HCH containing pesticide product contained more than 99% γ-HCH, the substance was termed as lindane. Additionally, varying mixtures of HCH isomers, termed technical grade HCH, were commercially sold for a variety of purposes with compositions of isomers in the following range of quantities (60–70% α-HCH; 5–12% β-HCH; 6–10% δ-HCH; 3–4% ε-HCH and 10–15% γ-HCH).[233] Technical grade HCH and α-HCH are considered to be probable human carcinogens while β-HCH is considered to be a possible human carcinogen. Current evidence suggests γ-HCH (lindane) may be carcinogenic but the evidence is considered insufficient for ascertaining the human carcinogenic potential of γ-HCH. Certainly, lindane has been restricted in the United States since 1977 due to safety concerns regarding its neurotoxicity[420] and has since been recognised as a persistent organic pollutant (POP) by the Stockholm Convention.[432] Furthermore, the United States Environmental Protection Agency and the International Agency for Research on Cancer have concluded that there are plausible indications that lindane may be a human carcinogen[187] (Fig. 6.4).

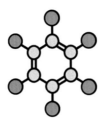

Figure 6.4 The basic chemical structure of HCHs.

Perfluoroalkylates

Perfluoroalkylated substances are a highly fluorinated class of chemicals which are regularly used in industry due to their advantageous surfactant properties and their ability to repel material, such as dirt, oil and water. Consequently, they are used in a multitude of applications, such as in packaging materials, textiles and paints, as well as in foams to extinguish fires. Perfluoroalkylated substances have been detected in a variety of seafood, including fish and are known to persist in the aquatic environment. Furthermore, they have been detected in birds, marine mammals and fish in remote locations, such as the Arctic[158] and in a large variety of aquatic organisms throughout the food web in the North American Great Lakes.[214] Additionally, they have been detected in 95% of the livers analysed from birds in Japan and Korea.[213] Following an assessment of 54,195 analytical results from 13 countries, the dietary exposure of perfluorooctanoic acid (PFOA) and perfluorooctane sulfonic acid (PFOS) was reported to be 2-3 higher in children than in adults.[121] In some animals, perfluoroalkylated substances have been reported to exhibit neurotoxic effects[271] (Figs. 6.5 and 6.6).

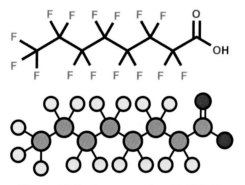

Figure 6.5 Perfluorooctanoic acid (PFOA).

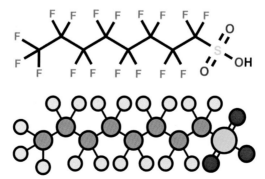

Figure 6.6 Perfluorooctane sulfonic acid (PFOS).

Phthalates

Phthalates are a class of chemicals which are added to plastics during manufacture to make them softer and more flexible (plasticise), and are predominantly used in polyvinyl chloride (PVC)[175] (Fig. 6.7). Phthalates incorporated into microplastics during manufacturing are not chemically bound to the plastic material and are therefore able to subsequently leach out.[405] Some phthalates, such as bis(2-ethylhexl) phthalate (DEHP) (see Fig. 6.8), have been detected in sewage sludge and effluents, as well as surface waters and sediments in Germany, while significant concentrations were detected in liquid manure, compost water and waste dump water.[144]

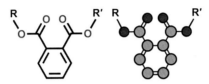

Figure 6.7 The basic chemical structure of orthophthalates (R and R′ represent placeholders).

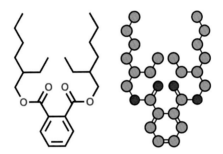

Figure 6.8 Bis(2-ethylhexl)phthalate (DEHP).

Interestingly, a study[221] which investigated the ability of Guar gum in removing DEHP from typical farm effluent reported that at pH 7, as little as 4.0 mg of Guar gum was capable of removing 99.99% of DEHP from 1 L of effluent. This is a promising area of research in removing DEHP from aqueous mediums since Guar gum, being the ground endosperm of guar beans, is biodegradable and non-toxic. Incidentally, a study[31] found an association between some phthalates, including DEHP, in house dust and the occurrence of asthma and allergic symptoms in children. Furthermore, a study[227] of 85 urine samples taken from the general German population calculated that the median intake of DEHP among the population sampled was 13.8 μg per kg of body weight per day. Phthalates are regularly used in many personal care products, such as perfumes, lotions and cosmetics as well as being used in varnishes, coatings and lacquers.[181] It has been reported that while humans can be exposed to phthalates from the use of consumer products, environmental contamination and leaching from other products, dietary intake is historically considered to be the main route of exposure.[373]

Polycyclic aromatic hydrocarbons (PAHs)

Polycyclic aromatic hydrocarbons (PAHs) are class of chemicals that can exist in more than 100 different combinations and are among the most ubiquitous pollutants in the natural environment. Many PAHs are considerably toxic to aquatic species, such as pyrene, which exhibits considerable toxicity even at low levels of exposure.[322] Furthermore, some are even carcinogenic, such as benzo[a]pyrene. Consequently, 16 PAH compounds have been listed by the United States Environmental Protection Agency as pollutants of concern and consequently a value of 200 ng/l has been set as the maximum allowable limit in drinking water[123] (Fig. 6.9).

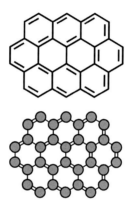

Figure 6.9 Ovalene, a PAH compound.

PAHs are readily produced as a consequence of incomplete combustion of wood, tobacco and other fuel sources composed of carbon compounds.[330] Consequently, it is recognised that some PAHs in the environment can originate from natural sources, such as forest fires and volcano eruptions.[111] However, their presence in the environment is mainly due to anthropogenic activities, such as coal burning power plants, shipping activities[442] and refuse dumping sites.[441] Furthermore, some PAHs are used in industry to produce plastics, pesticides and dyes.

Ocean-based industrial oil-extraction platforms regularly emit PAHs into the atmosphere as part of their manufacturing process,[363] while PAH by-products from combustion can be washed into marine habitats via rainfall and watercourses or settle from the atmosphere onto ocean surface waters.[444] Moreover, it has been determined that the burning of plastic refuse emits PAHs, with polystyrene producing the highest quantities.[450] Additionally, the manufacture of polystyrene can produce PAHs as an undesired consequence of incomplete polymerisation during processing in which the toxic PAH precursors, benzene and styrene, can become incorporated into the polymer matrix.[358]

The marine environment can also be directly polluted with PAHs due to the unintentional release of oil into seawater.[444] Interestingly, when analysing marine samples, the ratio between parent PAHs and alkylated PAHs can be used to determine whether the source of PAH contamination was of combustive origin or from petroleum-based fuels.[363] Accordingly, high levels of parent PAHs would equate to combustive emissions, such as the burning of oil and gas, while high levels of alkylated PAHs would represent contamination from petroleum products, such as fuel leakage during vessel activities in the region or from accidental spillages.

Polybrominated diphenyl ethers (PBDEs)

The polybrominated diphenyl ethers (PBDEs) class of compounds consists of 209 related molecules, known as congeners.[446] The extent of bromination of the diphenyl ring structure, combined with where the bromine atoms are situated in the ring structure, determines the congener identification number.[440] Importantly, highly brominated PBDEs (>5 bromine atoms) can undergo debromination in the environment to form lower brominated PBDEs (<5 bromine atoms) resulting in the production of congeners which exhibit increased toxicity[261] (Fig. 6.10).

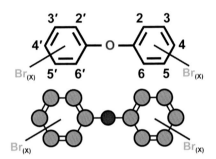

Figure 6.10 The basic chemical structure of PBDEs.

PBDEs, particularly deca-BDE and octa-BDE, were often added to plastic materials during manufacture to serve as flame retardants.[440] However, PBDEs represent a significant threat to marine biota due to their hormone disrupting abilities and other toxicological effects, and are recognised as tenacious aquatic pollutants of rising concern.[444] Consequently, in 2009 they were added to the list of pollutants in the Stockholm Convention.

Polychlorinated biphenyls (PCBs)

Polychlorinated biphenyls (PCBs) are a class of synthetic chemical compounds composed of a biphenyl ring with chlorine atoms attached at various possible locations around the benzene rings. These varying degrees of chlorination directly affect the half-life of the various congeners in the environment by reducing their susceptibility to photodegradation. Consequently, PCBs with a higher degree of chlorination tend to persist much longer in the environment, and the half-life of the various congeners has been estimated to range from 10 days to 548 days.[35] The PCB congeners with highest toxicological profile are 77, 81, 126 and 169 due to high similarities with 2,3,7,8-tetrachlorodibenzo-p-dioxin (TCDD).[446] Of all the polychlorinated dibenzodioxins, TCDD is considered to exhibit the highest toxicological potency to biological lifeforms and may be dispersed into the environment on account of industrial activities, such as combustion and refining.[318] Furthermore, PCBs are listed as probable human carcinogens and can suppress the human immune system (Fig. 6.11).

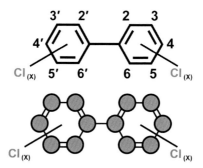

Figure 6.11 The basic chemical structure of PCBs.

The presence of PCBs in the environment is purely a result of anthropogenic activities[321] and these compounds were utilised in industry for around 70 years. Their significant proliferation into the aquatic environment has stemmed from sources such as municipal dumps, electronic components, transformer housings and incinerators.[35] Indeed, municipal dumps are responsible for the ingress of a large percentage of polychlorinated biphenyls (PCBs) into marine compartments, and PCBs are capable of travelling a significant distance from their point of origin via atmospheric mechanisms. Furthermore, the demolition of old industrial facilities, and their associated technology, still plays a significant part in PCB pollution of the aquatic environment despite cessation of manufacture of these compounds in the early 1970s.[444] Due to their highly resilient nature, and the way in which PCBs distribute and behave in ecosystems, it has been estimated that PCBs will remain as the most widespread contaminant of aquatic environments and organisms until at least 2050.[400]

Microplastics as vectors for chemical pollutants

Several decades ago, researchers found that polypropylene microplastics will readily adsorb hydrophobic organic compounds.[351] Since then, microplastics in the aquatic environment have been found to be contaminated with persistent organic pollutants (POPs), by way of sorptive processes, and consequently these interactions are coming under increasing scientific scrutiny.[212,279,354] For example, in a harbour in Sweden, near a polyethylene production plant, the abundance of microplastics in the seawater was reported to be as high as 100,000 per m3.[317] At the same time, microplastics in the marine environment have been reported to be capable of picking up and concentrating waterborne POPs at up to six orders of magnitude (1 million times) higher than the background concentration in the surrounding seawater.[191,279,353,403,460] Consequently, the readiness for microplastics to become contaminated with pollutants, combined with their ubiquity and capability of long distance travel, is a serious cause for concern.

The reason being that microplastics are often eaten by aquatic organisms. As such, the ingestion of contaminated microplastics by biological organisms represents a unique exposure route for the introduction of highly toxic chemical pollutants into the food web. However, there is a considerable knowledge gap with respect to these interactions and consequently, the way in which microplastics can act as vectors for the introduction of aquatic contaminants into the food web is considered by many as a research priority.[81] Thus, studies investigating the ingestion of microplastics by aquatic species are an important first step. However, it is of great importance that the complex interactions involving the ingestion of contaminated microplastics by organisms is systematically investigated to identify any serious risks and effects.[241]

The interaction of persistent organic pollutants with microplastics involves three distinct phenomena; desorption, absorption and adsorption.

Desorption

The term desorption essentially refers to a process in which a substance is released from the bulk or surface of another substance. For example, the pollutants which contaminate microplastics may be desorbed following ingestion of the microplastic by an aquatic organism. Thus, the processes of absorption and adsorption are important precursors to the desorption of chemicals from contaminated microplastics following ingestion, and the subsequent toxicity which may result from the diffusion of these chemical pollutants into the tissues. However, chemicals added to the plastics during manufacture, such as plasticisers and flame retardants, also have the potential to leach out from the plastics following ingestion and exhibit toxicological effects. Indeed, research has shown that chemicals added during the production of plastic artefacts, such as cookware, tableware and drinkware, can subsequently be transferred to humans.[172,226,238,287,289,398,405] Thus, it is plausible that following ingestion, these additive chemicals could leach out of the bulk of the plastic material and subsequently diffuse into biological tissues, thereby inducing toxicity.

Ultimately, there is limited evidence which relates to the effects of contaminated microplastics on organisms and further research is very much needed. However, some early studies have shown that chemical additives and sorbed chemical contaminants do indeed exhibit adverse effects on the health and fitness of aquatic organisms which have ingested contaminated microplastics.[28,42] Furthermore, a study[17] has demonstrated that the desorption of several persistent organic pollutants (POPs) from polyethylene microplastics could be up to 30 times greater than in seawater and of the POPs investigated in the study, polyethylene microplastics contaminated with the polycyclic aromatic hydrocarbon (PAH) phenanthrene exhibited the greatest potential for transfer from microplastics to organisms.

Absorption

The terms absorption and adsorption are often a source of confusion (Note that one term is spelled with a **b** while the other is spelled with a **d**). In fact, both terms relate to the action of sorption, but describe two entirely different sorptive processes. The main difference between the two processes is that absorption is a bulk phenomenon, while adsorption is a surface phenomenon (Fig. 6.12).

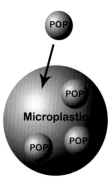

Figure 6.12 Absorption is a bulk phenomenon.

In the case of absorption, waterborne chemical pollutants diffuse into the bulk of the plastic material. For example, in the case of polypropylene, the PAH phenanthrene has been observed to undergo surface diffusion into the bulk of the polypropylene matrix. While deeper diffusion does not occur, it does take place in polyethylene which has a molecular geometry in which the polymer chains are less tightly packed than polypropylene thereby allowing phenanthrene to permeate through these gaps.[89] Similarly, in a study[358] which examined the levels of PAHs sorbed to microplastics recovered from beaches in San Diego, California, it was found that the concentrations of PAHs significantly differed between polystyrene, polyethylene, polyethylene terephthalate, polypropylene and polyvinyl chloride, and that polystyrene and polyethylene exhibited the greatest concentrations of PAHs, followed by polypropylene. Further investigations revealed that virgin polystyrene microplastics, which were deployed in San Diego Bay, California, exhibited a high sorption capacity for PAHs which was comparable to polyethylene, and which was significantly more than polypropylene, polyethylene terephthalate and polyvinyl chloride. Commonly, polystyrene is rigidified by inducing crosslinking during manufacture by introducing 0.1–1.0% p-divinylbenzene.[180] Thus, the researchers postulated that this crosslinking, which occurs via a carbocyclic arene ring,[180] results in greater spaces between the polymer chains.[358] Consequently, the gaps between the chains allow PAHs to permeate more efficiently into the polystyrene matrix, compared to a polypropylene matrix where the polymer chains are positioned closer together. Nevertheless, in some cases, chemical pollutants cannot penetrate into the bulk of the plastic material and adhere to the materials surface instead. This is known as adsorption.

Adsorption

Adsorption can be defined as a process in which material (adsorbate) travels from a gas or liquid phase and forms a superficial monomolecular layer on a solid or liquid condensed phase (substrate). Each phase is a distinct domain comprised of constant applicable physical properties, such as chemical composition, which is separated by a boundary from another part of a main-body system. Ultimately, the physiochemical properties of an adsorbate determine the degree as to which adsorption occurs.[11] In the context of adsorption of persistent organic pollutants (POPs) to microplastics, the POP would be considered as the adsorbate and the microplastic as the substrate. Furthermore, the water (liquid phase) and microplastic (solid phase) are considered immiscible, and the POP is considered as a solute since it is dissolved in a solvent (water) (Fig. 6.13).

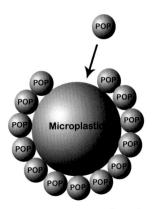

Figure 6.13 Adsorption is a surface phenomenon.

Some microplastics are composed of plastics which are non-polar, such as polyethylene, polypropylene and polystyrene, while some plastics are polar, such as polycarbonate, polyamide (nylon) and poly(methyl methacrylate). The more non-polar a plastic material is, the greater affinity it will have for hydrophobic (water repelling) POPs. Incidentally, the hydrophobicity of a substance is typically measured by its distribution at equilibrium between a partition of two immiscible phases of a hydrophobic substance (octanol) and water. Thus, when a substance which is considerably hydrophobic is introduced into the system, it will tend to disperse into the octanol layer and exhibit a greater concentration in this phase. Thus, the octanol-water coefficient (Kow) is a unitless expression of the ratio between the concentrations of a solute in known, equal and individual volumes of both octanol and water, as part of a two-phase system, while at equilibrium and a precise temperature.[328]

Similarly, when a hydrophobic POP molecule and a non-polar microplastic come into contact with one another, the POP will disperse between the two immiscible phases (water and microplastic) until equilibrium is reached and two distinct phases with differing concentrations of the solute are formed. The ratio between the concentration of the solute in the solid phase (microplastic) and the concentration of the solute in the liquid phase (water) is known as the partition coefficient (K_{pc}) (see Eq. 6.1).

$$\text{Kpc} = [\text{Concentration of POP (Solid phase)}] / [\text{Concentration of POP (Liquid phase)}]$$
(6.1)

The amount of solute (POP) adsorbed per unit weight of the adsorbent (microplastic) at equilibrium is represented by qe (mmol/g) and can be determined by Eq. (6.2).

$$\text{qe} = V\,(\text{Co} - \text{Ce})\,/\text{M}$$
(6.2)

where; qe = At equilibrium, the amount of POP adsorbed per unit weight of microplastic, V = Volume of the solution, Co = Initial concentration of the pollutant, Ce = Concentration of the pollutant at equilibrium, M = Mass of the microplastic.

Thus, adsorption can be thought of as a physical mass transfer process and in this case, is a process which involves the transfer of a substance (POP) from a liquid phase (water) to the surface of a solid (microplastic). For example, in terms of the adsorption of POPs to microplastics in an aqueous medium, water is considered as the solvent, the POP is considered as the adsorbate and the microplastic is considered as the adsorbent. In the absence of the POP, the microplastic is covered with water. However, when the POP is added to the system, the water is displaced into the bulk solution and a layer of adsorbed pollutants accumulates on the surface of the microplastic. Thus, a greater surface area results in a higher capacity for the adsorption of pollutants by microplastics and an increase in surface reactivity. The adsorption of POPs to microplastic in an aqueous medium can be described by Eq. (6.3).

$$H_2O\left(ad; X_1^{ad}\right) + POP\left(aq; X_{POP}\right) \rightarrow H_2O\left(aq; X_1\right) + POP\left(ad; X_{POP}^{ad}\right) \tag{6.3}$$

where; POP = The persistent organic pollutant, X_1^{ad} = The molar fraction of water in a thin solution next to the surface of the microplastic, X_{POP} = The molar fraction of the POP in the aqueous phase, X_1 = The molar fraction of water in the aqueous solution, X_{POP}^{ad} = The molar fraction of the POP in the adsorbed layer.

A study[248] which determined the partition coefficients determined for a variety of PAHs, hexachlorocyclohexanes (HCHs) and chlorinated benzenes between seawater and polyethylene, polypropylene and polystyrene found that all the plastics had a high capacity for the sorption of these hydrophobic organic chemicals (HOCs). Furthermore, it was determined that the two chlorinated benzenes and the PAHs exhibited strong and similar sorption characteristics, while the HCHs demonstrated poorer sorption due to their increased polarity. Importantly, microplastics with a high capacity for sorption of HOCs have an impact on the ability of HOCs to evaporate from aquatic environments, thereby facilitating their persistence in the environment and their transport to other regions.[248]

Furthermore, our own research involved a study of the interaction of 33 different polycyclic aromatic hydrocarbons (PAHs) with 3.8 mm spherical pristine polyethylene microplastics. It was found that, following exposure of the microplastics to the PAHs for a period of 24 h, the highly carcinogenic PAH compounds benzo[a]pyrene and dibenzo[a,h] pyrene demonstrated 100% sorption to the microplastics at concentrations between 5 and 37.5 ng/mL at 10°C; while at 50 ng/mL sorption was 95.58% (benzo[a]pyrene) and 98.18% (dibenzo[a,h]pyrene). Conversely, at 21°C, sorption for benzo[a]pyrene and dibenzo[a,h] pyrene was 100% across all concentrations. Furthermore, at both temperatures, naphthalene was the most poorly sorbed PAH at 89–92% (10°C) and 87–94% (21°C). Overall, the results of the study demonstrated that even at the greatest concentration of 1000 ppb, the sorption of all 33 PAHs to 3.8 mm pristine polyethylene microplastics was greater at 21°C than at 10°C (see Fig. 6.14). Thus, polyethylene microplastics in the environment likely have a greater affinity for picking up PAH pollutants in warm water regions, as opposed to cold water regions. As such, this is relevant in the levels of PAHs that different aquatic species may be exposed to when consuming PAH contaminated microplastics.

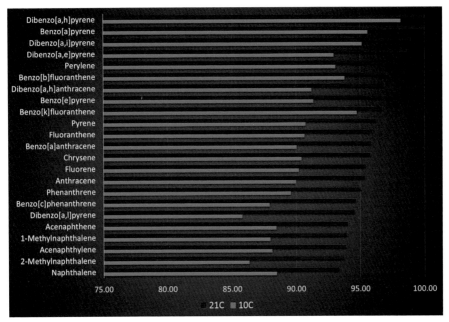

Figure 6.14 Adsorption of several PAHs at 1000 ppb to polyethylene microplastics was higher at 21°C than at 10°C.

The effect of the surface-area-to-volume ratio and weathering

Since microplastics in the marine environment are able to adsorb and transport POPs, they provide a mechanism for the widespread contamination of the aquatic environment with chemical pollutants.[460] Thus, there is a growing concern that microplastics in the aquatic environment are functioning as mobile reservoirs of toxic organic compounds.[229,403] This is of great concern since highly toxic pollutants, such as PCBs which exhibit a broad spectrum of hydrophobicities, are particularly ubiquitous in the aquatic environment[80] and are capable of adsorption to microplastics.[212,217]

Indeed, a study reported that 10–50 mm pieces of polypropylene (although not technically in the microplastic size range) were capable of adsorption of PCBs at significant concentrations of 4–117 ng/g.[279] However, the smaller a piece of plastic is, the greater the surface area and thus available sites for adsorption to take place. Indeed, it has been demonstrated that the smaller a piece of plastic is, the greater its affinity for POPs.[429] A study[429] which compared nano-sized pieces of polyethylene with microplastic-sized pieces of polyethylene established that smaller pieces had a considerably increased affinity for PCBs. The researchers ascribed the differences in affinity for PCBs as a result of differing surface-area-to-volume ratios between the nano-sized and microplastic-sized pieces of plastic. The surface-area-to-volume ratio of a piece of spherical plastic is the quantity of surface of that piece of plastic per unit of volume of that piece of plastic (see Eq. 6.2).

$$\text{Surface-Area-to-Volume ratio} = \left(\frac{4\pi r^2}{\frac{4}{3}\pi r^3} \right)$$

(6.4)

Importantly, the capacity for adsorption of persistent organic pollutants (POPs) becomes higher as the size of a piece of plastic decreases. This increase in adsorption capacity is a direct result of the increase in surface area and thus an increase in the available sites for chemical adsorption of POPs to take place.[403] As such, mini-microplastics have a much larger surface-area-to-volume ratio than microplastics (see Fig. 6.15) and consequently, a greater capacity to adsorb POPs. For this reason, and in terms of the transfer of chemicals to organisms which cannot penetrate into the bulk of the plastic material via absorption, smaller pieces of plastic represent a greater toxicological threat to marine species than larger pieces of plastic because they are easily consumed and can concentrate larger amounts of chemical pollutants onto their surface.

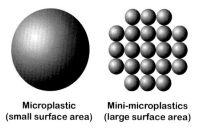

Microplastic   **Mini-microplastics**
(small surface area)   (large surface area)

Figure 6.15 Mini-microplastics have a greater surface-area-to-volume ratio than microplastics.

Microplastics which have been present in the environment for considerable time will tend to show some degree of weathering. Furthermore, some plastics are more prone to weathering than others, such as polystyrene and polyethylene.[85] Consequently, weathering of the surface of polymers causes smooth surfaces to become cracked, chipped and undulated,[9] thereby increasing the surface area (see Fig. 6.16). Indeed, samples of microplastic collected between 2003 and 2004 from various coasts of the North Pacific Ocean possessed a varying surface area.[354]

Importantly, a weathered plastic surface tends to become more reactive due to damage which increases the surface area and porosity of the material. This increase in surface area provides additional sites for adsorption of POPs, thereby increasing the adsorption capacity of the microplastic.[403] Certainly, in a study[89] which compared virgin microplastics and weathered microplastics collected from Greek beaches on Lesvos island, adsorption appeared to be the main sorptive process which occurred since the crystallisation of weathered microplastics decreased the ability of pollutants to diffuse into the bulk of the plastic material. This was especially true for polypropylene which only allowed surface diffusion due to its tightly packed molecular geometry. Furthermore, the surface of weathered microplastics will exhibit an increase in oxygen groups, thereby increasing the polarity[357] and thus the affinity for hydrophobic contaminants. Indeed, a study[112] concluded that

Virgin 3.8 mm polyethylene microplastic (pellet) (500x)

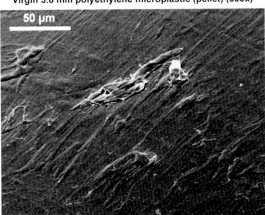

Weathered 3.8 mm polyethylene microplastic (pellet) (700x)

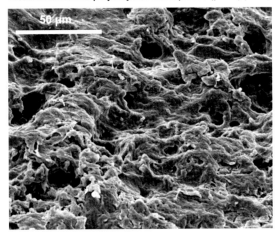

Figure 6.16 A comparison between the surface of a nonweathered virgin 3.8 mm
polyethylene microplastic pellet and a weathered 3.8 mm polyethylene microplastic pellet.

the two main factors which affect the degree to which polychlorinated biphenyls
(PCBs) adsorb onto polypropylene microplastics are the amount of degradation that
the microplastics have suffered and the length of time that they have spent in the
environment. For these reasons, weathered microplastics are able to transport higher
concentrations of POPs than non-weathered microplastics.[264] Thus, considering
the effects of the surface-area-to-volume ratio and weathering on increasing the
adsorption capacity for POPs by microplastics, small pieces of weathered plastic
have the greatest capacity for the adsorption of POPs which cannot penetrate into
the bulk of the plastic material via absorption.

Contaminated microplastics in the aquatic environment

Several environmental studies have been carried out that have established evidence that POPs readily adsorb to microplastics in the natural marine environment, as well as in a laboratory setting.[17,28,80,89,112,143,191,212,229,248,279,320,353,354,357,358,403,429,460] Furthermore, significant concentrations of POPs have been discovered in many lakes around the world, even in areas considered unspoiled,[83] and plastic debris recovered from the surface waters of Lake Erie, North America, has been found to be contaminated with polycyclic aromatic hydrocarbons (PAHs) and polychlorinated biphenyl ethers (PCBs).[104] Microplastics are also regularly detected in estuaries[152] and are transported there via river systems, especially during periods of heavy rainfall.[215,247,258,368] Similarly, estuarine regions tend to be more contaminated with POPs than neighbouring marine regions since the source of many POPs originate on land[207] and POPs regularly enter urban watercourses.[296] As a result of their dynamic nature, where currents consistently flow in and out creating turbulence, colloidal suspensions of microplastics coalesce into large masses that sink and become incorporated into POP contaminated sediment. Over time, the POPs become sorbed to the microplastics, and subsequent flushing of the estuary[368] or dredging activities[330] pulse release the contaminated microplastics into the wider aquatic environment where they can then be ingested by organisms. Contaminated microplastics also regularly wash up on beaches and are consumed by wildlife, such as birds.

Interestingly, a study[80] in Portugal found that black coloured microplastics collected from two Portuguese beaches (Fonte da Telha and Cresmina) appeared to be contaminated with the greatest concentration of pollutants while white microplastics appeared to have the lowest. While the black pellets were composed of polypropylene and polystyrene, it was found that if the pellets were aged, they were more likely to be composed of polypropylene and polyethylene. Similarly, several researchers[112,217,320] have reported that yellow microplastics collected from Japanese and Greek beaches consistently contained the highest levels of polychlorinated biphenyl (PCB) pollutants.

The yellow colour of the contaminated microplastics indicates the presence of quinoid based structures, produced from the oxidation of antioxidant additives added to the plastic during manufacture.[225] Thus, the yellowing of microplastics indicates that the microplastics have been in the aquatic environment for a considerable period of time.[320] Consequently, this longer exposure time increases the likelihood that the microplastics will become contaminated with pollutants present in the water and explains the higher levels of PCB pollutants detected in yellowed microplastics.[320] However, yellowing as an indicator of the amount of time that a microplastic has spent in the environment does not necessarily apply to polystyrene as it tends to quickly turn yellow when exposed to sunlight.[85]

Chlordane contaminated microplastics

There is very little research which has been carried out to ascertain the concentrations of chlordane on microplastics. However, a study[264] found the range of concentrations of chlordane adsorbed to microplastics collected from eight beaches in San Diego County, California (Mission beach, Mission Bay, Pacific beach, La Jolla Shores, Torrey Pines, Coronado and Imperial Beach) to be 1.8–170 ng/g.

Dichlorodiphenyltrichloroethane (DDT) contaminated microplastics

A study[279] reported that 10–50 mm pieces (although not microplastic-sized) of polypropylene were capable of sorption of DDE at concentrations of 0.16–3.1 ng/g. However, another study which collected microplastic-sized pieces of plastic between 2003 and 2004 from 10 industrial areas, as well as a river in San Gabriel, California, found that the concentration of DDTs sorbed to the plastics at the industrial sites were (<(0.03–2.03(LOD))-7100 ng/g), while the concentration in the San Gabriel River was 1100 ng/g. Similarly, a study[264] of microplastics collected from eight beaches in San Diego County, California (Mission beach, Mission Bay, Pacific beach, La Jolla Shores, Torrey Pines, Coronado and Imperial Beach) found that the range of concentration of DDTs sorbed to the microplastics was 0.56–64 ng/g. Another study[80] which collected microplastics from two Portuguese beaches (Fonte da Telha and Cresmina) determined that the range of concentrations of DDTs sorbed to the microplastics was 0.16–4.04 ng/g. A study[217] which collected microplastic pellets from four different Greek beaches (Aegena, Kato Achia, Vatera and Loutropyrgos) found levels of DDD, DDE and DDT sorbed to the plastic pellets to be 0.32–2.2 ng/g; 0.25–15.0 ng/g and 0.56–25.0 ng/g, respectively. Another study found DDTs sorbed to microplastics obtained from isolated islands (Fuerteventura, Cocos, Barbados, Hawaii and Oahu) at concentrations in the range (0.8–4.1 ng/g). Fig. 6.17 details the concentrations of DDTs (ng/g) reported as sorbed to microplastics at various regions worldwide.

Hexachlorocyclohexane (HCH) contaminated microplastics

A study[217] which collected microplastic pellets from four different Greek beaches (Aegena, Kato Achia, Vatera and Loutropyrgos) found that the average range of hexachlorocyclohexane (HCH) sorbed to the microplastic pellets, between all four beaches, was 1.05–3.5 ng/g. Another study[186] found HCH sorbed to microplastics collected from isolated islands (Fuerteventura, Cocos, Barbados, Hawaii and Oahu) to be 0.6–1.7 ng/g while the average HCH concentration sorbed to microplastics collected from St Helena was significantly higher at 19.3 ng/g. The researchers posited that increased levels of HCHs observed at St Helena may be attributed to use of γ-HCH (lindane) as an organochlorine insecticide. Indeed, the samples contained more than 80% lindane, as opposed to any of the other eight HCH isomers.

Perfluoroalkylate contaminated microplastics

There are a very few studies concerning the obscure perfluoroalkylated compounds. However, an unprecedented and extensive study[163] of the Atlantic Ocean, the Indian Ocean and the Pacific Ocean determined the levels of perfluoroalkylated substances in the surface waters of the Atlantic Ocean were in the range 131–10,900 pg/L. In the Indian Ocean, levels were 176–1980 pg/L, while in the Pacific Ocean the range was 344–2500 pg/L. In a study[435] of the interaction of perfluorooctane sulfonic acid (PFOS) and perfluorooctanesulfonamide (PFOSA) with polyethylene, polypropylene and polyvinyl

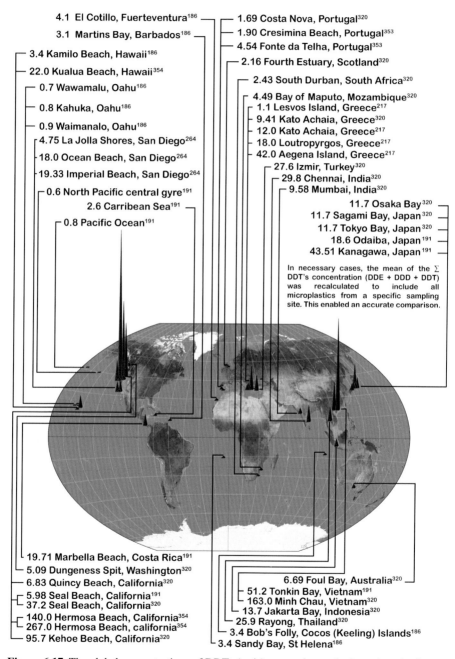

4.1 El Cotillo, Fuerteventura[186]
3.1 Martins Bay, Barbados[186]
3.4 Kamilo Beach, Hawaii[186]
22.0 Kualua Beach, Hawaii[354]
0.7 Wawamalu, Oahu[186]
0.8 Kahuka, Oahu[186]
0.9 Waimanalo, Oahu[186]
4.75 La Jolla Shores, San Diego[264]
18.0 Ocean Beach, San Diego[264]
19.33 Imperial Beach, San Diego[264]
0.6 North Pacific central gyre[191]
2.6 Carribean Sea[191]
0.8 Pacific Ocean[191]

1.69 Costa Nova, Portugal[320]
1.90 Cresimina Beach, Portugal[353]
4.54 Fonte da Telha, Portugal[353]
2.16 Fourth Estuary, Scotland[320]
2.43 South Durban, South Africa[320]
4.49 Bay of Maputo, Mozambique[320]
1.1 Lesvos Island, Greece[217]
9.41 Kato Achaia, Greece[320]
12.0 Kato Achaia, Greece[217]
18.0 Loutropyrgos, Greece[217]
42.0 Aegena Island, Greece[217]
27.6 Izmir, Turkey[320]
29.8 Chennai, India[320]
9.58 Mumbai, India[320]
11.7 Osaka Bay[320]
11.7 Sagami Bay, Japan[320]
11.7 Tokyo Bay, Japan[320]
18.6 Odaiba, Japan[191]
43.51 Kanagawa, Japan[191]

In necessary cases, the mean of the Σ DDT's concentration (DDE + DDD + DDT) was recalculated to include all microplastics from a specific sampling site. This enabled an accurate comparison.

19.71 Marbella Beach, Costa Rica[191]
5.09 Dungeness Spit, Washington[320]
6.83 Quincy Beach, California[320]
5.98 Seal Beach, California[191]
37.2 Seal Beach, California[320]
140.0 Hermosa Beach, California[354]
267.0 Hermosa Beach, california[354]
95.7 Kehoe Beach, California[320]

6.69 Foul Bay, Australia[320]
51.2 Tonkin Bay, Vietnam[191]
163.0 Minh Chau, Vietnam[320]
13.7 Jakarta Bay, Indonesia[320]
25.9 Rayong, Thailand[320]
3.4 Bob's Folly, Cocos (Keeling) Islands[186]
3.4 Sandy Bay, St Helena[186]

Figure 6.17 The global concentrations of DDTs (ng/g) reported as sorbed to microplastics.

chloride, it was found that polyethylene exhibited a greater adsorption capacity for PFOSA than for PFOS, while polystyrene did not exhibit any sorption for PFOS at all but did sorb PFOSA. However, levels of adsorption of both compounds were similar for PVC. Interestingly, the researchers suggest that the presence of a benzene ring in polystyrene may have the effect of increasing the polarity of this plastic, as opposed to polyethylene, thereby exhibiting a repulsive effect to the negatively charged PFOS molecule. Furthermore, the researchers surmise that alternatively, the benzene ring may be exhibiting a steric hindrance effect thereby preventing sorption of PFOS.

Phthalate contaminated microplastics

There are very few studies which describe the detection of phthalates in environmentally recovered microplastics. However, a study[136] of microplastics collected via neuston net from the surface waters of the Sardinian Sea found high concentrations of the phthalates bis(2-ethylhexyl) phthalate (DEHP) (23.42 ng/g) and mono(2-ethylhexyl) phthalate (MEHP) (40.30 ng/g). Furthermore, concentrations of DEHP and MEHP in the Ligurian Sea were found to be 18.38 ng/g and 61.93 ng/g, respectively.

PAH contaminated microplastics

In a study[354] which involved the collection of small plastic samples between 2003 and 2004 from beaches and industrial sites in California and Hawaii, 149 pieces were collected from 26 sites. However, from the data reported, 61 pieces were in the microplastic size range and microplastics were solely collected from 11 sites. Of these 11 sites, 10 were industrial areas and 1 was a river in San Gabriel, California. The range of concentrations of PAHs reported as sorbed to microplastics in the industrial sites was (74–12,000 ng/g), while the concentration in the San Gabriel river was (1200 ng/g).

A study[363] found that concentrations of PAHs in sediments in the East Shetland Basin in Scotland were almost 150 ng/g, and that the elevated levels of parent PAHs in comparison to alkylated PAHs indicate that the likely primary source of PAH contamination in these sediments is of combustive origin. However, the researchers found that levels of PAHs appear to be decreasing in this region, following a spike of high concentrations of PAHs in the late 1980s, which they attribute to drilling operations that involved the use of oil-based fluid to enhance lubricity and reduce interactions concerning shale and water. In a study[80] of microplastics on two Portuguese beaches (Fonte da Telha and Cresmina), it was determined that the concentration of PAHs sorbed to collected microplastics were 0.2–319.2 ng/g. Similarly, concentrations of PAHs were found to be between 18 and 210 ng/g on microplastics collected from eight beaches in San Diego County, California (Mission beach, Mission Bay, Pacific beach, La Jolla Shores, Torrey Pines, Coronado and Imperial Beach).[264] A study which analysed microplastic pellets collected from four Greek beaches (Aegena, Kato Achia, Vatera and Loutropyrgos) for the presence of PAHs found concentrations to be 180 ng/g; 160 ng/g; 100 ng/g and 500 ng/g, respectively. The researchers attribute the increased pollution at Loutropyrgos beach to emissions from vessels, cars and manufacturing works in the area.

Polybrominated diphenyl ether (PBDE) contaminated microplastics

There are very few studies relating to PBDEs sorbed to microplastics. However, the potential for sorption of PBDEs in the aquatic environment exists. For example, a study[467] found concentrations of PBDEs in high-altitude lakes in Tibet to be 0.09–4.32 ng/g. Furthermore, another study[261] found levels of PBDEs in various fish collected from the seas around China to be in the range 0.3–700 ng/g, which the researchers report as 'high levels'.

Polychlorinated biphenyl (PCB) contaminated microplastics

A study which investigated the concentration of PCBs on microplastic pellets, recovered from the high tide mark of 47 beaches around Japan, revealed a wide variety of concentrations of PCBs sorbed to the microplastics (6–18,700 ng/g). The second highest concentration in the range (18,600 ng/g) was measured at Tanegashima,[112] a long island which is positioned east of Takushima,[360] while the highest concentration (18,700 ng/g) was measured at Osaka Bay[112] and is the highest recorded level of reported in the literature thus far. Even a microplastic from the isolated island of Shikinejima, which is situated 160 km south of Tokyo,[360] contained PCBs at 1080 ng/g.[112] Furthermore, microplastics collected from Kasai beach in Japan were reported as having concentrations in the region of 28–2300 ng/g. However, the researchers stress that half of the total concentration at Kasai beach could be attributed to a single microplastic pellet. This study clearly demonstrates the large variation in the concentration of PCBs sorbed onto individual microplastics and the variation between different regions.

In a study of microplastics collected from industrial sites and beaches in California and Hawaii, none of the plastic samples recovered from terrestrial industrial sites contained any PCBs above the limit of detection (0.05–0.08 ng/g) and only microplastics recovered from beaches contained PCBs. This would seem to suggest that microplastics tend to accumulate PCBs via sorptive processes in the aquatic environment, as opposed to the terrestrial environment.

A study[80] which sampled two Portuguese beaches (Fonte da Telha and Cresmina) found concentrations of PCBs on microplastics to be in the range of 0.11–15.56 ng/g. Additionally, the PCB congener 2,30,5-trichlorobiphenyl (PCB-26) was found in the highest abundance (15.56 ng/g), while PCB congener 2,20,3,4,40,50-hexachlorobiphenyl (PCB-138) was discovered in every microplastic that was tested. In 1998, Pellet Watch International was setup by Takada and colleagues in Japan to monitor the levels of pollutants sorbed onto microplastic pellets which had been sent to them from different regions around the world. Consequently, the PCB concentration on contaminated microplastics collected from two nearby beaches (Guincho Beach and Costa Nova Beach) was 45 ng/g and 27 ng/g, respectively. Based on these findings, the concentration of PCBs sorbed to microplastics on Portuguese beaches were in the range 0.11–45 ng/g. Interestingly, a study[217] of four Greek beaches (Aegena, Kato Achia, Vatera and Loutropyrgos) found that of the 19 PCB congeners analysed, PCB-138 was found to be the most abundant congener sorbed to the microplastics. Additionally, the total concentration of PCB congeners on Aegena, Kato Achia, Vatera and Loutropyrgos was reported to be 230 ng/g; 5 ng/g; 6 n/g and 290 n/g, respectively. Loutropyrgos was identified as being the most polluted and the researchers explain that this is due to the large amount of industry in the area. Similarly, a study[212] of sediments obtained from Port Elizabeth harbour in South Africa found that PCB-138 was the most abundant PCB congener and represented 29% of the total PCB concentration measured, which were in the range 0.56–2.35 ng/g. Certainly, a study[279] reported that PCB-138 had an affinity for the particulate phase, and the researchers attribute this to the significant chlorination of PCB-138. As mentioned previously, increased chlorination of PCBs offers protection from photo-oxidative degradation. Consequently, this may offer an explanation as to the ubiquity of PCB-138 in the aquatic environment. Indeed, a study[122] ascertained that, of PCB-138 and PCB-153, PCB-138 was the most frequently detected PCB congener in foodstuffs; especially fish. Similarly, a study[212] ascertained that PCB-153 represented 24% of the total PCB concentration detected in sediment recovered from Port Elizabeth harbour in South Africa.

In a study which collected <5–50 mm pieces of plastic from eight beaches in San Diego County, California (Mission beach, Mission Bay, Pacific beach, La Jolla Shores, Torrey Pines, Coronado and Imperial Beach), 68% of the collected pieces were in the microplastic size range and had a concentration of sorbed PCBs in the range 3.8–42 ng/g. Furthermore, significant concentrations of PCBs sorbed to microplastics have been discovered on tropical islands. For example, microplastics recovered from the isolated islands (Fuerteventura, Cocos, Barbados, Hawaii and Oahu) were found to have concentrations of sorbed PCBs in the range (0.1–9.9 ng/g).[186]

At the present time, there does appear to be a consistent background level of PCBs in the aquatic background, and sporadic pockets of high concentrations of PCBs tend to be scattered around various regions of the world. Although these pockets of elevated levels are usually in areas of heavy industry, this is not always the case, as demonstrated by the high concentrations found in the remote Japanese island of Shikinejima.[112] Indeed, the researchers themselves cannot fully explain the elevated incidences of PCBs seen in these remote Japanese islands, but do hypothesise that perhaps the microplastics had previously travelled through a region of heavy industry and picked up PCBs there, before being deposited on the island. Certainly, industrial areas near coasts are commonly polluted regions of the marine environment. Furthermore, high levels of contamination with PCBs have been observed in areas of significant industrialisation, such as the Firth of Clyde, which is reported to possess the highest levels of pollution in Scotland.[439] Consequently, the deposition of toxic PCB contaminated microplastics onto remote regions of the world appears to be an important area which requires further investigation.

Metal and metalloid contaminated microplastics

The suspected human carcinogen antimony trioxide is often added to PET during production as a catalyst as well as being utilised as a flame retardant and smoke suppressant. There is the possibility that PET in contact with water can leach small amounts of the metalloid antimony. The amount of potential leaching is increased the longer the water is in contact with PET and the level of leaching is significantly increased at elevated temperatures. Thus, the United States Environmental Protection Agency (USEPA) has set the maximum contaminant level of antimony in bottled water at 6 ppb. One study[449] concluded that bottles of water stored at 22°C for a period of 3 months had very little change in antimony levels from the start of the study (0.195 ± 0.116 ppb) to the end of the study (0.226 ± 0.160 ppb) and thus were well below the USEPA maximum contaminant level. However, the study found that when the bottles were stored at higher temperatures, the number of days required to exceed the USEPA 6 pp maximum limit was decreased as shown in Table 6.2.

Table 6.2 **The effect of storage temperature on the days required for antimony-containing polyethylene terephthalate to leach levels of antimony that exceed the USEPA limit of 6 ppb**

Storage temperature (°C)	Days required to exceed 6 ppb limit
60	176
65	38
70	12
75	4.7
80	2.3
85	1.3

Testing of PET by the study[449] found that one sample of plastic used to produce a water bottle contained 213 ± 35 mg of antimony per kilogramme. Therefore, the study deduced that if all of the antimony was released into this 500 mL bottle of water, the level of antimony would equate to a concentration of 376 ppb. Thus, it is clear that only small amounts of antimony are potentially leached from the plastic. However, in terms of the aquatic environment, PET bottles are one of the most common types of litter. Thus, it could be hypothesised that if the conditions were right, antinomy-containing PET plastics may leach antimony into the aquatic environment, possibly adding to any background levels of antimony pollution in water, sediments and sands. Moreover, there may even be a risk of antimony bioaccumulation in marine biota that consumes antimony-containing PET plastics. However, for the time being, there is insufficient information upon which to form a cogent theory and accordingly, appropriate research would have to be undertaken to prove or disprove this hypothesis.

As a result of significant populations and intense industrial activity, the estuary of the River Seine is one of the most contaminated sea regions in Europe, for lead, mercury and cadmium.[401] However, highly populated and industrial regions are not the only areas to be contaminated with heavy metals. For example, large numbers of intact fluorescent lights were discovered on Saint Brandon's Rock, an isolated coral atoll within the Indian Ocean, as well as its various islets.[33] The fluorescent tubes contain mercury vapour and are easily shattered. Thus, their presence could potentially result in the long-range transport of this highly toxic heavy metal.

In an investigation[192] into the presence of trace metals on 26.4–32.5 mg randomly selected beached microplastics, collected from the coastline of South West England, it was discovered that high amounts of trace metals were found sorbed to the plastic pellets. Furthermore, it was established that lead and chromium exhibited the greatest level of adsorption to aged pellets, while cobalt and cadmium displayed the least. When the researchers suspended virgin polyethylene microplastics, with an average weight of 26.2 mg, in seawater containing lead, cadmium, nickel, cobalt, copper and chromium, a relatively rapid and considerable adsorption of these trace metals was observed. Interestingly, the researchers observed that during their exposures of microplastics to trace metals, approximately 5% of the chromium in the solution had adsorbed to the surface of the PTFE bottles that they had utilised for incubations. Based on this, it could be inferred that perhaps PTFE microplastics exhibit a particular affinity for chromium adsorption. In a further study,[193] the adsorption of the heavy metals lead, cadmium, nickel and cobalt, to virgin polyethylene microplastics, as well as beached pellets collected from Plymouth, United Kingdom, was investigated. It was found that an increase in pH, coupled with a decrease in salinity, increases adsorption of these metals. However, the opposite effect was observed with chromium (VI), while in the case of copper, pH and salinity did not appear to affect the degree of adsorption. Furthermore, it was observed that aged beached pellets, as opposed to virgin polyethylene microplastics, exhibited a greater affinity for adsorption of these metals.

In a study[357] of the relationship between five commodity plastics (LDPE, HDPE, PP, PET and PVC) and metals (Al, Cd, Co, Cr, Fe, Mn, Ni, Pb, Zn) present in seawater over 12 months, it was observed that HDPE exhibited the lowest accumulation of metals. Furthermore, it was observed that over the entire 12-month duration of the study, lead and cobalt did not reach equilibrium with 3 mm LDPE and HDPE polyethylene microplastics, thereby indicating that these types of plastics have a high adsorption capacity for these metals. At the same time, chromium took a period of 9 months to reach equilibrium. The researchers predicted that in the environment, a period of 22 months would be required for cobalt to reach equilibrium with polyethylene, while lead would take 64 months. Conversely, when the same metals were exposed to 3 mm polyethylene microplastics in the laboratory, equilibrium was reached within 100 h. Although these studies demonstrate rapid adsorption of trace metals in the laboratory setting, in the aquatic environment the process is expected to occur more slowly as a result of other contaminants and biofilms which form on the surfaces of plastics. Nevertheless, it has been suggested that biofilms may actually enhance adsorption of trace metals.[357] Ultimately, what is evident from previous research, as well as our own research, is that the longer microplastics remain in contaminated aquatic environments, the greater the concentration of contaminants that they will accumulate until equilibrium is reached.

The effect of salinity

An important aspect affecting the ability of POPs to adsorb to microplastics is salinity. Indeed, a study[89] found that an increase in salinity correlates with an increase in adsorptive behaviour of the polycyclic aromatic hydrocarbon (PAH) phenanthrene to polypropylene. In another study,[17] it was found that under simulated estuarine conditions, the salinity did not affect the capacity of polyvinyl chloride or polyethylene to adsorb phenanthrene. Furthermore, the researchers discovered that an increase in salinity hindered the capacity for adsorption of DDT by polyvinyl chloride or polyethylene. Nonetheless, the researchers concluded that the important factor in adsorption of phenanthrene and DDT to polyethylene and PVC is the concentration of phenanthrene and DDT in the surrounding medium and the amount of time the plastic materials spend there, as opposed to the salinity of the environment. Interestingly, a study[16] found that there was a competition between phenanthrene and DDT for adsorption sites on PVC and that DDT was preferentially adsorbed, even exhibiting an antagonistic effect that disrupted phenanthrene adsorption. Furthermore, a later study[17] found that polyethylene and PVC demonstrated a larger adsorption capacity for DDT than for phenanthrene. Nonetheless, it has been reported[403] that phenanthrene is capable of concentrating on polyvinylchloride, polypropylene and polyethylene microplastics at up to 10 times higher than concentrations found on sediments in the natural environment.

In terms of additives incorporated into plastic materials during manufacture, a study[395] reported that salinity has a negligible effect on the leaching of intrinsic plastic additives from polyethylene terephthalate, polyethylene, polyvinyl chloride and polystyrene. However, high levels of turbulence in the water were positively correlated with increased leaching of additives. Furthermore, additives such as bisphenol A and citrates have been shown to leach more readily than phthalates.

However, in the case of polychlorinated biphenyl (PCB) pollutants, salinity does appear to play a role. For example, a study[429] found that an increase in the salinity of an aqueous medium correlates with an increase in the capacity polyethylene microplastics to adsorb PCBs. Furthermore, an increase in salinity results in a decrease in adsorption of PCBs to organic sediments, while in freshwater the adsorption capacity of polyethylene and organic sediment was analogous. Thus, these findings would appear to suggest that in saltwater mediums, PCBs would preferentially adsorb to polyethylene microplastics, as opposed to organic sediments and adsorption of PCBs would be greater in regions of the world's oceans that possessed a high salinity, as opposed to low salinity regions. Furthermore, a study[112] found that polyethylene had a higher affinity for PCBs than polypropylene and that the density of polyethylene microplastics made no significant difference to the affinity for PCBs. Moreover, microplastics in the marine environment tend to form coatings of fouling material and it has been reported[112] that microplastics which have accumulated a coat of oil, or biofouling, have a greater potential to adsorb PCBs than microplastics free from any fouling material. As such, one could formulate a hypothesis. For example, ocean salinity data generated by the NASA Aquarius instrument located on-board the Aquarius/SAC-D spacecraft indicates that the Atlantic Ocean, and Mediterranean Sea have a very high salinity in comparison to the rest of the world's oceans (see Fig. 6.18). At the same time, the North Atlantic gyre has a high abundance of plastic pieces floating on the surface waters with a concentration of $20,328 \pm 2324$ pieces/km^2, while the Mediterranean Sea is considered to be one of the most polluted marine environments in the world.

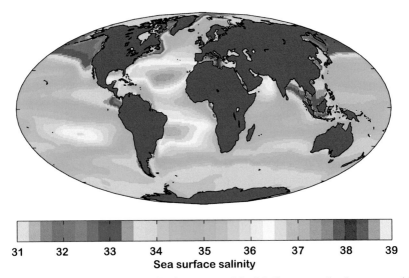

31   32   33   34   35   36   37   38   39
Sea surface salinity

Figure 6.18 The North and South Atlantic gyres, and the Mediterranean Sea have a very high salinity in comparison to the rest of the world's oceans.

Since it has been demonstrated that increased salinity results in an increase in the capacity polyethylene microplastics to adsorb PCBs[429] and that microplastics may accumulate PCBs while traversing industrial regions,[112] six factors could be relevant;

1. The North Atlantic, South Atlantic and Mediterranean Sea are regions of significantly high salinity.
2. The North and South Atlantic contain gyres with significant plastic accumulation, while the Mediterranean Sea is considered as one of the most polluted marine regions.
3. The North and South Atlantic gyres, as well as the Mediterranean Sea, are located between the coasts of many highly industrialised regions.
4. PCBs are among the most ubiquitous pollutants in the world's oceans.
5. Polyethylene is one of the most common plastics found in the marine environment.
6. Microplastics readily accumulate a layer of fouling material in the marine environment.

Thus, one could posit that the high salinity in the North and South Atlantic gyres, and the Mediterranean Sea, as well as their proximity to industrialised regions, may mean that a substantial number of polyethylene microplastics in these regions may be significantly contaminated with PCB pollutants, especially if they are fouled. Indeed, a study of plastic debris recovered from the North Pacific gyre found that 40% was contaminated with pesticides, 50% was contaminated with PCBs and 80% was contaminated with PAHs. Furthermore, 70% of the microplastic-sized pieces recovered were composed of polyethylene.[353] Consequently, the contamination of polyethylene microplastics in the North and South Atlantic gyres, and the Mediterranean Sea with PCB pollutants is an important area in which further work is required to test this hypothesis and if proved to be true, evaluate the risk to marine biota in these regions. Indeed, if polyethylene microplastics contaminated with PCBs subsequently sink to benthic regions as a result of fouling (see Chapter 5) and become incorporated into sediments, then they may be transporting PCB pollutants into these benthic regions. Furthermore, they may be ingested by benthic organisms thereby facilitating the introduction of toxic PCB contaminants into the food web.

However, another potential way in which microplastics floating on surface waters can become contaminated with toxic chemicals is via the deposition of pollutants from the atmosphere.[353] Interestingly, a study[143] which investigated differences in the adsorptive capacity for PAHs between different densities of polyethylene concluded that lower-density polyethylene demonstrates faster rates of adsorption and increased adsorption capacity. Thus, one could further hypothesise that lower-density plastics are likely to remain buoyant on surface waters for longer periods of time, despite the accumulation of fouling material, and are therefore more likely to undergo the deposition of PAHs onto their surface via atmospheric settling and subsequent adsorption. Indeed, one PAH which is particularly toxic to aquatic organisms, even at low levels of exposure, is pyrene.[322] Importantly, a study[14] was able to adsorb pyrene to polyethylene and polystyrene microplastics at a concentration range as high as 200–260 ng/g. Thus, the potential for contaminated microplastics to induce toxic effects in aquatic organisms, as well as enter the food chain as a result of ingestion by biota, is a serious issue that requires significant study to test these hypotheses and fully understand the risks posed to both the aquatic environment and, ultimately, ourselves.

The biological impacts and effects of contaminated microplastics

7

Ingestion of contaminated microplastics

The ingestion of contaminated microplastics by aquatic organisms provides a viable route for the transfer of toxic chemicals into the tissues of the organism, in which microplastics act as a vector for the transport of sorbed contaminants and chemical additives into organisms.[42,315] While the investigation of such phenomena and the mechanisms by which this occurs is very much in its infancy, there has been a significant garnering of evidence thus far. Certainly, microplastics can pick up waterborne chemical contaminants during their time spent in polluted waters and can concentrate these contaminants up to 1 million times greater than the surrounding water (see Chapter 6).[191,279,403,353,460] At the same time, fish regularly consume microplastics.[269,333,338, 359,371] Indeed, in the Gulf of Mexico, 10% of marine fish and 8% of freshwater fish were found to have consumed microplastics.[333]

However, the formation of a definitive link between the ingestion of contaminated microplastics and the body condition of pelagic fish is still in its infancy and requires further research.[411] Indeed, it has been suggested[107] that contaminated microplastics may not contribute significantly to the bioaccumulation of contaminants and that the levels of microplastics shown to have adverse effects on organisms in the laboratory are higher than microplastic concentrations measured in subtidal sediments, and similar to the maximum concentrations measured in beach sediments. Furthermore, based on the utilisation of model analyses, it has been suggested that the bioaccumulation of persistent organic pollutants (POPs) from the ingestion of contaminated plastics by organisms may be small due to the absence of a gradient between POPs and the fatty tissues of the aquatic organisms, and that some mechanism of removing POPs may take place.[165,229]

However, many models do not take into account the role of gut surfactants in the desorption of these contaminants, especially at different temperatures and pH. For example, in simulated physiological conditions, the desorption of several common pollutants in the aquatic environment from contaminated polyethylene microplastics (phenanthrene, DEHP, PFOA and DDT) was faster when gut surfactant was present and was further increased at elevated temperatures typically present in warm-blooded aquatic species. Furthermore, the desorption of contaminants from microplastics in the conditions typically found in the gut may be up to 30 times greater than in the conditions typically found in seawater.[17] Consequently, pollutants desorbed from ingested microplastics become freely available to diffuse into the tissues of the organisms. Indeed, the concentration of a chemical pollutants in an organism has been noted to increase, the greater the ability of that pollutant to dissolve in fats, lipids and oils (lipophilicity). Since many POPs are lipophilic, they tend bioaccumulate and readily distribute throughout

Microplastic Pollutants. http://dx.doi.org/10.1016/B978-0-12-809406-8.00007-4

the food web, ranging from small planktonic species to large air-breathers, such as whales.[127] Ultimately, the vast majority of researchers concede that drastic worldwide action must be taken since many aquatic regions have high levels of pollutants and global concentrations of microplastics in the aquatic environment are increasing (Fig. 7.1).

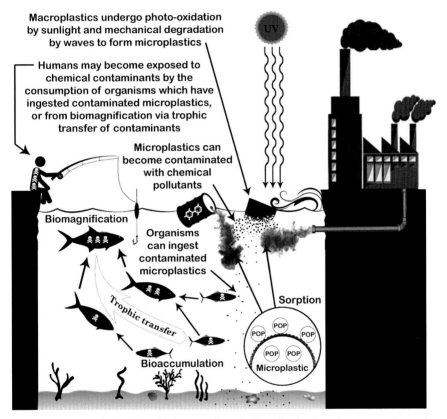

Figure 7.1 The mechanism by which contaminated microplastics introduce contaminants into the food web. (*POP*, persistent organic pollutant).

Aquatic invertebrates

In an attempt to combat plastic litter (see Chapters 2 and 3), the use of biodegradable plastics is slowly increasing. However, even biodegradable microplastics can be detrimental to the aquatic environment. For example, in one study[82] it was found that in natural settings, polylactic acid microplastics (average size $235.7 \pm 14.8\,\mu m$) exhibited minimal degradation over a period of 1 month in sediment and induced a significant reduction in microalgal biomass on the sediment surface. Moreover, the effects on the microalgae were comparable to non-biodegradable high-density polyethylene microplastics ($102.6 \pm 10.3\,\mu m$).

Furthermore, many plastic materials contain chemical additives which are incorporated into the plastics during manufacture (see Chapter 4), and which can subsequently leach out because they are not chemically bound to the plastic material,[405] although it has been suggested[228] that the leaching of chemical additives from ingested plastics may be a negligible route of exposure for the sediment dwelling lugworm (*Arenicola marina*) to these chemicals. However, in a study[42] which exposed lugworms (*A. marina*) to a mixture of 95% sand and 5% polyvinyl chloride (PVC) microplastics contaminated with the polycyclic aromatic hydrocarbon (PAH) phenanthrene, conclusive evidence was acquired for the first time in demonstrating that following ingestion by the lugworm, phenanthrene desorbed from the PVC microplastics and transferred into the worm's tissues. Furthermore, the same effect was observed for the flame retardant PBDE-47 and the anti-microbial triclosan, both of which are additives that can be used by industry in the manufacture of PVC. Indeed, in comparison to the concentrations of the additives in the sediments used in the experiment, the worms bioconcentrated the contaminants up to 950% higher in the walls of their body and up to 3500% higher in their gut.

In a study[82] carried out to compare the difference in biological effects on the lugworm (*A. marina*) between polyvinyl chloride (PVC) and high-density polyethylene (HDPE) (average size 130.6 ± 12.9 and 102.6 ± 10.3, respectively) and one biodegradable plastic (polylactic acid; average size $235.7 \pm 14.8\,\mu m$), it was found that the lugworm produced a reduced number of casts in the sediment that was contaminated with microplastics. Furthermore, by measuring the oxygen consumption, it was established that the metabolic rate of the lugworm was higher when exposed to the microplastics, indicating a stress response. The strongest effect was observed with PVC and the toxicity of PVC microplastics to the lugworm in the environment was hypothesised to be the result of leaching of intrinsic plasticisers and vinyl chloride monomers or adsorbed persistent organic pollutants (POPs). Furthermore, in a study[315] of the effects of virgin microplastics versus beach recovered microplastics on the development of green sea urchin (*Lytechinus variegatus*) embryos, it was observed that virgin microplastics exhibited the greatest toxicity due to the leaching of intrinsic additives. Furthermore, the degree of toxicity of the beach recovered microplastics was found to be highly variable and was postulated to result from differences in the levels of sorbed contaminants.

Thus, clear evidence exists that aquatic invertebrates are directly affected by the ingestion of contaminated microplastics, as well as intrinsic additives present within the bulk of the plastic material. One particular type of aquatic invertebrates that are particularly susceptible to microplastics and waterborne chemical pollutants are filter feeders,[14] such as mussels, due to their ingestion of small particles.[27] Indeed, a study[40] reported that detritovores, deposit feeders and filter feeders are quite capable of consuming polyvinyl chloride (PVC) mini-microplastics as small as $230\,\mu m$, while another study[437] found that bivalves were capable of ingesting polystyrene mini-microplastics as small as $10\,\mu m$ while undergoing feeding experiments. Moreover, it was reported[40] that the bioaccumulation of ingested $2\,\mu m$ spherical polystyrene mini-microplastics had been demonstrated in mussels during laboratory trials. Furthermore, a study[300] involved exposing around 500 blue mussels (*Mytilus edulis*) to <80 μm irregular shaped mini-microplastics composed of high-density polyethylene (HDPE) for up to 4 days. While none of the mussels died throughout the experiments, it was found that the mussels developed granulocytoma formations in their digestive glands following 6 h of exposure to the HDPE mini-microplastics and suffered destabilisation of their lysosomal membranes. However, the researchers were not able to determine whether these effects were the result of the physical attributes of the microplastics themselves or a toxic effect exhibited by the chemical composition of the plastics. For example, the toxic effect may have occurred as a result of leaching of chemical compounds incorporated during manufacture of the plastics, such as plasticisers, antioxidants or flame retardants.[84,175]

Nevertheless, tests in a laboratory are not always fully representative of the conditions and mechanisms inherent in an environmental system and thus experimental corroboration should be undertaken with organisms recovered from the environment.[339,340] Certainly, complementary approaches to monitoring marine contaminants have been suggested[411] which involve investigations into the biological impacts on aquatic organisms in their natural habitat, as opposed to solely monitoring of the contaminants themselves. For example, the environmental degradation of some persistent organic pollutants (POPs) can result in the formation of phenol, which has been shown to affect the motile dinoflagellate *Lingulodinium polyedrum* by altering metabolic systems and increasing antioxidant activity.[275] Thus, the reasoning behind these approaches is that if the presence of microplastics and chemical pollutants in the aquatic environment are having effects organisms, the presence of these contaminants would be easier to ascertain by looking for the characteristic effects in the organisms themselves. As such, there have been recent propositions[211] for the routine monitoring of microplastics and chemical pollutants by examining mussels in their natural habitat, to detect any risks and to provide an advance warning of the status of any environmental hazards in specific ecoregions.

However, despite the fact that mussels are vulnerable to bioaccumulation of waterborne pollutants and are therefore useful in exposure investigations, it is important to note that differences in biological responses and chemical accumulation have been observed between the three species (*Mytilus edulis*, *Mytilus galloprovincialis*, *Mytilus trossulus*).[37] Indeed, the mussel species typically found on the shores of the United Kingdom are *Mytilus galloprovincialis* and *Mytilus edulis*, with frequent breeding occurring between the two species.[374] Therefore, it is important to correctly identify the species being used to ensure that environmental assessments of marine pollution are accurately reported. However, it has been suggested that perhaps the physiological differences between species may be less pronounced as a result of adaption in situ to local environmental conditions.[37] Ultimately, further research is required to fully understand differences in biomarker responses and bioaccumulation between different *Mytilus* species from the same sampling area.

Interestingly, many bivalves, such as mussels, demonstrate material selection when feeding and consuming mini-microplastics, as opposed to simply engulfing all mini-microplastics that come within their proximity. Thus, any unwanted material is excreted as pseudofaeces.[25] Indeed, some kind of unknown sorting process takes place in the gut and it has been suggested that ciliary mechanisms are involved in catching, discerning and conveying material to the various digestive regions.[437] While there is some evidence that selective feeding can reduce the uptake and subsequent bioaccumulation of toxic metals, such as cadmium,[25] and that bivalve molluscs are capable of adapting to toxins present in their environment,[384] studies have shown that under certain conditions, material with no nutritive value can be favoured and ingested with detrimental effects on growth.[394] As such, this raises questions as to what effect this selective capability may have on the consumption of microplastics contaminated with pollutants. Certainly, a study which involved exposing free swimming crustaceans, bivalves and benthic deposit feeding biota to $10\,\mu m$ polystyrene mini-microplastics (density $\sim 1.05\,g/cm^3$), at concentrations of 5, 50 and 250 mini-microplastics per millilitre, concluded that the factors affecting mini-microplastic ingestion were the amount of mini-microplastics present, the rate at which the mini-microplastics were encountered in the environment and the way in which the animal fed.

Microplastics have been found in the tissues of mussels collected from the environment. For example, a study[53] of the blue mussel (*Mytilus edulis*) and the lugworm (*A. marina*) collected from several European coastlines (Belgium, the Netherlands and France) found microplastics in every organism that was collected, with the blue mussel containing approximately 0.2 microplastics per gram and the lugworm containing 6 times more at 1.2 microplastics per gram. In another study, the concentration of microplastics in the tissues of the blue mussel (*Mytilus edulis*) and the Pacific oyster (*Crassostrea gigas*) were approximately 0.36 and 0.47 microplastics per gram respectively.

At the same time, chemical pollutants have also been detected in Mussels. Indeed, a study[212] found the total concentration of PCBs in South African blue mussels (*Mytilus galloprovincialis*) to be in the range 14.48–21.37 ng/g, while another study[112] found the total concentration of polychlorinated biphenyls (PCBs) in blue mussels (*Mytilus galloprovincialis*), collected from 24 different regions around the Japanese coastline, to be between 11 and 1630 ng/g. Furthermore, a study[14] found an increase in the desorption of the polycyclic aromatic hydrocarbon (PAH) pyrene from polyethylene, polypropylene and polystyrene microplastics once the contaminated microplastics were ingested by blue mussels (*Mytilus galloprovincialis*). The researchers observed a large number of deleterious effects to the mussels, such as effects on DNA and neurotoxic symptoms. Furthermore, concentrations of pyrene in the gills of the mussels were three times higher than on the microplastics themselves, thereby demonstrating significant bioaccumulation. Another study[60] established that polybrominated diphenyl ethers (PBDEs) were capable of desorption from microplastics ingested by amphipods (*Allorchestes compressa*), resulting in the accumulation of PBDEs in the amphipods tissues.

A study of mercury[92] in blue mussels (*Mytilus edulis*) collected from the French coasts by the French Mussel Programme over 20 years, reported that the greatest levels of lead contamination occurred in mussels retrieved from the eastern areas of the Baie de la Seine, a bay and inlet of the English Channel, which itself is a $75{,}000\,km^2$ body of water that separates southern England and northern France. The researchers posit that coastal particles are the likely to be one of the main sources of mercury, as well as groundwater estuaries. While it was not specified whether these particles could be pieces of plastic, a review of contamination in the English Channel[401] asserted that as a result of significant population densities and intense industrial activity, the estuary of the River Seine and the Baie de la Seine are two of the most contaminated sea regions in Europe for heavy metals. Furthermore, high levels of heavy metals have been found sorbed to microplastics collected from beaches[192] and that microplastics are quite capable of accumulating heavy metals, such as lead, in the marine environment and that low-density and high-density polyethylene have a high sorption capacity for lead.[357]

As well as microplastics, nanoplastics can also have detrimental effects to organisms, although very little research has been carried out this far. However, nanoparticles of polystyrene are capable of distribution throughout the water column, and this can affect species of plankton and disturb the flow of energy within the marine ecosystem.[29] Indeed, following exposure of microalgae to uncharged and anionic polystyrene plasticles of 0.05, 0.5 and 6 μm diameter for a period of 72 h, no adverse effects on photosynthesis were observed. However, with a high concentration of uncharged polystyrene plasticles (250 mg/mL), up to a 45% reduction in growth of the microalgae was observed. Furthermore, a study[29] of the interactions of 40 nm anionic carboxylated and 50 nm cationic amino polystyrene nanoplastics with brine shrimp (Artemia franciscana) larvae found that after 48 h of exposure to both types of nanoplastic at a concentration of 5–100 μg/mL, a large accumulation of nanoplastics were discovered in the central cavity of the brine shrimp larvae digestive tract and subsequent excretion was limited.

Plasticles can also have physical effects on organisms. For example, 50 nm cationic amino polystyrene nanoplastics were found to increase the number of brine shrimp (Artemia franciscana) larvae moulting events and adsorbed to the surface of various appendages thereby likely hindering mobility. Consequently, increased moulting was observed which was attributed to activation of a detoxification mechanism in an attempt to remove the toxic cationic nanoplastics. In a study[241] of the physical effects of 1 and 100 μm high-density polyethylene microplastics (density 0.96 g/cm^3) at concentrations ranging from 12.5 to 400 mg/L on the small planktonic crustacean Daphnia magna for a duration of up to 96 h, it was observed that 1 μm microplastics resulted in immobilisation of the daphnids, which increased with time as more microplastics were ingested. Ultimately, the EC$_{50}$ (the effective concentration of microplastics at which 50% mortality of the daphnids occured) was 57.43 mg/L after 96 h. In the case of the 100 μm sized microplastics, the daphnids were unable to ingest them and consequently, no adverse effects were observed. Interestingly, Daphnia magna, has been found to exhibit acute toxicity to leachates from plastics.[22] Furthermore, it has been demonstrated that leachates from weathered plastics can become increasingly toxic to the small planktonic copepod Nitocra spinipes (Harpacticoida crustacea) when certain plastic materials are exposed to artificial sunlight, such as polyvinyl chloride (PVC) and polypropylene. Furthermore, the duration of irradiation demonstrated no particular correlation with respect to increasing toxicity.[22]

Thus, aquatic invertebrates are known to consume microplastics, and at the same time, they have been found to be contaminated with chemical pollutants. Indeed, the tissues of bivalves are often highly interspersed with ocean contaminants[239] and the blue mussel is often exposed to polychlorinated biphenyls (PCB) pollutants due to its constant filtering of seawater for food.[375] Furthermore, in some cases, very high concentrations of pollutants have been detected in the tissues of these organisms. However, what is not known thus far is whether the pollutants were simply accumulated by these organisms from the water in which they reside or whether the consumption of contaminated microplastics was the reason these organisms accumulated these pollutants. In reality, it may well be the case that a combination of both pathways resulted in the bioaccumulation. However, the main point here is that establishment of a definitive link between the accumulation of pollutants by aquatic invertebrates and the consumption of contaminated microplastics is still very much in its infancy and only limited evidence exists thus far. Nevertheless, the potential exists and the evidence is growing at a steady pace. Crucially, aquatic invertebrates tend to be located near the bottom of the food chain. Thus, the consumption of contaminated microplastics by aquatic invertebrates may well have the consequence of the bioaccumulation and biomagnification of contaminants throughout the food web. Ultimately, further scientific studies are required to fully understand the mechanisms involved and to assess the dangers that this may represent (Fig. 7.2).

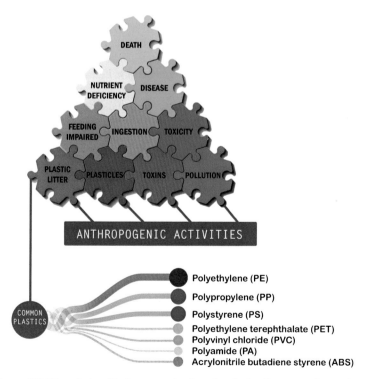

Figure 7.2 The relationship between plastics, chemical pollutants and organisms.

Humans

There is very little information in the literature thus far, with regard to effects of ingested microplastics on humans. However, a study[409] in the late 1980s fed healthy human volunteers with 15 g of microplastics, measuring less than 2 mm in diameter. While the specific type of plastic is not specified, the researchers noted that in comparison to the volume of ingested microplastics, the resulting faeces were typically 3 times greater in volume. Furthermore, there was a significant increase in transit through the gastrointestinal system, which they attributed to mechanical activation of mucosal receptors in the epithelial lining of the intestine. Another study[283] carried out in the late 1990s fed healthy human volunteers with 15 g of polyethylene microplastics measuring 1–2 mm in diameter. As with the previous study, this study also found that microplastics significantly accelerated the transit time of material through the human gastrointestinal system. Incidentally, increased gastrointestinal transit time from microplastic ingestion has also been demonstrated in pigs and dogs[58] indicating that the effect is not confined to humans. Certainly, any condition that increases transit time of food through the gastrointestinal system can reduce the absorption of vital nutrients and eventually result in nutritional deficits.[114] Furthermore, many plastics contain chemical additives, such as phthalates, which can be toxic (see Chapter 6) and the monomers of polyester and poly(methyl methacrylate) are mutagens and respiratory irritants.[333]

In terms of contaminated microplastics, conclusive evidence of the transfer of pollutants from contaminated microplastics to humans does not yet exist. However, the potential for microplastics as a vector for the introduction of waterborne contaminants into humans can be inferred from the research that has been undertaken thus far. For example, a recent study[151] concluded that humans are ingesting microplastics as a result of eating plastic-containing fish which are harvested for human consumption, such as the king mackerel (*Scomberomorus cavalla*) and the Brazilian sharpnose shark (*Rhizoprionodon lalandii*). Indeed, the colour of plastics ingested by these fish ranged from clear through white to yellowish, while yellow microplastics recovered from the environment are often contaminated with the highest levels of pollutants.[112,217,320] Furthermore, microplastics in the environment can be contaminated with polycyclic aromatic hydrocarbons (PAHs),[80] and levels have been detected as high as 500 ng/g.[354,363,427] Indeed, our own research has demonstrated (see Chapter 6) that 3.8 mm spherical pristine polyethylene microplastics, exposed to 33 different PAH compounds for 24 h in artificial seawater, exhibited a high affinity and sorption capacity for the carcinogenic compounds benzo[a]pyrene and dibenzo[a,h]pyrene, and to a greater extent at 21°C than at 10°C.

Importantly, some PAHs exhibit interactions with DNA resulting in genotoxic effects and cancer. One particularly deadly contaminant is benzo[a]pyrene.[291] Furthermore, the United States Environmental Protection Agency considers PAHs to be pollutants of concern.[123] While the general thinking is that fish can metabolise PAHs, thereby avoiding bioaccumulation, this may not entirely be the case. For example, in a study[99] which collected Indian salmon (*Eleutheronema tetradactylum*), Indian oil sardine (*Sardinella longiceps*), tigertooth croaker (*Otolithes ruber*), *mystus*

(*Mystus seenghala*) and goldspotted grenadier anchovy (*Coilia dussumieri*) between 2006 and 2008 from the harbour line in Mumbai, India, PAHs were detected in all of the species of fish studied. Indeed, the total levels of PAHs were ascertained to be 17.43–70.44 ng/g (wet weight), of which the level of carcinogenic PAHs detected was 9.49–31.23 ng/g (wet weight). The researchers estimated that the daily intake of carcinogenic PAHs for individuals consuming these species of fish is between 1.77 and 10.70 ng per kg of bodyweight per day. Another study[102] investigated the levels of PAHs in various fish collected from the Red Sea, a body of water which separates Africa and the Arabian Peninsula. The researchers found that the total levels of PAHs in the edible muscle of the fish was 422.1 ng/g (dry weight) and estimated that the daily intake of carcinogenic PAHs for individuals consuming these fish would be around 150 ng per person per day.

Similarly, a study found high levels of polybrominated diphenyl ethers (PBDEs) in fish collected from seas around China.[261] Incidentally, PBDEs are commonly added to plastics during manufacture to impart flame resistance.[440] However, the structurally analogous methoxylated (MeO-PBDE) and highly toxic hydroxylated (OH-PBDE) congeners commonly found in many biological organisms, including humans, may exist as a result of the metabolism of PBDEs or even be produced by algae (OH-PBDEs) or bacteria (MeO-PBDEs).[434] Thus, the presence of PBDEs in human blood may originate from various sources, including terrestrial origins. Nevertheless, MeO-PBDEs have been detected at higher levels in individuals who consume seafood.[376] Indeed, filter feeding organisms, such as mussels, are susceptible to the accumulation of pollutants and microplastics.[14] Furthermore, mussels are regularly consumed by humans worldwide on a daily basis,[206,454] and it is estimated that in Europe, humans consume up to 11,000 microplastics each year from eating shellfish.[54]

Certainly, a study estimated that humans consuming 300 g of shellfish meat from mussels exposed to polycyclic biphenyls (PCBs) could potentially be exposed to 288 ng of PCBs as a result of exposure of the organisms to contaminated water and sediment. However, it was estimated that only an additional 0.18 ng could be gained from microplastics contaminated with PCBs. This estimate is based on an average of 0.2 microplastics (of weight 5 μg) per gram of mussel tissue and with a maximum concentration of PCBs at 605 ng/g per microplastic.[425] Consumption of 0.18 ng of PCBs is considerably lower than the daily tolerable intake of PCBs, which has been suggested to be 20 ng/kg bodyweight.[453] However, while this calculation is based upon a study which found levels of sorbed PCBs as high as 605 ng/g,[320] other studies have found concentrations of PCBs on microplastics to be as high as 18,700 ng/g.[112] These high concentrations highlight the reality that the levels of pollutants sorbed to environmentally recovered microplastics can exhibit a high degree of variance.[130,315] Thus, if the exposure is recalculated using this higher sorbed concentration of 18,700 ng/g, then this could potentially amount to human ingestion of 5.61 ng of PCBs from microplastics (more than 30 times higher) from the same 300 g of shellfish.

A recent report found 105 mini-microplastics per gram (dry weight) in the tissue of blue mussels collected from the Dutch Coast and 19 mini-microplastics per gram (dry weight) in mussels collected from the North Sea coast.[255] Indeed, it was found that of all the organisms tested, blue mussels had the highest concentrations of mini-microplastics with 86% being 1–300 μm in size, in the case of the Dutch coast and 50% in the case of the North Sea coast. If the possible human exposure to PCBs is recalculated again, assuming only a single 5 μm microplastic per gram of mussel tissue, then the potential human exposure could increase to around 28 ng of PCBs, which is higher than the daily tolerable intake and more than 150 times higher than the original value. While it may be reasonably expected that the organism may die from ingesting such a high concentration of PCBs, mussels recovered from the natural habitat have been found to contain concentrations of PCBs in their tissues at levels as a high as 1630 ng/g.[112] Nevertheless, there is great difficulty in estimating the exposure levels of humans to aquatic contaminants via microplastics as there are many confounding factors and fundamental influences on the levels of contaminants sorbed, such as the degree of weathering, the surface area, they type of plastic and the time spent in the environment.[80,112,130,131,320] Ultimately, continued regular assessment of the situation is required due to increasing levels of microplastics in the aquatic environment.[255,425]

Seabirds

As early as 1988, researchers were drawing associations between persistent organic pollutants (POPs), seabirds and microplastics. For example, a study[366] found a positive correlation between the levels of polychlorinated biphenyls (PCBs) present in South African seabirds (*Puffinus gravis*) and the volume of small pieces of plastic that they had ingested and surmised that the seabirds were likely accumulating PCBs partly from plastic ingestion. Similarly, a later study[399] investigated the concentration of PBDEs in Northern Pacific seabirds (*Puffinus tenuirostris*) and found a relationship between ingested plastic and PBDE congeners 209 and 183 in 25% of the birds. Furthermore, the researchers emphasise that these two particular PBDE congeners are not found in the seabird's usual prey, thus pointing to plastic as the principle source.

Marine turtles

Studies of loggerhead turtles (*Caretta caretta*) and green turtles (*Chelonia mydas*) have indicated that the consumption of plastic litter by these species results in nutritional deficiencies since plastic material cannot be digested and has no nutritional value.[97,113] Furthermore, in a study[372] of 265 juvenile green turtles (*Chelonia mydas*) along the Brazilian coastline, it was found that only small amounts of ingested plastic litter were required to induce intestinal blockage and death. Furthermore, the researchers reported that a considerable portion of the ingested debris was disposable plastic products designed to be used and then discarded, such as plastic straws and cups, as well as products with a short duration of use, such as plastic bags. Furthermore, an investigation[167] into plastic ingestion by loggerhead turtles (*Caretta caretta*) along the Portuguese coast found that 56 of the 95 turtles examined (59%) had plastic in their gastrointestinal tract and 76.8% of turtles with plastic present had less than 10 plastic items. Furthermore, of all the ingested litter present in the gastrointestinal tract of the turtles, 56.8% was composed of plastic.

Seals and sea otters

It is entirely possible for higher organisms to be exposed to microplastics and contaminants purely as a result of their diet. Indeed, it has been suggested[129] that the exposure of seals to POPs is primarily via ingestion of contaminated food sources, such as fish, and not as a result of their contact with the aquatic environment. For example, in a study which examined the faeces of Antarctic fur seals on Macquarie Island, Australia, it was revealed that 164 microplastics were present in the faeces and 93% of the microplastics were composed of polyethylene.[118] Considering that the main diet of these fur seals was mesopelagic fish, such as lantern fish (*Electrona subaspera*), it is likely that the fish consumed the microplastics, after which they were then consumed by the seals, as opposed to the seals directly consuming the microplastics.

In a study[210] which analysed samples of blubber taken from various grey seal pups in the Farne Islands, United Kingdom between 1998 and 2000, polybrominated diphenyl ethers (PBDEs), Polychlorinated biphenyls (PCBs) and organochloride pesticides (DDE and DDT) were found in the blubber. These persistent organic pollutants (POPs) are passed from mother to pup in the womb and via milk following birth. However, upon maturation, dietary habits, such as the ingestion of fish, become the predominant exposure route to these pollutants.

While there are no reports thus far of sea otters ingesting plastics, it is expected that sea otter populations in the North Pacific will suffer increasingly detrimental effects from anthropogenic pollution of the aquatic environment. Populations in California in the United States have exhibited the highest effects from pollutants, suffering infectious diseases and increased morbidity and thus have been deemed as threatened by the United States Endangered Species Act.[18]

Whales

There have been several documented incidences of many species of whale ingesting plastic litter, such as the killer whale,[67] the beaked whale[378] and the sperm whale.[203,402] Furthermore, a dead sperm whale recovered from the Mediterranean Sea was ascertained to have died as a result of a ruptured stomach, which contained 7.6 kg of plastic material.[93] However, microplastics were directly identified for the first time in a cetacean species during a study[268] of True's beaked whale carcasses discovered on the North and West coasts of Ireland. In the carcass examined, the average size of the microplastics was reported to be 2.16 mm while the greatest accumulation of microplastics (38%) was in the whale's main stomach. Furthermore, the main type of plastic recovered overall was rayon, which comprised 53% of the recovered microplastics. This was significantly higher than other types of plastic and the next most abundant, polyester at 16%. Indeed, the common plastics polyethylene and polypropylene were reported to only amount to 4% and 6% respectively. In another whale, a 7 cm polyethylene shotgun cartridge was found in the accessory main stomach of the whale, however, the researchers stress that ingestion of large plastic debris by these whales is quite common.

However, as the largest filter feeder in the ocean, baleen whales (such as the blue whale) were expected to be exposed to large quantities of microplastics on a daily basis via there filtering activities.[7,134,286] Indeed, in a recent study,[27] microplastics were directly identified for the first time in a baleen whale (*Megaptera novaeangliae*) carcass located on a sandbank in the Netherlands. The plastic materials identified were mainly fragments, sheets and threads ranging in size from 1 to 17 mm. Furthermore, it was reported that the abundance and types of plastic found were polyethylene (55.01%), polyamide (37.64%), polypropylene (5.61%), polyvinyl chloride (0.97%) and polyethylene terephthalate (0.77%).

In a study[136] which analysed the blubber of stranded baleen whales along the Italian coast, a significantly high concentration (57.97 ng/g) of mono-(2-ethylhexl) phthalate (MEHP), a metabolite of di-(2-ethylhexyl) phthalate DEHP, was detected in 80% of the blubber. The researchers concluded that whales are chronically exposed to phthalates, and possibly other persistent organic pollutants (POPs), from the regular consumption of plastic litter and have stressed that this population of whales is now in sharp decline. Indeed, the researchers have proposed that phthalates may possibly be used as a marker for the ingestion of plastics by baleen whales. Furthermore, studies of beluga whale carcasses have revealed high levels of POPs upon toxicological examination[90,109,273] and reproductive orders exposed animals.[23] Furthermore, examinations of five species of toothed whale found high concentrations of polybrominated diphenyl ethers (PBDEs) and polychlorinated biphenyls (PCBs) in their blubber, with main factors affecting levels of pollutants in these animals considered to be the local environment, food sources and trophic level.[290] Certainly, the levels of POPs within the tissues of organisms tend to increase in concentration at consecutively higher levels in the food chain, thereby demonstrating biomagnification.[127]

While POPs which exhibit considerable hydrophobicity, such as PCBs, build up in the tissues of organisms to concentrations that exceed that of the surrounding water (bioconcentrate), models suggest that compounds such as hexachlorocyclohexane (HCH), which exhibit a low octanol–water partition coefficient (K_{OW}) (see Chapter 6), tend not to bioconcentrate in aquatic species. However, biomagnification of β-HCH in the fatty tissues of beluga whales was higher than the biomagnification of the PCB congener 2,2′,3,4,4′,5,5′-heptachlorobiphenyl (PCB-180), which was attributed to an increased digestive capability of air breathers to assimilate their food sources, coupled with decreased urinary elimination of POPs with a log K_{OW} greater than 2. For water-respiring creatures, POPs require a log K_{OW} greater than five before bioconcentration occurs.[223] Thus, the bioconcentration of POPs may be substantially greater in air breathing animals than water-respiring animals.

Fish

Microplastics are ingested by fish,[269,333,338,359,371] and the consumption of microplastics by fish can interfere with biological processes, such as the inhibition of gastrointestinal function, as well as causing blockages and inducing feeding impairment.[269,460] Furthermore, fish are often found to have bioaccumulated waterborne pollutants to various degrees. For example, a study[439] found high levels of polychlorinated biphenyls (PCBs) of 1066–3112 ng/g in the liver of fish collected from an abandoned metropolitan dumping area for sewage sediment in Garroch Head, a Scottish coastal region in Argyll and Bute. A further study[445] found concentrations of PCBs in North

Atlantic fish, collected off the West coast of Scotland, to be in excess of 500 ng/g. An examination[348] of fish products sold in the Canadian retail market during 2002 found total concentrations of PCBs to be on average 12.9 ng/g (net weight) in the salmon they tested, with PCB 101, 118, 138 and 153 being the most abundant congeners observed. The human daily tolerable intake of PCBs has been suggested to be 20 ng per kg of bodyweight.[453] In a study[157] assessing the levels of perfluoroalkylated substances (PFAS) in wild eels in Comacchio Lagoon and Po River (located on the Northwest Adriatic coast), it was found that concentrations were in the ranges <0.4–92.77 ng/g w.w. for perfluorooctanoic acid (PFOA) and <0.4–6.28 ng/g w.w for perfluorooctane sulfonic acid (PFOS). The highest accumulations were found in blood while the lowest was in muscle tissue. Furthermore, the researchers reported that lesions in the liver of the eels were observed and PFAS levels were found to be on par, or lower, than levels observed in other European fish. A study[261] of fish collected from seas around China found concentrations of polybrominated diphenyl ethers (PBDEs) in the tissues of the fish to be 0.3–700 ng/g, which was reported as high concentrations.

Incidentally, fish, by their very nature, are highly mobile creatures and many travel great distances. Thus, if fish are ingesting plastic then they may be unwittingly serving as a vessel for the long-range distribution of microplastics and persistent organic pollutants (POPs). For example, large abundances of microplastics have been reported in deep ocean sediments[458] and a study has identified plastic ingestion by deep sea fish.[6] Furthermore, deep ocean fish are more prone to bioaccumulation of high levels of pollutants, such as polychlorinated biphenyls (PCBs) and polybrominated diphenyl ethers (PBDEs) pollutants, on account of their typical extended lifespans and their propensity to consume organisms situated further up the food web, in contrast to their epipelagic counterparts.[446] Certainly, PCBs and PBDEs have been detected in deep sea fish in Northwest Scotland.[443]

It is likely that the capability of microplastics to become contaminated with hydrophobic pollutants via sorptive processes facilitates the transfer of these hydrophobic pollutants to fish if the contaminated microplastics are ingested,[250] thereby inducing toxicological effects. For example, it was determined[201] that when European seabass (*Dicentrarchus labrax*) were fed <0.3 mm of polyvinylchloride (PVC) microplastics, which had been exposed to seawater in Milazzo Harbour, Italy for 3 months to adsorb contaminants. Upon examination, it was discovered that there were significant pathological alterations to the distal part of the intestine of the fish. Exposure to >0.3 mm virgin PVC microplastics resulted in a similar effect. However, the polluted PVC microplastics exhibited more damage and it was evident that the intestinal function was capable of being completely compromised by PVC microplastics. Furthermore, as exposure times of the fish to PVC microplastics was extended, the pathological alterations increased from moderate to severe. In another study,[322] which involved exposing juveniles of the common goby (*Pomatoschistus microps*) to the polycyclic aromatic hydrocarbon (PAH) pyrene (20 and 200 μg/L) and 1–5 μm polyethylene microplastics simultaneously, it was observed that the microplastics delayed mortality of the fish from pyrene. However, it was found that the mixture of microplastics and pyrene resulted in a considerable reduction in acetylcholinesterase and isocitrate dehydrogenase activity within the fish. Consequently, the researchers posited that the combined effects of microplastics and pyrene could potentially increase fish mortality in populations within the natural environment.

Additionally, the interactions of microplastics with heavy metals and their effect on these fish have been studied. For example, in a study[266] of the effects of a mixture of chromium (VI) and 1–5 µm polyethylene mini-microplastics on early juveniles of the common goby (*P. microps*), obtained from the estuaries of the Lima and Minho rivers in the Northwest Iberian Peninsula, it was observed that toxicological interactions between the microplastics and chromium (VI) occurred. The researchers established that in the fish collected from the Lima River, concentrations of chromium (VI) of more than 3.9 mg/L significantly increased levels of lipid peroxidation. Furthermore, the researchers observed up to 31% inhibition of the activity of acetylcholinesterase. These are important biomarkers which are part of mechanisms that have a direct impact on the performance and survival of these fish in the natural habitat. Accordingly, a 67% decrease in predatory performance of the fish was observed from the mixture. Additionally, the researchers found that in the mixture, there was a decrease in the nominal concentration of chromium (VI) in the aqueous medium, in the presence of microplastics, thereby suggesting that chromium (VI) sorbed to the microplastics when the two substances were mixed. Conversely, in another study[78] which involved gold, rather than chromium IV, it was observed that when the juvenile common goby (*P. microps*) was exposed to ≈5 nm gold nanoparticles for 96 h, there was a reduction in predatory performance of approximately 39%, while an increase in water temperature from 20 to 25°C resulted in a 2.3-fold increase from 0.129 to 0.129 g/g w.w. in the weight of gold nanoparticles incorporated within their body. However, the presence of 1–5 µm polyethylene mini-microplastics, during the exposure, had no effect on the toxicity of gold nanoparticles to the organism.

The evidence that fish regularly consume microplastics in their natural habitat is growing rapidly. Indeed, fish recovered from the English Channel in the United Kingdom contained microplastics in their gastrointestinal system[269] while a study[151] carried out along the eastern coast of Brazil found between two and six microplastic pellets (ranging in size from 1 to 5 mm) in the stomach of every king mackerel (*Scomberomorus cavalla*) and Brazilian sharpnose shark (*Rhizoprionodon lalandii*) examined. The rate of plastic ingestion of each species was 62.5% and 33% respectively. Furthermore, a study[371] of wild gudgeon freshwater fish (*Gobio gobio*) collected from French rivers found microplastics in 12% of the fish. In another study[359] which involved investigating albacore (*Thunnus alalunga*), bluefin tuna (*Thunnus thynnus*) and swordfish (*Xiphias gladius*) collected from the Mediterranean Sea over a period of 1 year for evidence of plastic consumption, it was found that 18.2% of the fish had ingested plastics, with Bluefin tuna having had ingested the greatest amounts. Furthermore, 75% of the plastics were in the microplastic size range (<5 mm), as opposed to mesoplastics (5–25 mm) or macroplastics (>25 mm) (see Chapter 5).

An examination of the gut tract of 535 marine and freshwater fish in the Gulf of Mexico revealed that 10% of marine fish and 8% of freshwater fish had ingested microplastics, as well as other sized plastics. Although the plastics found had a maximum size of 14.3 mm, the most common sizes were 1–2 mm. the types of plastics found were polystyrene, poly(methyl methacrylate), nylon, polyester and polypropylene. Interestingly, the researchers did not report finding any polyethylene microplastics. Bearing in mind that 535 fish were examined and that polyethylene is typically one of the most common types of plastic reported in many other studies, this is quite remarkable. Consequently, other factors may be at play, such as colour and buoyancy. Indeed, a study of the common goby (*P. microps*) found that the juvenile fish, as visual predators, were confused by polyethylene mini-microplastics in the size range 420–500 μm and misidentified them as their natural prey (*Artemia nauplii*) thereby reducing their predatory abilities drastically. Interestingly, the researchers speculate that mini-microplastics moving in the aquatic environment as a result of turbulence would likely resemble the movements of natural prey and thus the mini-microplastics would be more susceptible to misidentification and consumption. Certainly, it has been reported[50] that many plastic items collected from marine environments show signs of bite marks by fish and it has been estimated that 1.3 tonnes of plastic were attacked in a 15 km area near Hawaii. Indeed, microplastics collected from the Goiana Estuary in Brazil had a colour and shape that was similar in appearance to the local species of zooplankton.[258]

The effect of colour on the consumption of microplastics

Microplastics exist in a vast array of colours (see Chapter 5). However, whether the colour of microplastics is a major factor in their misidentification as prey by aquatic organisms and thus affects the likelihood of ingestion of these plastic materials is still a matter that is up for debate. However, based on the fact that many aquatic species are visual predators, this would seem intuitive. Indeed, a study[47] concluded that fish in the North Atlantic preferentially consumed white polystyrene microplastics, as opposed to transparent microplastics, despite the identical abundance of both colours. However, it may well be that white microplastics are considerably easier to recognise by foraging fish than transparent microplastics, which would be expected to be far less visible.[237] However, a study[359] of albacore (*Thunnus alalunga*), bluefin tuna (*Thunnus thynnus*) and swordfish (*X. gladius*) collected from the Mediterranean Sea revealed that white and transparent plastics were the most commonly ingested. Aside from these pale plastics, bluefin tuna ingested blue and yellow plastics while albacore ingested only ingested blue plastics. Conversely, swordfish only ingested yellow plastics. Importantly, yellow and black coloured microplastics have consistently been found to be contaminated with the greatest concentrations of pollutants.[80,112,217,320]

However, the current evidence appears to suggest that black and red coloured microplastics are the least likely to be ingested by aquatic species. For example, a study[92] investigated the effects of 420–500 μm polyethylene spherical mini-microplastics on early juveniles of the common goby (*P. microps*) obtained from the estuaries of the Lima and Minho rivers in the NorthWest Iberian Peninsula. The study was carried out in a salt water medium and with, and without, the presence of brine shrimp (*Artemia nauplii*). It was revealed that fish from the Lima river ingested a greater amount of white microplastics (density: 1.2 g/cm³), as opposed to black (density: 1.15 g/cm³) and red (density: 0.98 g/cm³) microplastics. The researchers posited that since brine shrimp tend to be pale in colour, combined with the fact that juveniles of the common goby are visual predators, the white microplastics had an appearance most like their natural prey. Furthermore, it was hypothesised[92] that based on experimental research, the ability of fish in discriminating between microplastics and their natural prey may be influenced by the environmental conditions during development of the fish, such as prey availability and the levels of contaminants.

Interestingly, microplastics recovered from the faeces of Antarctic fur seals contained plastics which were white, brown, blue, green and yellow in colour.[118] From the details of the study, it appears that red and black microplastics were absent from the faeces. However, the researchers reported that red microplastics were common in flotsam.[118] Importantly, fur seals are not considered to be consuming microplastics directly.[129] Thus, the presence of microplastics in their faeces is highly likely to be the result of the consumption of plastic-containing fish, and consequently it can be inferred that these fish are not consuming red and black microplastics. Incidentally, microplastics recovered from True's beaked whale carcasses, discovered on the North and West coasts of Ireland are hypothesised to have originated from fish and squid ingested by the whales.[268]

Importantly, benthic prey tends to have a red appearance[92] and many species of seal feed at considerable depth, such as the Southern elephant seal which feeds in the mesopelagic zone[44] and the Australian fur seal which forages in deep benthic habitats.[76] Furthermore, analysis of the diet of Antarctic fur seals has found their diet to consist of 94% mesopelagic lantern fish.[59] Thus, one could hypothesise that the reason for the absence of red-coloured microplastics in seal faeces may be a result of the inability of low-energy red wavelengths of light to penetrate deep water, thereby rendering the microplastics as black-coloured and less visible to foraging fish. This hypothesis could be tested by investigating whether microplastics recovered from deep sea sediments contain higher abundances of red and black plastics.

For example, red light is attenuated in the deep ocean, where the frequency of light inclines to be in the blue 475 nm region of the visible spectrum.[256] Furthermore, a lack of red wavelengths in a blue aquatic environment would render a red-coloured object as an imperceptible black to observing fish from above.[237] Thus, the lack of red perception by mesopelagic fish, and the consequent decrease in the misidentification of red and black microplastics as prey, could permit red microplastics to distribute throughout the aquatic environment or sink to benthic regions unimpeded. Certainly, many deep sea organisms exploit the attenuation of red light by adopting red pigmentation in an attempt to camouflage themselves from visual predators.[423] Thus, it appears from the current evidence garnered that the colour of microplastics does indeed play a major role in the probability of ingestion of microplastics by aquatic organisms.

Translocation and trophic transfer of microplastics

The health implications for humans consuming organisms which contain microplastics are undetermined. However, it has been suggested that mini-microplastics can traverse the gut wall and enter the circulatory system of humans, rodents and invertebrates,[421] although conclusive evidence is not yet forthcoming. However, due to their small size, nanoplastics could potentially be distributed via the circulatory system, ultimately passing the human blood–brain barrier and placenta, or becoming deposited in vital organs where they may accumulate. While there is insufficient evidence at present to ascertain if this is occurring, or even to form a cogent theory, there are some studies of other nanoplastic-sized particles in humans which may shed some light on the way in which nanoplastics may behave in the body. For example, the ingestion or overexposure to colloidal solutions containing silver nanoparticles, with a size of 5–50 nm, results in translocation of the nanoparticles to various areas of the human body where they tend to accumulate in the liver, kidney, spleen, skin, eyes and testes, leading to toxic effects.[326] Furthermore, nanoparticles of 70 nm in size, and partly composed of polyethylene glycol (PEG), have been systematically administered to humans for the purposes of targeted drug delivery and have been demonstrated to circulate in the bloodstream and accumulate and permeate within solid tumours.[86]

In aquatic organisms, some limited evidence exists. For example, a study[39] demonstrated that 3.0 and 9.6 μm fluorescent spherical polystyrene mini-microplastics were able to translocate from the gut cavity to the circulatory system of the blue mussel (*Mytilus edulis*). Another study[124] found that 500 nm spherical polystyrene fluorescent nanoplastics were ingested by mussels (*Mytilus edulis*). Upon feeding the nanoplastic-containing tissues of the muscles to a common crab (*Carcinus maenas*), subsequent testing of the haemolymph of the crabs revealed that the nanoplastics had undergone translocation into the tissues and haemolymph of the crab and thus undergone trophic transfer. Furthermore, the nanoplastics remained in the tissues of the crabs for up to 21 days. In a study[379] using 10 μm spherical fluorescent polystyrene mini-microplastics, it was demonstrated that the mini-microplastics were capable of undergoing trophic transfer from a lower trophic level (mesozooplankton) to a higher trophic level (macrozooplankton). In another study,[36] the translocation of 180–250 μm polluted polystyrene mini-microplastics to the stomach, hepatopancreas and gills of the fiddler crab (*Uca rapax*) was observed following exposure of the crabs to the mini-microplastics for a period of 2 months. Importantly, the study demonstrated that as well as the 500 nm nanoplastics demonstrated to translocate to the hepatopancreas in a previous study,[124] mini-microplastics up to 500 times larger have been shown to have a similar effect. Thus, both studies demonstrated that nanoplastics and mini-microplastics have the potential to block glands and gills as well as disrupting respiratory functions, nutrient absorption, secretion of enzymes and the storage and maintenance of energy reserves.

Microplastics as vectors for microorganisms and invasive species

Invasive species can use microplastics as a floating raft that enables them to traverse great distances across the world's oceans to new habitats.[20,21] However, one distinct possibility is that microplastics may be acting as a vector for the introduction of pathogens into organisms. As the size of a microplastic decreases, the surface area-to-volume ratio increases (see Chapter 6). For this reason, large concentrations of mini-microplastics can provide a considerable surface area for microorganisms to adhere to. Indeed, it is often the case that microplastics in the environment will become fouled with organic material. Thus, if bacteria are feeding on this material, the potential exists for microplastics to support colonies of pathogenic bacteria. Furthermore, the interactions between microorganisms and microplastics needs to be further investigated to ascertain what role and extent, if any, those specific microbes may play in facilitating the degradation of microplastics. As such, it is recommended[211] that there is an increase in investigations into the types of microorganisms capable of colonising microplastics, especially within the aquatic environment. Certainly, in our own investigations of microplastics using a scanning electron microscope (SEM), we have found various organisms colonising fouled primary and secondary microplastics collected from the marine environment, including bacteria, fungi and diatoms (see Figs. 7.3 and 7.4).

Primary microplastics

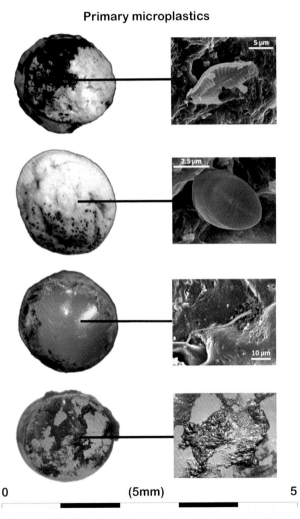

Figure 7.3 Organisms and fouling detected on primary microplastics using a scanning electron microscope (SEM).

Secondary microplastics

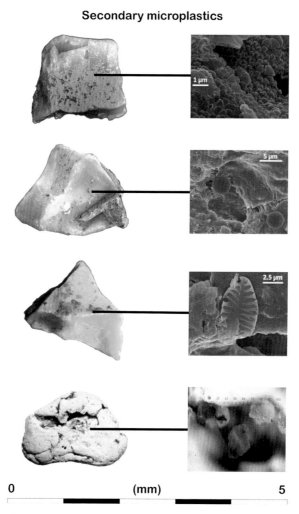

Figure 7.4 Organisms and fouling detected on secondary microplastics using a scanning electron microscope (SEM).

Furthermore, we detected the structure in Fig. 7.5 on a 2 mm spherical polystyrene microplastic collected from the marine environment, using a scanning electron microscope (SEM). Despite several attempts by many individuals, the structure has not been able to be identified, and thus its origin remains a mystery.

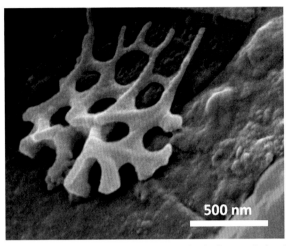

Figure 7.5 An unknown structure detected on a 2 mm spherical expanded polystyrene microplastic recovered from the marine environment.

Ultimately, microplastics may be a vector for the introduction of unknown pathogens into organisms by way of ingestion or even by direct contact with skin and mucous membranes. This may be particularly relevant to humans. For example, many people visit the beach and children often enjoy playing on the beach and building sand castles. During these activities, sand is inadvertently handled which may contain microplastics. If these microplastics are contaminated with pathogens, then the potential exists for these pathogens to be transferred to the hands, foodstuffs and the mouth. Thus, research is required to determine the survivability and adeptness of pathogens on microplastics in the environment. Certainly, weathered microplastics are often pitted and undulated (see Chapters 5 and 6). As such, it may well be the case that these indentations offer refuse to pathogenic organisms that might otherwise be washed off on smoother surfaces. This raises an interesting, though somewhat disconcerting, hypothesis that perhaps microplastics may become more susceptible to pathogen accumulation as they become more weathered. If this is the case, then the large quantities of microplastics currently present in the environment are potentially floating reservoirs harbouring ever increasing colonies of pathogenic organisms. The need for research here seems intuitive and one could postulate that there may be competition among specific pathogenic organisms for plastic surface area, or that certain species exhibit a predilection towards specific types of plastic. Ultimately, to investigate the types of species colonising microplastics, it is necessary to collect samples from the environment. Since the way in which microplastics are collected can have a bearing on the results obtained, it is important to choose a suitable and effective collection technique.

Microplastic collection techniques

Collection of microplastics from the environment

The analysis of microplastics in the environment starts with sample collection. While this first step is often given less consideration and described in less detail in the literature, selection of an appropriate technique is essential as it will determine the types of microplastics that are collected, separated, identified and subsequently reported. The method of sample collection is influenced by many factors. However, primarily the matrix to be sampled (water, sediment, soil, air or biota) will determine the abundance, size and shape of the microplastics obtained. There are no universally accepted methods for sampling any of these matrices and the methods available all have potential bias.[156] However, a carefully designed and implemented sampling strategy is the key to obtaining reliable and representative results for international monitoring programmes. Importantly, with any sampling strategy, the cost benefit is an essential consideration. Thus, the methods used should be simple enough to allow for replication and reproducibility, as well as being cheap enough to be accessible, while ensuring precision, accuracy and minimal contamination.[110]

In terms of an atmospheric matrix, only one study has investigated atmospheric samples for microplastics thus far.[105] At various frequencies during the monitoring period, the study used a relatively simple technique of collecting atmospheric fallout in stainless steel funnels with a sampling surface area of $0.325\,m^2$. However, in the terrestrial environment, relatively few studies have been undertaken thus far, despite the assumption that soils are considered to be acting as a sink for microplastics.[107] One reason for this is that soils are typically complex organo-mineral matrices and as a consequence, extraction of microplastics from soil is not as straightforward as extracting from water or sediment.[352] Furthermore, the identification of sites suitable for sample collection is less obvious than in the aquatic environment. For example, shorelines exhibit a pattern of accumulation, such as that seen on the high tide mark, which is not present in the terrestrial environment. Nevertheless, synthetic fibres have been measured on land and used as an indicator of municipal sewage sludge.[415] Ultimately, the vast majority of studies have involved the investigation of microplastics in water and sediment, as opposed to terrestrial and atmospheric environments. For these reasons, this chapter shall focus on the collection of microplastics from water and sediment matrices.

Standardisation of collection techniques

One of the predominant recurring themes throughout the published literature on microplastics research is the lack of standardisation in the collection of microplastics from aquatic, solid fraction or biological samples.[107,188,334,355,428,448] However, while there is no standardisation for collecting microplastics, the standardised size and

Microplastic Pollutants. http://dx.doi.org/10.1016/B978-0-12-809406-8.00008-6

colour sorting (SCS) system for the effective categorising of microplastics, based on their size and appearance, was introduced with this book (see Chapters 5 and 10). To effectively monitor microplastics in the environment, other steps in the monitoring process, such as the collection of microplastics from the environment, shall require standardisation through the development of standard operating procedures (SOPs) and rigorous quality control measures.

Furthermore, the abundance of microplastics in environmental samples, or the concentrations used in laboratory based experiments, are often expressed with differing units of measurement, which in some cases can be incomparable.[53,107,156,334] Indeed, the abundance of microplastics is commonly presented as a numerical or mass concentration.[107] However, with water samples this is expressed as the weight or number of microplastics per area (such as km^2 for sea surface samples) or per volume (such as m^3 for water columns). In the case of sediments, the abundance of microplastics is recorded as the weight or number of microplastics per sediment area or weight [wet weight (ww) or dry weight (dw)], as well as volume (mL or L). Consequently, this wide variation in the way in which the abundance of microplastics are quantified means that comparisons between studies can often be very difficult.[53,107] For that reason, it has been suggested that environmental studies should provide sufficient information to allow unit conversion[188] and that preferably both numerical and mass concentrations are provided.[41,116,156]

Since the issue of microplastics in the environment is gaining increased media and political attention, it is imperative that regulators and other interested parties have the necessary data to make large scale spatial and temporal comparisons. Certainly, the lack of standardisation not only in sample collection, but also in the units used to report the abundance of microplastics in various mediums is hampering progress in the monitoring of this emerging contaminant. However, the issue is beginning to be addressed. For example, the multinational BASEMAN project, which is funded under the EU Joint Programme Initiative (JPI) Ocean, has been tasked with the specific goal of method standardisation in microplastic analysis. However, this project has only recently been established, and it will take some time before standard operating procedures (SOPs) are developed. Furthermore, laboratory methods have been developed by the National Oceanic and Atmospheric Administration (NOAA) in the United States, such as the separation of microplastics from sediment, as well as recommendations for quantifying microplastics in waters and sediments.[276] Ultimately, the goal of standardisation is to overcome the barriers hindering the monitoring of microplastics, such as the requirement for the mastery and execution of several different techniques,[334] and to developing a cohesive microplastic monitoring strategy that facilitates a clear understanding of what is happening in the environment.

Sampling methods

It is inevitable that the sampling method adopted shall greatly influence the type of sample collected. Certainly, the lower size limit of any microplastics collected is determined by the sample collection technique, such as the size of mesh used, and the processing method.[107] There are three main sampling methods[188] used for recovering microplastics from the environment and while each method has its own advantages and disadvantages, sampling methods will always change, develop and adapt over time. Ultimately, the sampling method to be employed will largely depend upon the environment which is being sampled and the size limitations of microplastics to be collected. Thus, in many cases it may be appropriate to use more than one sampling method, particularly where both water and sediment samples are required. The three main methods used for sampling are selective sampling, volume-reduced sampling and bulk sampling.[188]

Selective sampling

In selective sampling, items visible to the naked eye are directly extracted from the environment, such as on the surface of the water or sediment. This collection method is adequate in situations where different microplastics of similar morphology and of a size greater than 1 mm are present, such as primary microplastic pellets and similarly shaped secondary microplastics. However, the main disadvantage of this technique is that the less obvious, more heterogeneous items are often overlooked, particularly when they are mixed with other beach debris.[156] Despite this, selective sampling has been extensively used and is reported in 24 of the 44 studies involving the extraction of microplastics from sediment.[188]

Volume reduced sampling

In volume reduced sampling, the volume of the bulk sample is reduced until only the specific items of interest for further analysis remains. Thus, the majority of the sample is discarded. Consequently, this method is typically utilised to collect samples from surface water because it has the advantage that large areas or quantities of water can be sampled. However, the disadvantage of volume reduced sampling is that discarding the vast majority of the sample introduces the risk of under-representation of the abundance of microplastics in the sample due to the potential loss of microplastics.

Bulk sampling

In bulk sampling, the entire sample is taken without reducing its volume. Although there are practical limitations to the amount of sample that can be collected, stored and processed, the advantage of this method is that in theory, all the microplastics in the sample can be collected, regardless of their size or visibility. Furthermore, processing the full sample prevents any microplastics from being lost or overlooked during the sampling process, as can happen in selective sampling or volume-reduced sampling. Additionally, the reduction in handling of the sample can also help to decrease any contamination by reducing the amount of time that the sample is exposed to the surrounding environment. While this method was undertaken in 1 seawater study and 18 sediment studies,[188] it has been used less frequently in more recent studies.

Environmental parameters

When collecting samples in the environment it is important to take into consideration and to record the prevailing weather conditions, not only on the day of sampling but in the period leading up to sampling. On the day of sampling, it is necessary to note the wind direction as this may influence any potential contamination from the person carrying out the sampling, as well as from others nearby. Furthermore, the time of sampling in relation to the tidal height (be it diurnal or semi-diurnal) should also be noted, as should the stage in the tidal cycle.[302] These will impact the sampling of both water and sediment from marine and estuarine environments. Interestingly, it has been reported in several studies that increased rainfall prior to sampling can have a significant positive impact on the amount of plastic debris observed, particularly in tropical areas,[133,344] and where there is a freshwater influence. Certainly, the quantity of debris entering the marine environment is reported to increase with rainfall,[240] and a significantly greater abundance of neustonic plastic litter has been observed within the California coastal surface water following storm events.[117,297] Thus, documentation of the prevailing weather conditions is necessary when sampling for microplastics and is particularly important when sampling in the marine environment.

Contamination mitigation

Due to their ubiquitous nature, the contamination of a solid or liquid environmental sample with microplastics that were not originally part of that sample is one of the major issues involving the examination of samples for microplastics and consequently it has been raised by several researchers.[4,96,107,118,198,316,319,355,459] Indeed, the processes involved in the collection, separation and identification of samples for microplastics often result in the inadvertent introduction of microplastics that would not otherwise be found in the sample. Mini-microplastics (particularly microfibers) can be introduced from the ambient air, but also via the use of sampling or laboratory equipment, improper storage of samples or even from the clothing of the researchers themselves. In many cases this contamination can compromise the analysis, leading

$\longrightarrow$

Figure 8.1 A contamination prevention protocol for the processing of microplastics in the laboratory.
Adapted from Wheeler J, Stancliffe J. Comparison of methods for monitoring solid particulate surface contamination in the workplace. Annals of Occupational Hygiene 1998;42:477–88. Woodall LC, Gwinnett C, Packer M, Thompson RC, Robinson LF, Paterson GLJ. Using a forensic science approach to minimize environmental contamination and to identify microfibres in marine sediments. Marine Pollution Bulletin 2015;95:40–6. Alomar C, Estarellas F, Deudero S. Microplastics in the Mediterranean Sea: deposition in coastal shallow sediments, spatial variation and preferential grain size. Marine Environmental Research 2016;115:1–10; Murphy F, Ewins C, Carbonnier F, Quinn B. Wastewater Treatment Works (WwTW) as a source of microplastics in the aquatic environment. Environmental Science and Technology 2016;50: 5800–8; De Wael K, Gason, FG, Baes CA. Selection of an adhesive tape suitable for forensic fiber sampling. Journal of Forensic Sciences 2008;53(1):168–71.

Step 1:Preparation

- A clean white cotton laboratory coat and nitrile gloves should be worn at all times.

- Only clothing composed of natural fibers should be worn, synthetic garments should be avoided, even if worn underneath a laboratory coat.

- Air movement in the laboratory should be minimised by closing all windows and doors.

Step 2:Cleaning

- Ensure that the laboratory is kept clean and free from dust. Avoid working below overhead fixtures which may have accumulated settling dust.

- Clean all equipment with 70% ethanol and then rinse 3 times with distilled water.

- After cleaning, cover all equipment with aluminium foil.

- Wipe all work surfaces with 70% ethanol 3 times prior to the commencement of work.

- Examine all petri dishes, filter papers and forceps with a dissection microscope before use.

Step 3: Solid particulate surface monitoring

- This process is carried out before and after all analyses of microplastics in samples.

- While ensuring that gloves are worn at all times, the adhesive side of fresh 5 cm^2 sections of transparent high tack adhesive tape is pressed 3 times onto the work surface and then lifted. Any solid particulates present should adhere to the adhesive on the tape.

- Each 5 cm^2 section of adhesive tape is then adhered to a clean piece of cellulose acetate film.

- The date and time that each 5 cm^2 section of adhesive tape was used is written next to it on the cellulose acetate film with a permanent marker pen.

- The 5 cm^2 sections of adhesive tape on the cellulose acetate film are then examined under a microscope for the presence of mini-microplastics, such as microfibers and microfragments.

- Further analysis, and positive identification, of any mini-microplastics found adhered to the 5 cm^2 sections of adhesive tape can be undertaken by infrared spectroscopy and subsequently excluded from the sample of interest.

Step 4: Solid airborne particulate monitoring

- Before work commences, clean pieces of dampened filter paper are placed in 9 cm standard glass petri dishes, ensuring that the filter paper covers the entire internal area of the petri dish.

- The petri dishes are then placed around the work surface, where they remain for the duration of the laboratory work.

- Upon completion of the work, the filter paper is then examined for the presence of mini-microplastics using a microscope, or a glass lid is placed on the petri dish and labelled with the date and time for subsequent microscopic analysis at a later date.

- Further analysis, and positive identification, of any mini-microplastics found on the filter paper can be undertaken by infrared spectroscopy and subsequently excluded from the sample of interest.

to overestimations of the abundance of microplastics in the sample.[263] For this reason, several methods have been suggested to help reduce this type of contamination. For example, during sample collection, samples should always be collected downwind to prevent air borne contamination and collected, transported and stored using non-plastic tools or containers, such as aluminium trays. When handling samples, synthetic clothing should always be avoided and natural fibres, such as cotton should be worn wherever possible. One type of clothing that should be expressly avoided is fleece, due to the significant release of plastic microfibers from this material.

Samples should also be kept covered to reduce exposure to the ambient air and the processing of samples in a clean room or sterile laminar flow hood may be particularly effective,[96,317,345,428,459] although this is not likely to be practical in many instances. Furthermore, particular attention needs to be paid in the laboratory during sample handling or processing and all equipment and laboratory surfaces should be cleaned with alcohol and then rinsed with distilled water before use.[4,267,344,345,459] Inspired by the techniques employed in the field of forensic science, such as the detection of solid particulate matter on surfaces using adhesive tapes,[451,94] contamination reduction and monitoring approaches have been proposed.[459] Furthermore, the concentration of solid airborne particulates during the analysis of microplastics can be monitored using dampened filter paper and glass petri dishes.[4,307,459] Ultimately a strict contamination mitigation protocol, such as in Fig. 8.1, should be adhered to when collecting and handling samples for microplastic analysis. Indeed, it has been demonstrated that when adequate cleaning is undertaken, the abundance of background fibres is considerably reduced.[459] Furthermore, it has been suggested that developed protocols are standardised and cross-calibrations are run between laboratories.[188]

Aquatic sampling

Microplastics are commonly found in most bodies of water and the strategy used to collect water samples for the examination of microplastics depends upon the type of aquatic environment that is to be sampled. In terms of their distribution in aqueous environments and owing to their physiochemical properties, such as variations in density, shape and chemical composition, as well as the extent of biofouling, microplastics can be found floating on surface waters, suspended in the water column or in the depths of the ocean (see Chapter 5). The specific location of the microplastics in the water shall influence whether horizontal sampling along the water's surface is required, or whether vertical sampling through the water column is needed. As with all environmental sampling methods, it is key to establish the parameters of the sampling method to be undertaken. Thus, consideration needs to be given to the physical environment to be sampled (e.g., area, depth, flow), the type of sample to be collected (positively buoyant upon the water surface, negatively buoyant at depth or neutrally buoyant in between) and the sampling method to be employed (vertical or horizontal towing or pumping from the water column). Since the sampling method used shall greatly influence the results obtained, these questions should be addressed early in the planning stages of sample collection, but shall often be dictated by logistical restrictions, such as the weather and the availability of equipment and boats.

Furthermore, concentrations of microplastics in the aquatic environment can vary dramatically between regions and typically, large volumes of water need to be examined. Consequently some form of sample reduction or filtration is needed, and while there have been some cases where bulk water samples have been taken,[98] the majority of water samples analysed for microplastics have been reduced in volume. This is usually carried out by the use of neuston nets for positively buoyant plastics on the sea surface, and zooplankton nets for sub-surface areas, with a typical mesh width of 300–390 µm. Ultimately, the sampling strategy adopted depends upon the purpose of the study (see Table 8.1).

Table 8.1 The choice of sampling strategy depends upon the purpose of the study

Study purpose	Objective
Qualitative	To rapidly gather non-numerical information about the types of microplastics present in the environment
Quantitative	To gather numerical data regarding the distribution and abundance of microplastics in the environment

Freshwater

Both lentic (relatively still) and lotic (flowing) freshwater systems are influenced by many of the same physical forces, such as wind and water currents, which can influence the can transport and accumulation of microplastics in the marine environment. However, since bodies of freshwater are generally smaller in size, the influence of these forces may be greater, thereby resulting in potentially larger spatial and temporal differences in the mixing and transport of microplastics.[110] Furthermore, site-specific physical drivers, such as advective transport (influenced by velocity) and diffusive/dispersive transport (influenced by turbulence), can have an impact on the distribution and concentrations of microplastics in the fresh water environment and will greatly be influenced by geology and relief.

This is apparent in Lake Erie, North America where higher concentrations of microplastics were found in sites of converging currents and areas closer to the shore.[116] Certainly along flowing rivers, higher concentrations of microplastics were found in samples taken from the mid-channel, as opposed to samples taken closer to the river bank or riverbed.[297] The residence time of lake waters can also influence the abundance of microplastics,[138] and seasonal variations in rainfall can affect the abundance in rivers.[297,342,345] Indeed, rainfall can substantially increase the amount of microplastics found in freshwater environments,[297,344] and this is particularly relevant in tropical areas where the rainy season can positively impact the concentrations of microplastics found[345] (see Chapter 5). Furthermore, as in the marine environment, powerful upward and downward movements of water can occur in freshwater bodies of water as result of temperature differences at different depths (vertical mixing). These movements of water can disperse microplastics throughout the water column. Importantly, an increased influence of the wind has been observed in freshwater environments, resulting in greater vertical mixing within the water column.[232,349] Freshwater environments also tend to be in close proximity to point sources of contamination, such as waste water treatment plants, and are generally associated with large urban areas where plastic rubbish is regularly concentrated.

There are relatively few studies of microplastics in freshwater systems and it has been reported[110] that the surface waters of rivers and lakes were only sampled in six studies, with one in Asia,[138] three in Europe[125,246,368] and two in North America.[116,297] Since then, more recent work was undertaken in rivers in China.[342] The studies all used nets with mesh sizes of 300–800 μm for collecting samples and the collection techniques were similar to those used in the marine environment. When sampling freshwater, it is particularly important to provide a good characterisation of both the water body studied and the surrounding area, describing land use and possible sources of pollution.[138,246,368] Furthermore, in freshwater environments there may be a larger amount of organic debris in comparison to marine waters, particularly from vegetation. Consequently, a potential issue with sampling is the way in which this organic debris may influence sampling in these areas.

Manta trawls were used in several freshwater studies[116] with mesh sizes varying from 333 to 500 μm.[138,297,344,368] Additionally, stationary conical driftnets (500 μm mesh, 50 cm diameter, 150 cm long), attached to iron rods driven into the riverbed, were used to sample the top 0.5 m of the water at various stages along the River Danube in Austria and at different times of the day.[246] One study[297] used several different types of net with a mesh size of less than 1 mm, such as hand nets, a manta trawl, a heavy streambed sampler and a rectangular net to sample the surface, edge, middle, bottom and middle sub-surface to bottom of the Los Angeles River and San Gabriel River in California, United States. More recently, sampling has also taken place in the streams of treated effluent from municipal waste water treatment plants in Southern California using sieving methods.[49]

Estuarine

Estuaries are the transitional zone between freshwater rivers and the marine environment. They are subject to both marine influences (tides, waves and saline water) and riverine influences (flowing freshwater), which form a partially enclosed body of brackish water that is connected to the open sea. Estuaries are typically close to urban areas, where they are exposed to contaminants, and have been identified as microplastic hotspots[460] (see Chapter 5). However, little is known regarding the characterisation of microplastics in estuarine ecosystems,[41,202,346,368,468] or on the influence of estuarine salinity gradients on the transportation and deposition of microplastics. As with freshwater sites, the prevailing weather, such as wind and rainfall, has been proven to play a significant role in influencing the distributions and abundance patterns of microplastics within the estuarine environment.[110,345]

However, the responses of microplastics in estuaries to meteorological events are sparsely investigated.[468]

Indeed, it is not known if microplastics suspended in estuaries occur in comparatively greater abundance under the influence of weather phenomena, such as that which has been witnessed on the open sea surface.[240,297] Interestingly, in contrast to the upwind sites, higher concentrations of microplastics were found in the downwind sites within the Tamar Estuary in the United Kingdom.[41] The tidal cycle is particularly important when sampling this environment and may greatly influence the amount of microplastics found. In a study[368] which sampled the Tamar Estuary in the United Kingdom near the mouth of the river, during both spring and neap tides, a manta net was towed against the tidal flow at a speed of 4 knots for 30 min during the maximum flow period.

Marine

The vast majority of sampling for microplastics has occurred in the marine environment, which ranges from the relatively shallow coastal littoral zone that can be heavily influenced by tides, to the abyssal oceanic pelagic environment. Due to the vastness of the oceans and the wide variation in the abundance of microplastics in these waters, volume reduced techniques are generally practiced. As with freshwater systems, samples from marine waters can either taken from the surface of the water or from within the water column.

Surface water sampling

The most common method for sampling microplastics in surface waters is to use established methods for plankton sampling which utilise monofilament nylon mesh plankton nets of various designs. Occasionally, alternative methods have also been used, such as a rotating drum sampler.[314] There are numerous standard operating procedures (SOPs) available from different monitoring agencies, such as the National Oceanic and Atmospheric Administration (NOAA) Ecosystem Survey Programs in North America,[277] which outline methods to sample surface waters using various plankton nets. This approach allows for the sampling of large volumes and surface areas of water relatively quickly in a volume-reduced method resulting in a relatively small, concentrated final sample. There is no defined definition of what constitutes the surface water layer in microplastic sampling. Nevertheless, it has been described as the water surface layer less than 15 cm deep, which is where 95% of small plastic debris is concentrated.[51] However, in most studies the depth of the surface layer is not specified and where it has been, the specified depth varied from between 50 and 60 μm to 25 cm deep.[103,314]

The most common type of net used for microplastic sampling in surface waters is the neuston net (Fig. 8.2), which was used in 28 of 33 studies.[188] Neuston nets are commonly used for the horizontal sampling of both the epineuston (organisms which live in the air on the surface film of the water) and the hyponeuston (organisms which live just beneath the water's surface). Certainly surface sampling works best in calm flat waters, but in more open waters where the depth of the net may vary considerably, the use of a more stable manta trawl (Fig. 8.3)[71,103,116] or catamaran (Fig. 8.4) is recommended. The manta trawl has wing-like structures on each side of the net to maintain stability and buoyancy in the water, whereas the catamaran has two runners on each side of the net to provide stability and buoyancy.

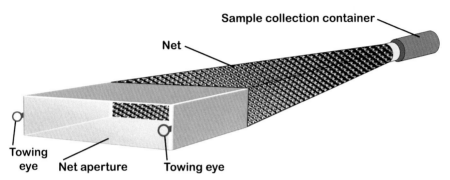

Figure 8.2 A neuston net can be used to sample surface waters.

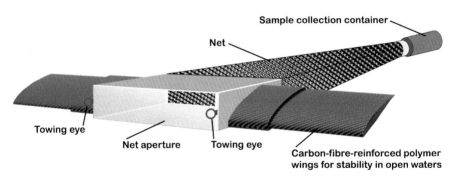

Figure 8.3 A manta trawl can be used to sample surface waters.

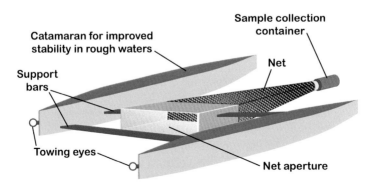

Figure 8.4 A catamaran can be used to sample surface waters.

The design of the net, its opening area or aperture, its length and the mesh size all have an influence on the sample of microplastics obtained. The typical mesh sizes used for the collection of microplastics range from 53 μm to 3 mm with the most frequent size being 333 μm, which is also the most common size used for plankton studies.[71,107,188,355] The choice of mesh size is commonly a trade-off between the mesh being small enough to catch the microplastics of interest and being large enough to prevent rapid clogging.[334] Since fine mesh will clog quickly, it will need to be towed slowly. The net aperture is often not provided in publications but where reported, ranges were between 3 and 200 cm for rectangular neuston nets.[188] Similarly, net length was provided in only half the studies investigated and was commonly 350–400 cm long. Flow rate can also greatly influence the sample and to aid more reliable quantification, a flow meter should be fitted to the net so that distance travelled and the flow rate of water through the net can be measured. The flow rate together with the area of the net aperture can be converted into the volume of water filtered (V) using Eq. (8.1).

$$V = \pi r^2 d \qquad\qquad (8.1)$$

where V = the volume of water filtered by the net; r = the radius of the net aperture; and d = the distance that the net is towed.

Importantly, net clogging can introduce errors and the use of two flow meters is strongly recommended. The best place to position the first flow meter is midway between the centre and the net rim, whereas the second meter should be positioned outside the net to estimate the nets speed. In this way, the combination of readings from both flow meters can indicate the filtration efficiency of the net and clogging.[154] Although the speed of trawling can be influenced by many factors, such as the weather, the sea state and marine currents, it usually takes place between 1 and 5 knots. While the trawling time can vary, and may be dependent on how quickly the net clogs from the concentration of seston (living and non-living matter) in the water, the time that the net was trawled should always be recorded. Ultimately by taking all these factors into account, a reliable quantification of microplastics should be able to be obtained from the water sampled (Fig. 8.5).

There are many good examples,[71,103,135] and a review[107] of the collection of microplastics using this method. However, since there is no standardisation of this sampling procedure, differences in techniques used can produce large variations in the quantity of microplastics collected.[65] As 95% of microplastics are thought to be located less than 15 cm below the water's surface,[51] horizontal towing is the most common method undertaken for sampling of microplastics in water.[51,334] Nonetheless, samples can also be collected from surface waters using several other alternative methods, such as a manta trawl, a hand net or the sieving of bulk water samples.[388]

- Clean the net and check for any contamination before use. Steps to avoid microplastic contamination should be undertaken, such as avoiding wearing synthetic garments, especially fleece fabrics.

- Record the net aperture, the length and the mesh size.

- Record the date, time, location, weather conditions, vessel course and speed.

- Securely attach the first flowmeter midway between the centre and the net rim.

- Securely attach the second flow meter on the outside of the net.

- Lower the net into the water to the correct depth. Wherever possible, a boom should be used to collect samples from the side of the boat in order to avoid sample disturbance by the bow wave.[71]

- In open waters, a catamaran or manta trawl may be used to ensure the correct surface depth is maintained.

- Tow the net at a constant speed (1-5 knots) for a set period of time, such as 20 minutes per sample.[71,135]

- Care should be taken to ensure that the net has not become clogged. If this has occurred, reduce the trawl time/speed.

- Once the tow is complete, retrieve the net and allow the water to drain away. Wash the debris collected by the net to the cod end of the net using a hose.[103]

- Transfer the collected sample to an appropriate container for storage and preserve it where necessary, such as with 4% formaldehyde,[135] 70% ethanol or freezing.[138]

- The net is then cleaned and checked for any contamination, after which it is ready for reuse.

- Appropriate replication should be undertaken.

- It is often a good idea to get a small sample of the materials used for sampling, such as the net and rope, as these are often composed of plastic materials and can be used to exclude sampling induced contamination.

- Using the data from the flow meters, the measurement of the net aperture, and the abundance of microplastics in the sample, calculate the results.

- The results can then be reported as the number of microplastics per m^3 of water.

Figure 8.5 A protocol for the sampling of microplastics in marine surface waters using a neuston net.

Water column sampling

In a similar way to surface water sampling, water column sampling for microplastics uses techniques and equipment developed for zooplankton sampling in both the marine and freshwater environments.[103] However, the mesh size of zooplankton nets is typically larger than plankton nets and the typical size of the nets used for the collection of microplastics is 500 μm[188] and at a depth of 1–212 m.[314,103] Importantly, the water column can be sampled both horizontally and vertically.

Horizontal sampling of the water column

For sub-surface horizontal sampling of microplastics, a bongo net (see Fig. 8.6) is used.[263] This type of net is generally comprised of a pair of circular aluminium frames connected to a central axle to which a flow meter and pair of nylon plankton nets are attached.[103] The net is lowered to a chosen depth, often just above the bottom, and towed at this depth at a set speed for a set amount of time and then recovered. For example, lowering the net to 212 m depth at a rate of 50 m per minute, then trawling for 30 seconds, and then recovering the net at a rate of 20 m per minute. The net aperture is often not provided but where reported, ranges were between 79 and 158 cm for circular bongo nets.[188] Various depths can be trawled to establish where in the water column microplastics of different densities are located. This allows for sample quantification similar to that used for surface sampling using the relevant parameters that have been recorded (see Eq. 8.1).

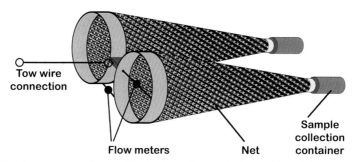

Figure 8.6 A bongo net can be used for horizontal or vertical sampling of the water column.

Vertical sampling of the water column

For vertical sampling of microplastics, the net is either pulled up towards the surface from a specified depth to give an overview of the entire water column. This can be undertaken by pulling a bongo net upwards. However, in rough waters, the vessel will pitch and roll thereby pulling on the net briefly and then releasing it, resulting in slack in the towing wire. To compensate for this, a spring mechanism can be fitted between the towing wire and the bongo net to compensate for the motion of the vessel and ensure the net is pulled upwards at constant speed. The water column can also be sampled vertically by opening and closing at discrete depth layers. For example, this can be undertaken using a modified multiple-net tucker trawl, which is a messenger operated closing net that allows specific depth layers to be sampled without contamination.[232] Other variations, such as Hensen, Apstein or Juday nets, have a reducing cone forward on the net aperture and a simple closure system to collect samples at discrete depths.

Other techniques can also be used for the sampling of microplastics from the water column, such as direct in situ filtration[317] and bulk sampling with subsequent filtration.[106,314] Furthermore, the use of a Continuous Plankton Recorder (CPR) can be used to sample from 10 m depth.[406] Indeed, in the European Union, the use of a CPR for the monitoring of microplastics is a technique that has been reported as 'in development' under descriptor 10 of the Marine Strategy Framework Directive (MSFD), relating to marine litter.[174] Water pumps can also be used to collect large volumes of water, which may be advantageous in areas where the density of microplastics is suspected to be low. However, while using pumps can provide representative samples, the main disadvantage is that the samples are only obtained from a single sampling point[334] and the definition of the sampling point is imprecise, owing to the fact that the pump can draw water from different layers in the water column.

There are also other techniques developed for sampling plankton and zooplankton that can be used for sampling microplastics but have not been reported thus far. For example, plankton traps, such as a Schindler-Patalas trap (see Fig. 8.7), and a variety of water collection bottles designed for the removal of zooplankton (primarily in lakes), such as the Ruttner bottle, the Friedinger bottle and the Bernatowicz bottle. These devices work by lowering the container to the desired depth in the water column. The device can then be closed under the control of the user to allow sample collection at a chosen depth. Consequently, quantification is precise as a known volume of water is sampled. However, due to the small volume of the containers, they may require a large number of replicates to give a more precise estimate of abundance.

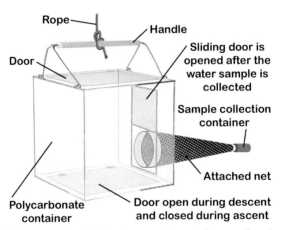

Figure 8.7 The Schindler-Patalas plankton trap can be used for sampling the water column for microplastics.

The Clarke–Bumpus Plankton Sampler can also sample at discrete depths or multiple depths and has the advantage of ASTM International standard practice guidance (ASTM E1199-87). A simple method for sampling the entire water column for microplastics involves vertically lowering a tube to a desired depth and collecting a water sample using stoppers (see Fig. 8.8). While this method is precise because it collects an exact volume of water, it is only effective at relatively shallow depths and this narrow sampling device may need to be used many times to obtain a representative sample.

- Cut a piece of rigid tube (such as 40 mm internal diameter white PVC pipe) at a length of 230 cm.

- Using a permanent marker, draw a line around the circumference of the tube at a point 30 cm from one end.

- Obtain two suitably sized rubber stoppers which will effectively seal both ends of the tube.

- Rinse the 210 cm tube with clean water and check for any contamination before use. Steps to avoid microplastic contamination should be undertaken, such as avoiding wearing synthetic garments, especially fleece fabrics.

- At the sample site, record the date, time, location and weather conditions.

- Very slowly lower the tube vertically into the water until 30 cm of the tube remains visible above the water surface (use the line drawn earlier to gauge this).

- Seal off the top of the tube with one of the rubber stoppers.

- Very slowly start to raise the tube back out of the water and before completely pulling the tube out of the water, seal the bottom of the tube with the other rubber stopper to avoid loss of the water sample.

- Carefully invert the tube 3 times to ensure the contents are thoroughly mixed.

- Decide if a bulk sample or a volume reduced sample is required.

- For a bulk sample: Hold the tube vertically, position the bottom end of the tube above a suitable collection vessel and remove the bottom rubber stopper. Next, remove the rubber stopper from the top end of the tube and allow the contents to drain into the collection vessel for analysis later. Label the vessel with the date, time, depth and location.

- For a volume-reduced sample: Hold the tube vertically, position the bottom end of the tube above a clean contamination-free net of chosen mesh size (such as 333 μm) and remove the bottom rubber stopper. Next, remove the rubber stopper from the top end of the tube and allow the contents to drain through the net. Transfer the net contents into a suitable container for analysis later and label with the date, time, depth and location.

- Thoroughly rinse all equipment with clean water in preparation for the next sample to be taken.

- When storing the tube after sampling has finished, ensure the rubber stoppers are removed to facilitate air drying.

- For the analysis of the samples, first work out the volume of water sampled using the following equation;

$$V = \pi r^2 h$$

  Where;
  V = The volume of water sampled.
  r = The radius of the internal diameter of the tube.
  h = The length of the section of tube that was filled with water. (In this case 200 cm)

- Next, analyse each sample collected for the presence of microplastics.

- The results can then be reported as the number of microplastics per m^3 of water.

Figure 8.8 A protocol for constructing and using a device to sample the top 2 m of the water column for microplastics.

Sediment sampling

Sediments are considered to be ideal mediums for environmental pollutants and tend to experience long-term contamination,[55] including contamination with microplastics. In terms of the global sampling of sediments and sands for microplastics, the vast majority of studies in Europe have reported on sediments, whereas studies in Asia have predominantly reported on sands and North America has reported on both equally.[334] However, as is the case with water sampling, there tends to be a lack of specific protocols for the sampling of sediments for microplastics.[355] Furthermore, the technique that is chosen for the collection of sediment will have a significant bearing on the type of microplastics which are collected and the way in which the results are expressed. Consequently, the technique needs to be relevant and aligned with the purpose of the study (quantitative or qualitative), and ultimately the sampling to be carried out will depend upon the question that requires to be answered. For this reason, clearly defined goals need to be established which take into account any possible limitations and which ensure that the sampling regime is practical, repeatable and reproducible. For example, larger scale studies, particularly involving volunteers or individuals located in highly polluted areas, may only be interested in obtaining a general understanding of the extent of microplastic pollution over a large area, as opposed to highly precise results. On the other hand, smaller scale studies may choose to undertake a detailed investigation of a small area, thereby generating reduced sample volumes but providing exact information pertaining to the abundance of all types of microplastics (see Chapter 5). Nevertheless, in both scenarios it is important to ensure that suitable methods are employed to prevent contamination of the recovered samples with microplastics which were not part of the original sample, such as microfibers from clothing.

Importantly, as the size of microplastics decrease, they become progressively more difficult to collect from the environment. For this reason, some studies[53] have chosen to collect and report only microplastics (<5–1 mm) in size and thus have excluded mini-microplastics (<1 mm to 1 μm) in size (see Chapter 5 for more information on the size categories of microplastics). This selective sampling approach, particularly when excluding mini-microplastics, has been commonly used on coastal sediments when using visual identification and sorting by hand.[107] While this approach can be useful for sampling larger pieces of plastic, the exclusion of mini-microplastics can result in highly underestimated concentrations of microplastics being reported. This is especially important since mini-microplastics are considered to represent 35–90% of all the microplastics present in the marine environment.[116,346,388]

The lower size limit of microplastics reported is highly dependent on the sampling and separation methods used. For example, for small irregularly shaped microplastics, such as microfragments (see Chapter 5), which are intermixed with other debris, bulk sampling is the favoured method to obtain a representative sample.[32,188] However, bulk sampling tends to result in the collection of large amounts of unwanted material. Thus, for reasons of practicality, it is generally beneficial to reduce down the mass of the sample. This can be undertaken by sieving it dry on the beach or wet in the water, thereby facilitating the collection of microplastics within a particular range of sizes without retaining the additional unwanted material. The distribution of microplastics in a dynamically shifting sediment environment (coastal or sublittoral) is often heterogeneous; so to obtain a representative sample, attention needs to be paid to the sampling location, the sample size and the number of replicates to be collected.[41,156,414] As with the aquatic environment, the distribution of microplastics in sediment varies over time and location as microplastics can become buried or exposed according to the prevailing sediment depositional regime.[156,414] However, the size of the grains in sediment or sand does not appear to have any relationship with the abundance of different-sized microplastics in those sediments or sands.[41,345] Nevertheless, the characteristics of the substrate to be sampled (sediment, sand, silt) should be recorded where possible as it may influence the degree of microplastic deposition and retention, as well as the collection methods used.

In terms of reporting the results, one particular difficulty is that variations in the sampling techniques employed for the collection of sediment samples has led to differences in the quantification units reported,[53,334] thereby hampering the comparison of results. For example, studies which sample many random 1 m^2 plots of a nominated area will report the number of microplastics per m^2. Conversely, studies which obtain bulk samples from a specific depth of sediment at random locations in a nominated area will report the number of microplastics per m^3.[53] Since the sampling depth can range from the sediment surface to 50 cm deep and is often not reported, conversion between both methods can be problematic. Furthermore, sediment samples may also be reported as units of weight (g or kg) in both wet and dry weight, as well as volume (mL or L), thereby further complicating comparisons[53,188] and highlighting the lack of standardisation in sample collection protocols.

Intertidal and estuarine

Due to the extremely dynamic nature of intertidal and estuarine zones, one of the key issues to be addressed when sampling for microplastics in the sediment or sand in these areas is the location to be sampled. For example, microplastics which have been transported by the tide to coastal areas, can reach different parts of the shoreline as a result of tidal movements, drift currents and the wind, thereby influencing their pattern of distribution.[302] Indeed, the variation in the abundance of microplastics between beaches may have a direct relation to the orientation of the beach and the wind regime.[209] Initially, microplastics are deposited in the intertidal zone, commonly along the strandline defined by the high tide mark[406] (see Fig. 8.9).

The deposition of microplastics across the beach profile, including depth, is heterogeneous, and microplastics are typically found at a greater depth towards the backshore.[414] This indicates that the intertidal zone actively transfers microplastics from the sea to the backshore, where the majority of microplastics accumulate.[302]

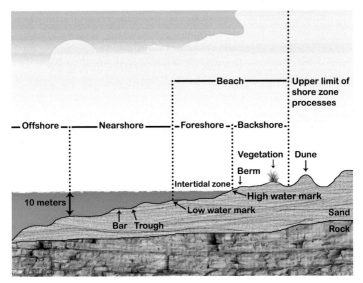

Figure 8.9 The beach profile.

The dynamic movement of microplastics along the shore will greatly impact their distribution and abundance at different locations along the shore. Thus, samples taken from the intertidal surface sediments represent the load of microplastics to the beach system, whereas samples taken from the backshore represent the standing stock of microplastic accumulation.[302] Consequently, this needs to be taken into account when devising the sampling strategy. Furthermore, the geographical location of the area studied appears to be a major factor that influences the distribution of microplastics.[334] Consequently, potential sources of microplastic pollution need to be taken into account, such as waste water treatment plants, ports and industrial areas to obtain a representative sample.

Of the studies reported thus far, there appears to be considerable variation in the specific tidal zones sampled. For example, a review[188] of 44 sampling studies reported that sites ranged from the sublittoral zone to the supralittoral zone (splash zone), while some studies investigated several littoral zones.[406] Furthermore, some studies combined samples from across several different zones,[74,413] whereas others did not mention the specific zone sampled. Nevertheless, most studies sampled the most recent flotsam deposited on the horizontal high tide line on the upper beach zone where large macroplastic litter is typically found.[263,302,344] However, the drift line which was considered often varied among studies.[38,96,188,189,274] Although some studies have shown an increased abundance of microplastics in the upper strandline, as opposed to the cross sectional line,[185] other studies found no significant difference in the abundance of microplastics between the upper and lower drift lines.[96,302] Ultimately, the most appropriate drift line for obtaining a representative sample is an issue that has yet to be fully resolved (Fig. 8.10).

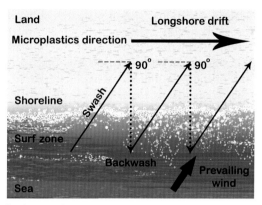

Figure 8.10 Microplastics may be transported along the coastline by waves approaching the shoreline at an angle (longshore drift). The microplastics will be transported up the beach at the same angle as the waves (swash) and then carried back down again at right angles (backwash).

There have been some vertical studies undertaken from the edge of the water to the backshore of the beach by testing several drift lines along the way. This method has the added advantage of providing a better understanding of the pattern of microplastic deposition. However, this pattern will often be very much dependent on the physical aspects of the shore, such as its gradient and reach. Furthermore, the considerable effort required in obtaining such a large number of samples, as well as processing them, has meant that most studies concentrate on the most recent drift line.

Often upon arrival at the shore, it is difficult to distinguish the most recent tidal drift line from the other drift lines that have been left by previous tidal events. For this reason, it is particularly useful to know which point in the tidal cycle (spring or neap) and tidal stage (ebbing or flooding) that sampling is taking place since these parameters will influence tidal height and have a large influence on one's ability to find the most recent high tide mark. This is important to avoid the effect of overlapping strandlines.[274,302] Furthermore, recent weather conditions should also be taken into account since high winds,[367] heavy rainfall and recent flooding will influence the location of the drift line and the rate at which microplastics are deposited. Consequently these sources of variability need to be addressed and may partly offer an explanation as to why there is a high level of variation between the data reported thus far on the abundance of microplastics in these environments.[302]

Importantly, the number of replicates and the frequency of sampling to be undertaken need careful consideration due to the heterogeneity of this environment. Most studies report one off sampling events,[96,185,189] which are relevant in obtaining a snap shot of the level of microplastic contamination at that particular period in time. However, if longer term spatiotemporal trends are of interest, then repeated sampling is necessary. While a number of studies[1,302,424] have investigated the distribution of microplastics in the same location over time and have found large temporal variation, the overall necessary long-term data does not yet exist.

Most studies tend to sample along a transect, such as horizontally along the most recent drift line, or vertically from the edge of the water to the backshore of the beach,[188,414] although there are some cases where random sampling was undertaken.[342] The advantage of using a transect is that spatial variability can be accounted for by collecting a number of replicates from the same general conditions. This makes comparisons easier and conclusions more appropriate in a standardised spatially integrating sampling design.[263] It is recommended that a minimum of five replicates, separated by 5 m, are taken from each site.[174] However, for the replicates to be comparable, they need to be taken from relatively similar locations. Thus, they should all be obtained from locations with similar grain size and topography. Furthermore, there should be no features which are specific to a particular sampling location, such as a stone, a large mass of seaweed or a freshwater influence. For this reason, an effective sampling method is selective random sampling. This involves randomly choosing sampling locations along the drift line (potentially using a random number generator), but only choosing sampling locations which have similar characteristics to facilitate a better comparison. While this is not truly random sampling, it is effective in overcoming unknown variables which can affect the reproducibility of the results and that only become apparent once actual sampling of the area takes place. For example, the location of sand dunes and walls or individuals walking on the beach may result in the mixing of different zones.

In order that the sampling method can be replicated and the reported results can be applied in the appropriate context, it is important to ensure that sufficient details are provided about the sampling method used. Furthermore, the chosen method will determine the quantification units that shall be used to present the results. In a review[188] of sediment studies, the tools and specific methods used for sampling sediments were reported in 31 out of 44 studies. Selective sediment samples were obtained by using either tweezers, metal spoons or by hand. However, the sampling technique used along the high tide line varied from linear sampling along strandline using a spoon or trowel with areas randomly marked out with a 25 cm^2 or 4 m^2 frame (quadrat)[147,205,344] to vertical stratified sampling schemes which sampled layers at different depths using corers.[51,61,431] While superficial surface samples were generally taken, over half of the studies did not report the sample depth. However, in the studies that did, sample depths were taken from between the surface and up to 32 cm deep, with the majority taken up to 5 cm deep.[188] While it has been reported[414] that microplastic pellets were only found on the surface sediment of the intertidal zone on sandy beaches, the sample depth can still be a considerable source of variability and is required to be normalised.

While there is no standardised method for collecting sediment samples for the analysis of microplastics, the European Commission Joint Research Centre has developed recommendations for the Marine Strategy Framework Directive Technical Sub-Group on Marine Litter.[174] The recommendations suggest sampling the strand line for microplastics and mini-microplastics by taking five replicate samples separated by 5 m.[263] Importantly, when devising a sampling protocol, the aim is to keep the protocol as simple as possible to enable comparisons of the results and to facilitate the large scale sampling of several locations as quickly and as cost effective as possible. Ultimately, the sampling protocol should be developed around the specific sampling goals and limitations of the study. An intertidal and estuarine sediment or sand sampling protocol is provided in Fig. 8.11.

- First, undertake preliminarily characterisation of the sampling location by analysing maps or images.

- Gather information about the tidal cycle, the moon phase and the time of the high tide, as well as information on the weather, such as wind and rainfall.

- Once on site, identify the relevant pre-defined sampling location, such as a horizontal transect along the most recent high tide drift line.

- Make a note of which tide line was used, since at certain times, the highest tide line is not always suitable.

- Follow a relevant contamination prevention protocol, such as always collecting samples downwind and avoid wearing synthetic fabrics.

- Randomly select at least 5 specific locations along this transect, separated by at least 5 m, and marked out with a 20 cm² quadrat. (If necessary, a quadrat can easily be constructed by cutting a piece of white 20 mm diameter PVC pipe into four 20 cm lengths. These are then joined together using four 20 mm 90° pipe connectors).

- Ensure that each selected location has no features which are specific to that particular sampling location, such as a stone or a large mass of seaweed (selective random sampling).

- Take photographs of the study area and of each specific location that is to be sampled.

- Record a GPS reading of the location and make a note of the weather conditions at time of sampling.

- In the case of sampling for microplastics (<5 mm to 1 mm), take a volume reduced sample by collecting all of the sediment within the area of the quadrat to a depth of 5 cm and sieve this material on site using a sieve with a mesh size of 1 mm. Collect the material left in the sieve. Pass the collected material which was left in the sieve through a second sieve with a mesh size of 5 mm. Collect the material which has passed through the second sieve in an appropriately labelled aluminium container. Responsibly discard any material that did not pass through the second sieve.

- In the case of sampling for mini-microplastics (<1 mm to 1 μm), take a bulk sample by collecting all of the sediment within the area of the quadrat to a depth of 5 cm using a metal scoop and store it an appropriately labelled aluminium container.

- Where possible, collect an additional sediment sample for grain size analysis.

- If desired, grade the location in relation to the amount of large macroplastic litter or other debris present by using a simple 1-10 scale.

- In the laboratory, analyse each sample collected for the presence of microplastics.

Figure 8.11 A protocol for the sampling of microplastics in intertidal and estuarine sediment or sand.

As per the protocol in Fig. 8.11, sediment or sand can be sampled for microplastics (<5 mm–1 mm) by collecting a volume reduced sample. This is undertaken by collecting the material within the area of a quadrat of chosen size to a depth of 5 cm and then passing the material through a sieve with a 1 mm mesh size. The material left in the sieve is then passed through a second sieve with a mesh size of 5 mm and any material which passes through is collected in an appropriately labelled aluminium container. Any material which did not pass through is responsibly discarded. In the case of sampling sediment or sand for mini-microplastics (<1 mm to 1 μm), a bulk sample is collected. This can be undertaken by collecting all of the material within the area of a quadrat to a depth of 5 cm using a metal scoop and then storing the material in an appropriately labelled aluminium container. Alternatively, the sediment or sand can be sampled to a depth of 5 cm at each location by using a metal scoop to take several scoops at arm's length in an arc-shaped area and then combining the scoops in an appropriately labelled aluminium container until about 250 g of sediment is collected. The collected material from all of these methods is then taken to the laboratory for the separation and identification of microplastics.

Sub-tidal

Sampling sediments for microplastics below water requires considerably more effort and expense than intertidal sampling, which is perhaps the main reason why there has been fewer studies undertaken in this environment. As a general rule, the deeper you go, the more expense you shall incur. In the coastal sublittoral zone, samples have been collected for microplastic analysis from boats using Ekman and Van Veen grabs,[38,61,406] bottom trawling and corers.[426] Ultimately, the method that is used will be influenced by the type of sediment sample that is required. For example, grabs tend to disturb the sediment and are therefore most suited for sediment surface or bulk sampling. In contrast, corers produce smaller samples but allow sampling of the sediment surface and depth layers to investigate the deposition of microplastics over time.[263] More recently in the Mediterranean, samples have been obtained by using scuba divers to collect superficial sediment samples along a transect similar to the intertidal method[140] or randomly using core tubes from sand patches.[4]

When sampling from the deep sea at depths of more than 1 km, more specialised equipment is needed which greatly increases the resources required and the cost of sampling. While several studies have sampled macroplastic pollution in the deep sea using several techniques, such as deep sea trawls,[347,390] submersibles,[438] remotely operated vehicles (ROVs)[149] or combinations of these,[128,332,457] fewer studies have investigated microplastics in deep sea sediments as a result of the difficulties in sampling such small items at considerable depths. Nevertheless, a study[128] primarily utilised a box corer (see Fig. 8.12), as well as an Agassiz trawl and a Camera epibenthic sledge (see Fig. 8.13) for the collection of samples at depths between 4869 and 5766 m for the analysis of macroplastics and mini-microplastics.

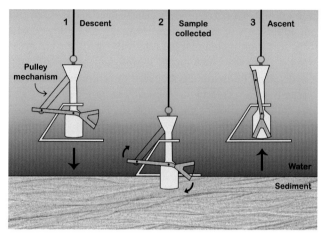

Figure 8.12 The box corer is a sampling tool which is used to collect sediment from the deep sea.

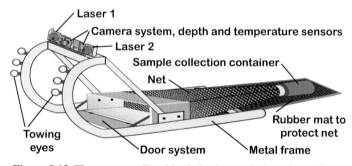

Figure 8.13 The camera epibenthic sledge is towed along the seafloor.

The camera epibenthic sledge is a sledge which is towed along the seafloor at depths of up to 4 km. The device is composed of a stainless steel frame with two doors on the front and a camera mounted on top. The steel frame is designed to withstand impacts with rocks or other solid items on the seafloor. The doors can be controlled by the operator on the vessel, thereby ensuring that anything collected by the sledge is obtained from the seafloor and not from the water column during descent and ascent. When towed just above the seafloor, the bow wave created by the sledge pulls up items from seafloor and into the net. The camera allows visualisation of the tow, thereby enabling calculation of the tow time and identification of the substrate on which towing took place. Furthermore, the camera enables identification of any macro fauna and plastic litter not collected by the sledge. Additionally, twin lasers are fitted to the sledge to allow the determination of the size of any items of interest, and the sledge is equipped with sensors which provide precise information about the water temperature and depth. Other devices which have been used to sample the deep sea are sediment megacorers and box corers,[344] which have been used for sampling from 1000 to 3500 m depth with ROVs or push-corers. However, in a study[426] which employed deep sea corers, it was highlighted that the relatively small surface area (25 cm^2) that was sampled by the deep sea corer meant it was difficult to collect representative samples at these depths. Following the recovery of sediment samples from the deep sea, storage is usually undertaken by drying or freezing the samples and then storing them in darkness until separation of the microplastics from the samples is ready to be carried out.

Microplastic separation techniques

9

Separation of microplastics from samples

Upon collection of a water, sediment or biological sample from the environment, the microplastics contained within that sample need to be separated from all the other organic and inorganic material which is present. This ensures that the microplastics can be quantified by counting or weighing and positively identified. In many cases, the sample will have been subjected to some form of separation or sample reduction on-site, such as filtration through a net or sieve. In other cases, a bulk sample will be taken back to the laboratory for the separation of microplastics, particularly in the case of sediment samples. However, due to the rapid emergence and development of the field of microplastic research, there is a general lack of standardisation and consistency in the extraction techniques used to remove microplastics from organic and inorganic matter, especially with regard to sediments.[53,448] Nevertheless, there are several techniques commonly used in the laboratory for the separation of microplastics from organic and inorganic matter, including visual sorting, filtration, sieving, density separation, elutriation, flotation and chemical digestion.[107,156,188] While there tends to be an absence of harmonisation in the way in which these techniques are reported in the literature and regulatory bodies, such as the National Oceanic and Atmospheric Administration (NOAA) in the United States, are producing guidelines and protocols in an attempt to standardise the way in which microplastics are separated from environmental samples.[156,208,276] However, detailed cross-calibration of equipment costs, extraction efficiency, sampling time and health and safety procedures are yet to be undertaken amongst the majority of methods used.[147]

Visual detection and separation

Visual detection and separation is a mandatory step in the analysis of microplastics, regardless of whether the sample is water, sediment or biological. Thus, no matter where the sample came from, visual detection and separation is necessary in every study for the removal of debris, including naturally occurring organic fragments, such as seaweed, wood, shell fragments and anthropogenic contaminants, such as metals, paints and oil residues. These unwanted contaminants are almost always present in bulk samples and are commonly found in volume reduced samples, even after the use of laboratory based separation techniques. For this reason, visual detection and separation remains an obligatory step in microplastic research.

Visual detection can be carried out by direct examination with the naked eye,[208,389] for microplastics (<5–1 mm) or with the use of binocular microscopy[135,355] or high

Microplastic Pollutants. http://dx.doi.org/10.1016/B978-0-12-809406-8.00009-8

magnification fluorescence microscope[389] for mini-microplastics (<1 mm to 1 μm). Suspected microplastics are detected based upon their physical characteristics, such as their size, shape, texture, colour and lack of biological structures. These optically detected items suspected of being microplastics are then separated from the rest of the material using tweezers. These microplastics are then visually sorted and categorised using the standardised size and colour sorting (SCS) system (see Chapter 10) and further identification can be carried out using spectroscopic techniques, such as Raman spectroscopy, to positively confirm the type of plastic.

While the use of visual detection and separation in microplastic research can be controversial due to its subjective nature and the potential for bias, there is no better method of detecting and separating plastic materials from other debris than with a human operator. Certainly there is the risk that microplastics in the sample may be overlooked due to the accumulation of organic material on their surface or as a result of their transparency, and this can lead to underrepresentation of the abundance of microplastics in the sample. However, this can be markedly reduced by undertaking a careful and methodical examination, leaning towards preliminarily selecting doubtful items as being microplastics. The reason for this approach is that while underrepresentation is difficult to detect, overrepresentation can be eliminated by the subsequent use of spectroscopic techniques to positively confirm selected items are indeed composed of plastic (see Chapter 10). Ultimately, the visual detection, separation, sorting and identification of microplastics from samples are a well-established method that represents the cornerstone of microplastic research.

Water samples

Most water samples taken from the environment for microplastic analysis are subjected to some form of volume reduction at the time the sample was taken. This is usually undertaken by filtration with a net or sieving (see Chapter 8). In other cases, the collection of bulk water samples may have been undertaken. However, once the samples arrive at the laboratory, microplastics are routinely removed from both volume reduced and bulk water samples by the standard laboratory techniques of sieving and filtration, using various mesh or pore sizes.[188]

Filtration

Filtration is an effective physical or mechanical method used to separate solids from fluids (in this case liquid) by use of a medium through which only the fluid (filtrate) can pass. The size of the solids retained shall depend upon the pore size of the filter medium utilised. In most cases, the process of separating microplastics from water is normally undertaken physically using a funnel, a filter medium (paper) and a vacuum system[188,406] in which the microplastics are retained in the filter paper.

While the pore size of the filter paper used can vary between studies, it is generally 1–2 μm in size and importantly, qualitative filters should be utilised, such as VWR Grade 310[188] (Fig. 9.1).

However, while filtration can be a relatively simple process, complications often arise because the water sample is contaminated with particulate matter or debris. This rapidly clogs the filter paper, markedly reducing the effectiveness of filtration in separating the microplastics from the water and often results in a complete blockage. There are generally two ways to prevent this from happening:

1. Reduce the volume of liquid to be filtered.
2. Add in a pre-filter clean-up step.

Typical examples of pre-filter clean-up steps include

- Settling the liquid for a period of time to allow the separation of the heavier solid fraction to leave a purer liquid fraction.
- Sieving or pre-filtering using a mesh or filter with a larger pore size.
- Employing the density separation technique (See Density separation).
- Adding chemicals to the liquid, such as Iron(II) sulphate ($FeSO_4$) to encourage flocculation or coagulation of the solid fraction.

Importantly, caution should be exercised with all of these clean-up steps since they can result in the loss of microplastics from the liquid fraction. Consequently, these lost microplastics escape observation, thereby resulting in underrepresentation of the abundance of microplastics in the sample.

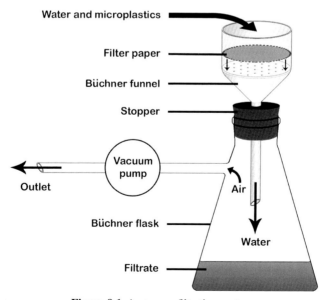

Figure 9.1 A vacuum filtration system.

Sieving

The process of sieving water samples is commonly applied in microplastic research. The sieve physically captures the microplastics and allows the water to be lost from the sample. This method is well documented and has been used in numerous studies to separate microplastics from water[98,188] as well as tertiary treated municipal effluent.[49,307] The size of the mesh used in the sieve depends upon the size of the microplastics that are desired to be collected. While the mesh size used in most studies ranged from 38 μm to 4.75 mm,[188] it is important to note that the mesh size must not exceed 5 mm, otherwise the size of the pieces of plastic collected would be outside the size-range of microplastics (see Chapter 5).

The most commonly used method to sieve water and sediment samples for microplastics is multi-tier sieving. This involves separating material of different sizes by passing the sample through a series of sieves with a decreasing mesh size (see Fig. 9.2). This approach has also been successful for recovering microplastics from tertiary treated municipal effluents using mesh sizes which decreased from 400 to 20 μm. However, the technique was unsuccessful for dirtier raw effluent as the sieves rapidly became clogged.[49] Following sieving, the microplastics are then collected by rinsing them from the sieve and concentrating them into a smaller volume. Alternatively, they can be picked out of the sieve using tweezers. However, removing the microplastics from the sieve with tweezers is more subjective and thus may be prone to greater error.

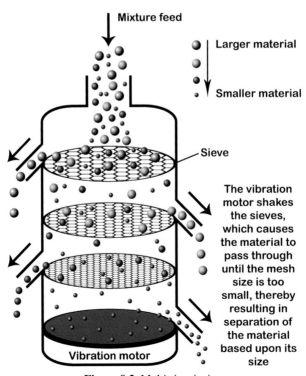

Figure 9.2 Multi-tier sieving.

Sediment samples

The type of separation method used to extract microplastics from sediment is influenced by the physical characteristics, such as size, shape and density, of both the sediment and the microplastics of interest. The sediment grain sizes applicable to microplastic research are listed in Table 9.1.

Table 9.1 Sediment grain sizes of interest in microplastic research

Aggregate	Grain diameter
Fine gravel	8–4 mm
Very fine gravel	4–2 mm
Very coarse sand	2–1 mm
Coarse sand	1 mm–500 µm
Medium sand	500–250 µm
Fine sand	250–125 µm
Very fine sand	125–62.5 µm
Silt	62.5–3.9 µm
Clay	3.9–1 µm
Colloid	<1 µm

As a result of their physical characteristics, it is very difficult to separate small microplastics from sediment consisting of silt, clay or colloid. Indeed, the finer the microplastics and sediment are, the more difficult it becomes to separate them. Furthermore, the shape of the microplastics can also greatly influence their ability to be separated from sediments, with fibres being more difficult to separate than spherical shapes. However, sieving or filtration (as described previously) of finer sediments for larger microplastics is an appropriate technique that is commonly practiced,[4,188] and several methods can be utilised to aid in the separation of smaller microplastics from smaller grain diameters, such as density separation.

Density separation

Density separation is a process that exploits the actuality that plastic materials have a range of different densities. As such, when a mixture of materials with varying densities, such as sediment and microplastics, is placed in a liquid of intermediate density (water or a salt solution), the material with a lower density than that of the liquid will float, while material with a density greater than that of the liquid will sink (see Fig. 9.3). The floating microplastics can then be collected by decanting the liquid that lies above the layer of sediment (supernatant).

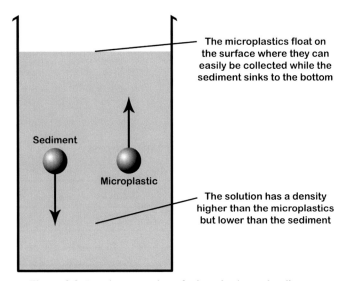

Figure 9.3 Density separation of microplastics and sediment.

In the literature, density separation has become the most reliable and commonly used method for the separation of microplastics from sediment, or sand, because the density of the liquid (typically a sodium chloride or other salt solution) can be adjusted to allow certain plastic materials to float on the surface. This facilitates faster separation and collection of the material of interest. Consequently, density separation was applied in 65% of studies undertaken to extract microplastics from sediments[188] most commonly using a saturated sodium chloride (NaCl) solution,[61,156,406] which has a density of 1.202 g/cm^3. Thus, any particles with a density lighter than this shall float in the liquid, thereby separating it from the denser sediment (typically 2.65 g/cm^3), which sinks to the bottom. The plastic debris floating on the surface of the saturated NaCl solution is then separated for later identification by filtration, usually facilitated by a vacuum.[188] However, there are a large number of variations on how this extraction method is undertaken, each exhibiting different extraction efficiencies. A highly effective protocol is provided in Fig. 9.4 which describes procedure which is carried out on a bulk sample of sediment that has been taken back to the laboratory for the separation of microplastics. Incidentally, for a procedure on how to avoid contamination while following this protocol, see the microplastic contamination prevention protocol described in Chapter 8. Importantly, unnecessary contact of the sample with the atmosphere should be avoided and as a general rule, the more steps there are in a protocol, the more chance there is for contamination to occur.

While many studies have followed a protocol similar to that which is described in Fig. 9.4, there have been variations in the salt solutions utilised, the sample mixing time (ranging from 30 s to 2 h), the settling time (ranging from 2 min to 6 h) and the number of replicates.[188] Nevertheless, this method of density separation is particularly advantageous in that NaCl is cheap to procure, widely available and environmentally friendly. However, one disadvantage of the method is that a saturated NaCl solution has a specific gravity of 1.202 g/cm^3 and will only allow items to float that possess a density below that value. However, the protocol in Fig. 9.4 is appropriate for many microplastics, such as polyethylene, foamed polystyrene and some blends of polypropylene, which are typically the most common type of plastic found in the aquatic environment, as well as being adaptable. For example, if a low-density solution is desired, perhaps for the extraction of specific low-density plastics (see Table 9.2), the saturated NaCl solution can be substituted with a mixture of ethanol and distilled water. By changing the ratio of ethanol to water, solutions with different densities which are lower than that of water can be produced. Similarly, the saturated NaCl solution can also be substituted for a higher density solution in the case of recovering microplastics with a high density.

- Clean all the laboratory surfaces and adhere to the microplastic contamination prevention protocol.
- Thoroughly mix the bulk sediment sample with a metal spatula for 2 min, taking care to avoid contamination.
- Optional: oven dry the sediment at around 60°C. Once dry, sieve the sediment to separate the appropriate sizes of microplastics of interest. However, caution should be exercised in this step. While many studies do indeed dry and sieve the sediment sample prior to the process of density separation, this can potentially remove microplastics and result in a sample which is less representative.
- Prepare a concentrated solution of sodium chloride (NaCl) (density 1.2 g/cm^3).
- Weigh a subsample of the homogenous sediment using an appropriate balance. Note: while the weight of the sediment subsample is optional, ideally the ratio of NaCl solution to sediment is three parts NaCl solution (210 mL) to one part sediment (70 g).
- Pour the 70 g of sediment into a 400 mL glass beaker.
- Add 210 mL of the saturated NaCl solution to the 400 mL beaker containing the sediment. Note: alternatively, higher density solutions may be used in place of NaCl, such as zinc chloride (ZnCl$_2$).
- Suspend the solution/sediment mixture in the 400 ml flask using an overhead stirrer for 3 min at 300 rpm.
- Rinse the stirrer paddle using deionised water to remove any attached material.
- Allow the solution/sediment mixture to settle for 10 min.
- Secure a 1 L three-necked round-bottomed flask with a retort stand.
- Fit a stopper to the middle neck and connect a vacuum pump to the left neck. To the right neck, connect a length of tubing with a glass tube connected at the other end, such as a Pasteur pipette (see Fig. 9.5)
- Switch on the vacuum pump and use the glass tube connected to the rubber hose to collect all the material floating on top of the solution/sediment mixture in the 400 mL beaker.
- Once all the floating material is collected in the round-bottomed flask, set up a filtration apparatus (see Fig 9.6) by connecting a Buchner funnel to a 500 ml Buchner flask, ensuring an airtight seal.
- Switch on the vacuum pump and place a piece of filter paper in the Buchner funnel. Pour a small amount of NaCl solution into the Buchner funnel to wet the filter paper and flush the system.
- Slowly pour the collected material in the round-bottomed flask into the Buchner funnel in order to collect any solid material. Carefully rinse the flask with deionised water and pour the rinsing into the Buchner funnel to ensure all the material in the round-bottomed flask is collected.
- Flush some deionised water through the filtration system to rinse any salt off the material collected by the filter paper.
- Carefully remove the filter paper from the Buchner funnel, ensuring that no material is lost from the filter paper, and place it on a watch glass. Place the watch glass in an oven to dry at 60–70°C for 10 min.
- Once the filter paper is dry, carefully remove the material from the filter paper and weigh it.
- It is advisable to carry out the entire procedure in triplicate to ensure that all microplastics are collected from the sediment.

Figure 9.4 A density separation protocol using a saturated sodium chloride (NaCl) solution. Adapted from Thompson RC, Olsen Y, Mitchell RP, Davis A, Rowland SJ, John AWG, McGonigle D, Russell AE. Lost at sea: where is all the plastic? Science 2004;304:838; Claessens M, Cauwenberghe LV, Vandegehuchte MB, Janssen CR. New techniques for the detection of microplastics in sediments and field collected organisms. Marine Pollution Bulletin 2013;70:227–33.

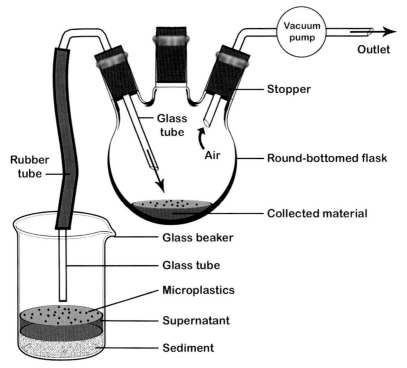

Figure 9.5 The experimental setup of the three-necked round-bottomed flask.

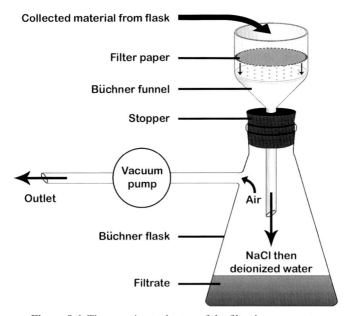

Figure 9.6 The experimental setup of the filtration apparatus.

Importantly, the vast majority of plastic materials are seldom exclusively composed of the pure polymer and are normally blended with glass fibre, or other materials, as well as being plasticised and impact modified (See Chapters 1, 2 and 4). Consequently, these additives alter the density of the plastic materials, often increasing it. The minimum and maximum density values for common plastics are listed in Table 9.2.

Table 9.2 **The density ranges of common plastics**

Plastic	Abbreviation	Density (g/cm³)
Polystyrene (expanded foam)	EPS	0.01–0.05
Polystyrene (extruded foam)	XPS	0.03–0.05
Polychloroprene (neoprene) (foamed)	CR	0.11–0.56
Polypropylene (impact modified)	PP	0.88–0.91
Polypropylene (homopolymer)	PP	0.90–0.91
Polypropylene (copolymer)	PP	0.90–0.91
Low-density polyethylene	LDPE	0.92–0.94
Linear low-density polyethylene	LLDPE	0.92–0.95
High-density polyethylene	HDPE	0.94–0.97
Polypropylene (10–20% glass fibre)	PP	0.97–1.05
Polystyrene (crystal)	PS	1.04–1.05
Polystyrene (high heat)	PS	1.04–1.05
Acrylonitrile butadiene styrene (high impact)	ABS	1.00–1.10
Acrylonitrile butadiene styrene (high heat)	ABS	1.00–1.15
Nylon 6,6 (impact modified)	PA	1.05–1.10
Polypropylene (10–40% mineral filled)	PP	0.97–1.25
Polypropylene (10–40% talc)	PP	0.97–1.25
Acrylonitrile butadiene styrene	ABS	1.03–1.21
Nylon 6	PA	1.12–1.14
Nylon 6,6	PA	1.13–1.15
Poly(methyl methacrylate) (impact modified)	PMA	1.10–1.20
Polypropylene (30–40% glass fibre)	PP	1.10–1.23
Polycarbonate (high heat)	PC	1.15–1.20
Acrylonitrile butadiene styrene (flame retardant)	ABS	1.15–1.20
Poly(methyl methacrylate)	PMA	1.17–1.20
Poly(methyl methacrylate) (high heat)	PMA	1.15–1.25
Polychloroprene (neoprene) (solid)	CR	1.20–1.24
Polyvinyl chloride (plasticised and filled)	PVC	1.15–1.35
Nylon 6,6 (impact modified and 15–30% glass fibre)	PA	1.25–1.35
Polyethylene terephthalate	PET	1.30–1.40
Nylon 6,6 (30% mineral filled)	PA	1.35–1.38

Continued

Table 9.2 The density ranges of common plastics—cont'd

Plastic	Abbreviation	Density (g/cm³)
Nylon 6,6 (30% glass fibre)	PA	1.37–1.38
Polyvinyl chloride (rigid)	PVC	1.35–1.50
Polycarbonate (20–40% glass fibre)	PC	1.35–1.52
Polyethylene terephthalate (30% glass fibre and impact modified)	PET	1.40–1.50
Polystyrene (30% glass fibre)	PS	1.40–1.50
Polycarbonate (20–40% glass fibre and flame retardant)	PC	1.40–1.50
Polyvinyl chloride (20% glass fibre)	PVC	1.45–1.50
Polyvinyl chloride (plasticised)	PVC	1.30–1.70
Polyethylene terephthalate (30% glass fibre)	PET	1.50–1.60
Polytetrafluoroethylene	PTFE	2.10–2.20
Polytetrafluoroethylene (25% glass fibre)	PTFE	2.20–2.30

For the separation of dense microplastics, a $1.202\,g/cm^3$ saturated NaCl solution is unsuitable. Indeed, dense microplastics, such as polyvinyl chloride and polyethylene terephthalate, make up over 17% of the European plastic demand.[337] If a saturated NaCl solution is used, then these heavier plastics will sink with the sediment resulting in an underrepresentation when determining the amount of microplastics present in the sediment samples. This is particularly important since seawater has a density of $1.025\,g/cm^3$. Consequently, dense microplastics are often the first to settle to the benthos and be incorporated into marine sediments. The Marine Strategy Framework Directive (MSFD) Technical Subgroup on Marine Litter have recommended the use of a saturated NaCl solution for the density separation of microplastic from sediments.[208] However, if it is desired to include dense microplastics in the quantification of microplastics in sediment samples, it is recommended that an alternative method is used to ensure that any dense microplastics present in the sample are accounted for.

One way in achieving this is to use heavy liquids to create high density solutions. The addition of distilled water to heavy liquids can produce a solution of the required density. Zinc chloride,[198,257] sodium polytungstate[74,344] and sodium iodide[62,96,319]; solutions have all been used successfully. Furthermore, a publication by the National Oceanic and Atmospheric Administration (NOAA) in the United States recommends the use of 5.4M lithium metatungstate ($1.6\,g/cm^3$) for sediment density separations.[276] Importantly, solutions with a density of more than $1.4\,g/cm^3$ are recommended to ensure the efficient separation of all plastic materials from the sediment.[53,198] Although high density solutions will extract plastic materials more efficiently, if the density is too high then other debris, or even the sediment itself, shall float to the surface with the consequence of making the separation ineffective. This is particularly apparent with sediments with a smaller grain size and a balance must be found by experimental trial and error (Table 9.3).

Table 9.3 Heavy liquids that have successfully been used to create high density solutions for the separation of microplastics from sediment

Salt solution	Chemical formula	Density used (g/cm³)
Sodium polytungstate	$Na_6(H_2W_{12}O_{40})$	1.4
Zinc chloride	$ZnCl_2$	1.5–1.7
5.4 M lithium metatungstate	$Li_6(H_2W_{12}O_{40})$	1.6
Sodium iodide	NaI	1.8

While heavy liquids aid in attaining better recoveries of microplastics from sediments, the use of different salt solutions has resulted in variations in the reported extraction efficiency of microplastics, which introduces difficulties in comparing results between studies.[334] Moreover, while several studies developing extraction methods have reported high percentage recoveries of microplastics from sediments using various salt solutions, the microplastics used were generally larger than 1 mm in size, were spherical in shape and were present in relatively coarse sediments. Consequently, these may not reflect samples taken directly from the environment.

In many cases, the extraction process has to be repeated three times to achieve high efficiencies. For example, with a sodium chloride (NaCl) solution, the extraction efficiency of polyethylene microplastics increased from 61% for the first extraction to 83% for the second and 93% for the third.[406] Subsequent minor modifications to this method resulted in up to a 4% improvement in extraction efficiency with recoveries of up to 97%.[61] However, even higher recoveries have been achieved with other salt solutions. For example, 68–99%[319] recovery and 98–100%[62] recovery was attained with a sodium iodide solution, while 96–100%[199] recovery was achieved with a zinc chloride solution. However, the recovery of fibres from sediments samples obtained directly from the environment has proven to be far more difficult. Indeed, the recovery efficiency has been reported to be over a considerably wider range of 0–98%.[319] Similarly, the recovery of 40–309 μm mini-microplastics from sediments was reported to be insufficient at around 40% recovery,[198] thereby offering a greater challenge.

Nevertheless, a novel technique based on density separation has been developed and is termed as the Munich Plastic Sediment Separator (MPSS).[198] Utilising a solution of zinc chloride, the MPSS is reported to reliably separate mini-microplastics from sediments with an improvement in the recovery rate from 40% to 95.5%. The rate of recovery for microplastics greater than 1 mm in size is even higher at 100%. An added advantage of the MPSS is that only a single extraction step is required, as opposed to three with the traditional method, thereby reducing the time taken to carry out the extraction process. However, the disadvantage is that the MPSS is a highly specialised piece of equipment that is not widely available at the present time.

Ultimately, an important aspect to consider when undertaking density separation is that while heavy liquids aid in better recoveries of microplastics from sediment samples, they are generally very expensive to procure and some are even considered to be toxic to the environment. Consequently, this often precludes their use in large scale studies for reasons of practicality. As a result of these cost implications, several researchers have developed methods to reduce the volume of a sample by introducing a reduction step before density separation, such as elutriation or froth flotation.[62,319]

Elutriation

Elutriation is a process in which particles are separated based upon their size, shape and density by using a stream of gas or liquid flowing in a direction opposite to the direction of sedimentation (see Fig. 9.7). In terms of microplastics, the technique was first used[62] to separate microplastics from sediment by directing an upward flow of water through a column, thereby inducing fluidisation of the sediment. Ultimately, the technique was used a pre-treatment step to reduce the sample volume before density separation with a sodium iodide (NaI) solution and demonstrated an excellent recovery efficiency for PVC particles at 100% and fibres at 98%.

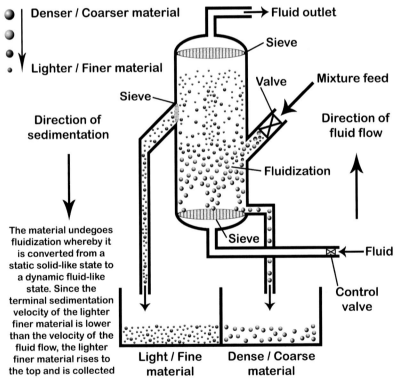

Figure 9.7 The process of elutriation.

Elutriation was also the basis of the Air-Induced Overflow (AIO) technique, which involved the fluidisation of sediment in a sodium chloride (NaCl) solution.[319] In the AIO technique, air-generated turbulent gas bubbles induced a stochastic process that forced specific lighter particles to move more frequently and quickly to the top layer of the solution than heavier ones. This allowed the sediment mass to be reduced by up to 80% before density separation with a sodium iodide solution. The combination of the elutriation based AIO technique and density separation with sodium iodide resulted in the recovery of 91–99% of ~1 mm polypropylene, polyvinyl chloride, polyethylene terephthalate, polystyrene and polyurethane microplastics.[319] Other elutriation devices have successfully achieved recoveries of 50%[381] indicating that some optimisation of these techniques may still be required.

Froth flotation

Froth flotation is a process that selectively separates materials based upon whether they are water repelling (hydrophobic) or have an affinity for water (hydrophilic). Importantly, the word flotation is also used in the literature to describe the process in density separation in which lighter microplastics float to the surface of a salt solution. However, this process is based upon density alone and should not be confused with the process of froth flotation. Thus, the process of froth flotation is not solely dependent upon the density of the material; it is also dependent upon its hydrophobic nature. For example, froth flotation is a technique commonly used in the mining industry. In this technique, particles of interest are physically separated from a liquid phase as a result of differences in the ability of air bubbles to selectively adhere to the surface of the particles, based upon their hydrophobicity. The hydrophobic particles with the air bubbles attached are carried to the surface, thereby forming a froth which can be removed, while hydrophilic materials stay in the liquid phase[272] (Fig. 9.8).

Since plastics are generally hydrophobic materials, froth flotation has successfully been used for the separation of plastic materials.[5,137] For example, two plastic materials which are buoyant in a specific liquid phase can be separated from one another by the addition of a wetting agent which selectively adsorbs to one of the plastics and not the other. Thus, the wetting agent acts as a flotation depressant by selectively adsorbing to the surface of a specific plastic, thereby rendering it hydrophilic. However, adsorption to the other plastic is far less pronounced. Consequently, the hydrophobic plastic will continue to float while hydrophilic plastic will sink to the bottom as a result of depressed flotation. The floating plastic can then be recovered from the surface of the mixture. However, there are many factors to take into account in the process of froth flotation, such as the surface free energy of the microplastics and the surface tension of the liquid in the flotation bath, as well as the critical surface tension, which defines the surface tension at which the liquid completely wets the solid microplastics. Ultimately, the selective separation of inherently hydrophobic microplastics requires that the microplastics are only partially wetted by the liquid in the flotation bath, thereby allowing the bubble to adhere to the surface of the solid phase and bring the microplastics to the surface of the liquid phase to be collected, while the sediment particles are completely wetted and sink. Nevertheless, the technique has rarely been used for the separation of microplastics from sediments.[198,436] Perhaps the reason for this is that while recovery rates of up to 93% have been reported,[436] the technique was found to be negatively influenced by the presence of wetting agents or additives in the plastics.

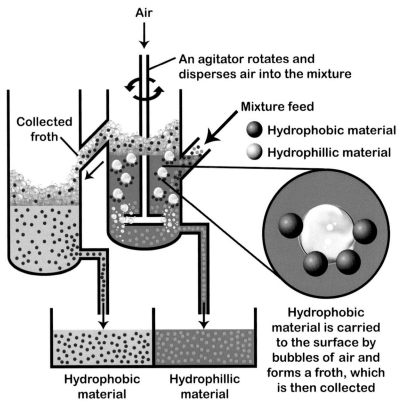

Figure 9.8 The process of froth flotation.

Biological samples

The uptake and ingestion of microplastics by many marine and freshwater organisms has been widely reported (see Chapter 7). Indeed, invertebrates (especially bivalves and crustaceans), fish, marine mammals and birds are known to regularly ingest these materials, and consequently, there has been a general focus on the analysis of gut contents. However, in terms of the presence of microplastics in the tissues of organisms, only very small mini-microplastics tend to be found as these are more likely to be translocated from the gastrointestinal tract.[156] Consequently several techniques have been developed for the separation of microplastics from biological material and the appropriate technique depends upon the size of the organism and the microplastics of interest, as well as the type of plastic that is to be recovered.

Visual detection and separation from biological material

Visual detection and separation is the most common technique used for the detection and separation of microplastics from biological material as the procedure is relatively straightforward and does not require any specialist equipment, other than a binocular microscope. Indeed, this method has been particularly popular for the isolation of microplastics from the gastrointestinal tract of fish.[15,30,69,87,269,361] The procedure typically involves cutting open the gastrointestinal tract using scissors and removing any detected microplastics with tweezers under observation with a binocular microscope.[269] Alternatively, the microplastics are flushed from the tissue by distilled water and suspended in a petri dish for observation.[361] For example, in a study[331] of freshwater fish, the stomach contents of the bluegill (*Lepomis macrochirus*) and longear sunfish (*Lepomis megalotis*) were washed with distilled deionised water through four filters (1000, 243, 118 and 53 μm) to separate individual ingested items into unique size populations.

This method has also been popular for the investigation of the gastrointestinal tract of birds for microplastics. For example, the carcasses of the northern fulmar (*Fulmarus glacialis*) seabird are routinely examined for plastics and microplastics as part of a monitoring programme operating in the North Sea.[427] Nevertheless, the drawback of this technique is that there is the potential for bias when undertaking this potentially subjective method due to difficulties in visually detecting very small items, such as mini-microplastics, which may be transparent or fouled with material and consequently overlooked by the observer.

Chemical and enzymatic digestion

While visual detection and separation is a mandatory step for the removal of debris and naturally occurring organic fragments, in many cases it may be impractical to completely rely on visual detection and separation to remove microplastics from biological material. This is especially the case when dealing with dense amounts of material or very small organisms retrieved from the environment, such as zooplankton. For this reason and to overcome the potential for bias, techniques have been developed that allow faster separation of plastic materials from organic material by digesting away the organic material to leave behind the plastic material for quantification. This is typically accomplished with the use of acids, bases and enzymes. Ultimately, this approach is particularly appropriate for smaller organisms where the entire body or visceral mass may be digested, as well as with fish stomachs, and ensures that all of the plastic materials present are collected. Nonetheless, care has to be exercised when utilising this approach since although the reagent may successfully digest away the biological tissue, it may also have a chemical impact upon the microplastics themselves, especially with small items such as fibres.[355] As such, a delicate balance needs to be found through experimental trial and error.

One of the most extensively studied methods is based upon the principle of wet digestion of biological tissues using acids. Indeed, the most successful is the acid destruction method and is recommended by the International Council for the Exploration of the Sea (ICES) as part of a preliminary protocol which was introduced for the monitoring of plastics in fish stomachs and shellfish. The acid destruction method (also known as the acid mix method) uses a mixture of 65% nitric acid (HNO_3) and 68% perchloric acid ($HClO_4$) in a 4:1 ratio (HNO_3:$HClO_4$ 4:1 v:v) and completely digests the tissues and removes other organic material, leaving behind only silica and plastic.[428] This method has the added advantage that it also removes rayon fibres, which are common fibres composed of regenerated cellulose and are not considered to be microplastics, and which have been known to skew results. However, the technique is still under development and variations on the concentrations of the acids may be required since there have been some reports of detrimental effects on nylon fibres,[428] which are known to be sensitive to acids and alkalis (see Chapter 4).

Other methods which have been developed involved the use of nitric acid, hydrogen peroxide and sodium hydroxide and have demonstrated effective rates of tissue digestion (particularly in mussels), with high recovery yields of polystyrene microbeads at 94–98%, but highly variable results for nylon fibres at 0–98% recovery.[62] Furthermore, considerable variation was found based upon the size and type of the microplastics extracted.[334] A review[428] of the presence of microplastics in organisms from natural habitats reported that crustaceans, fish, molluscs (mostly *Mytilus edulis* mussels) and polychaetes (lugworms) were assessed using visual detection and separation, as well as with tissue dissociation methods utilising potassium hydroxide (KOH), hydrogen peroxide (H_2O_2) and various acid destruction methods. Treatment with H_2O_2 was undertaken in various studies[132,278,448] but was demonstrated in other studies to result in incomplete tissue dissociation and a significant loss of microplastics of specific sizes from the sample.[319] Thus, to avoid the prospect of dissociation of the microplastics themselves, the use of enzymes, such as proteinase, lipase, cellulase and chitinase, have been recommended[107] for use in tissue dissociation as an alternative to acids and alkalis. Indeed, a study[66] which investigated the ingestion of microplastics in zooplankton, the enzyme proteinase-K was used in a sample clean-up step to remove large amounts of biogenic material (97% by weight) which was successfully filtered from water samples without destroying any microplastics present.[66] While the technique is yet to be more widely used, the results look promising.

Ultimately, continued research is needed to develop more efficient methods in separating microplastics from different mediums. Furthermore, while it is important to develop improved extraction techniques, there should also be a focus on developing standardised separation protocols that will allow the comparison of results over time, as well as between different locations and species. Once microplastics have been successfully separated from a sample medium, the final stage is to sort, categorise and identify the type of plastic materials present in order to report the abundance and type of microplastics in the sample.

Microplastic identification techniques

10

Identification of microplastics

Following sample collection and separation, the final stage in the assessment of microplastics in the environment is the positive identification of those items suspected to be composed of plastic. However, the precise chemical composition of suspected microplastics is potentially very broad. Since many plastics are blended with fillers and additives during manufacture, or exist as copolymers, the range of different types of microplastics recovered from an environmental sample can be quite extensive. This gives each specific plastic its own characteristic physiochemical properties (see Chapter 4), creating challenges in accurate identification.[156] Further complications arise from the size and shape of the microplastics, which can influence the methods used for identification. This is particularly the case for mini-microplastics (<1 mm–1 µm) where the limits of detection of the instrument are tested, as well as the practical difficulties encountered in handling items of such a small size. However, several techniques are available for microplastic identification, ranging from a simple visual identification to analytical techniques based on the chemical composition of the polymer. This is typically carried out on specialist instruments, such as Fourier transform infrared spectroscopy (FTIR), Raman spectroscopy and pyrolysis–gas chromatography–mass spectrometry (Pyr-GC–MS), all of which have been reviewed in the literature.[334] Incidentally, there has been a call for the inclusion of spectroscopic techniques in standardised microplastic monitoring schemes to ensure the accurate identification of microplastics recovered from environmental samples.[448]

Sample purification

Any surface exposed to the marine or freshwater environment will be subject to some form of fouling, particularly biofouling where a layer of microorganisms (bacteria, algae, fungi or plankton) accumulates on the surface. The presence of biofouling can be determined from the detection of nitrogen during elemental analysis and has been demonstrated successfully with virgin polypropylene and polyethylene microplastics.[303] However, the change in surface appearance can greatly impact how the item is identified and may result in the underrepresentation of microplastics in a sample. The most obvious example of this is with visual identification since the appearance (colour, texture and shape) of the item can change, depending on the extent of the fouling, thereby resulting in misidentification. Furthermore, several surface based techniques which are frequently employed to identify microplastics, such as FTIR or Raman spectroscopy, may also be negatively impacted.

Microplastic Pollutants. http://dx.doi.org/10.1016/B978-0-12-809406-8.00010-4

For this reason, various methods have been proposed to purify suspected microplastic items and clean the surface of any biofouling organisms. This typically takes place after the microplastic separation step. The methods proposed include freshwater rinsing,[282] ultrasonic cleaning[73] and the treatment with various chemicals including hydrochloric acid, sodium hydroxide and hydrogen peroxide.[198,257,319] However, for the same reason that care is taken when using chemicals for the separation of microplastics from biological material, similar caution must be exercised when attempting to remove biogenic material since the chemicals could potentially alter the characteristics of the microplastic, leading to misidentification, or may even completely dissolve the sample. For example, treatment with hydrogen peroxide (30%) was found to be an effective method in removing biogenic organic matter from microplastics. However, it was found to alter the characteristics of the polymer making it difficult to identify or in some cases even dissolving it, particularly with regard to fibres.[61,319] Consequently, it has been suggested that this purification step should only be undertaken if there is a high degree of organic matter present on the samples which is hindering visual selection of the microplastics.[319] Nevertheless, in a protocol recently developed by the National Oceanic and Atmospheric Administration (NOAA) in the United States, which concerns the separation of microplastics from sediment, solids are subjected to wet peroxide oxidation (WPO) in the presence of a Fe(II) catalyst to digest liable organic matter.[276] If it is decided to carry out a sample purification step, it is vital that the impact of that step on a selection of representative microplastics is undertaken. This is especially important with regard to fibres, which due to their large surface area, may be subject to greater change. In many cases, where surface based identification techniques are to be undertaken, simply scraping the surface of the plastic with a scalpel may be sufficient. However, this is unlikely to be practical for smaller items. As such, one interesting possibility is that there may be potential in the use of enzymes to remove biological material, as has occurred with the separation of samples from biological material.[66] However, as yet, this has not been fully explored in terms of cleaning samples for microplastic identification.

Visual identification

The simplest method in identification of microplastics is visual identification. Although the technique is more time consuming, it is often the most appropriate, especially for high volume samples where limited resources are available and access to expensive analytical instruments may not be possible. To aid in the positive identification of microplastics by visual examination, the following strict criteria have been suggested[316] to identify microplastics, which work best for microplastics in the size range 0.5–5 mm.[188]

Selection criteria to identify microplastics

- The particle or fibre in question has no observable organic or cellular structures.
- In the case of fibres, the diameter should be consistent along the length with no evidence of tapering or bending in three-dimensional space. If the fibre is not straight, biological origin is suspected.

- In the case of red coloured fibres, additional scrutinisation with high-magnification microscopic examination, fluorescence microscopy and staining of chloroplasts is required to preclude algal sprouts.
- Particles should be clear and unvaryingly coloured.
- In the case of transparent, opaque or white particles, further high-magnification microscopic examination, as well as fluorescence microscopy, should be undertaken to preclude the possibility of biological origin.

Note: The absence of any criteria for transparent and white coloured fibres is a result of their typical confusion with the antennae of organisms and plant fibres.[316] Thus, overrepresentation from the misidentification of these organic structures as microplastics is avoided. However, the disadvantage in excluding these types of fibres is that it can lead to underrepresentation of the amount of microplastics in an environmental sample. Thus, if it is desired to include such fibres, a highly detailed examination is required using high-magnification microscopic examination, fluorescence microscopy and staining techniques to confirm the items are microplastics. In some cases, it may even be necessary to utilise a scanning electron microscope (SEM) to be confident that the identification is accurate.

These selection criteria help to exclude biogenic materials and aid in the identification of microplastics. Indeed, visual methods as the soul means of microplastic identification have been used in several recent publications, based upon the items shape and colour, as well as in the study of fraying (splitting, breakdown, wear) of microplastic fibres.[32,173,331] In cases where the identification of large numbers of items may not be practical, representative sub-groups of plastics, based on visual examination, should be randomly chosen for identification using spectroscopy.[344,345]

Importantly, there is always an element of subjectivity and potential bias during visual identification which results from individual interpretation by the examiner. Furthermore, if the microplastic is weathered (see Chapters 4 and 5), then there may be changes in morphology which can make identification more difficult. In a comparison between microscopic and spectroscopic (FTIR) identification, fragmented microplastics were significantly underestimated and fibres, which can be of natural origin, were significantly overestimated using a stereomicroscope on the surface microlayer and on samples derived from beaches.[389] In another comparison, a 70% overestimation was reported for items visually classified as microplastics when compared to FTIR.[188] The effectiveness of visual identification is dependent on the size of the item being scrutinised and becomes considerable more difficult as the size of the item reduces, with error rates ranging between 20%[116] and 70%.[188] Thus, the accuracy in visually identifying very small items as microplastics is less reliable than with larger items.[107,188,389] Consequently, spectroscopic conformation has been recommended for particles smaller than 1 mm in size,[188] while others have suggested a cut-off of less than $500\,\mu m^{263}$ or less than $100\,\mu m.$[120] Nevertheless, techniques such as high-magnification fluorescence microscopy[389] or scanning electron microscopy can be utilised to confirm the absence of cellular structures and improve visual identification. Ultimately, visual identification is an important technique, and the selection of an appropriate identification method is determined by the number of samples to be analysed and the size of the microplastics of interest. Once suspected items have been visually identified as microplastics, they can then be sorted using the standardised size and colour sorting (SCS) system.

The standardised size and colour sorting (SCS) system for categorising microplastics

The standardised size and colour sorting (SCS) system (see Fig. 10.1) effectively categorises pieces of plastic based upon their size and appearance. The SCS system is capable of categorising any piece of plastic.

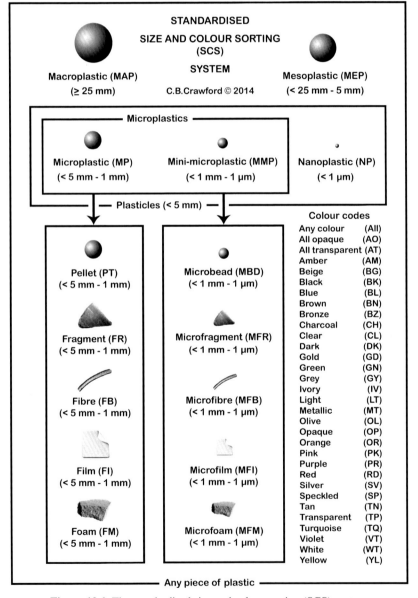

Figure 10.1 The standardised size and colour sorting (SCS) system.

The size and colour sorting (SCS) system procedure

As well as providing an effective categorising system, the SCS system generates unique codes for dealing with microplastic abundance data. In the following five-step procedure, an example will be used to demonstrate how the SCS system works. Thus, we shall assume that a sample has been collected from the environment and the detected pieces of plastic have been separated from the sample. The next step would be to use the SCS system to sort and categorise those pieces of plastic.

Example
In this example, we shall explain how to categorise a mini-microplastic measuring 0.8 mm in size of the microfragment type which is light-blue in colour and is composed of polyethylene.

Step 1: category

First, the plastics should be sorted based upon their size. This is determined by measuring them along their longest dimension and then allocating them with the appropriate category abbreviation by referring to Table 10.1. For example, (MP) = microplastic and (MMP) = mini-microplastic.

For macroplastics and mesoplastics, allocate each piece of plastic with the appropriate category abbreviation (MAP) or (MEP) and size, then go to **Step 3**.

For nanoplastics, use the abbreviation code (NP) and the colour abbreviation code (ALL) and then go to **Step 4**.

Table 10.1 Categorisation of pieces of plastic based on size

Category	Abbreviation	Size	Size definition
Macroplastic	MAP	≥25 mm	Any piece of plastic equal to or larger than 25 mm in size along its longest dimension
Mesoplastic	MEP	<25 mm–5 mm	Any piece of plastic less than 25 mm–5 mm in size along its longest dimension
Plasticle	PLT	<5 mm	All pieces of plastic less than 5 mm in size along their longest dimension
Microplastic	MP	<5 mm–1 mm	Any piece of plastic less than 5 mm–1 mm in size along its longest dimension
Mini-microplastic	MMP	<1 mm–1 μm	Any piece of plastic less than 1 mm–1 μm in size along its longest dimension
Nanoplastic	NP	<1 μm	Any piece of plastic less than 1 μm in size along its longest dimension

Example
The piece of plastic used in the example is 0.8 mm in size along its longest dimension. By referring to Table 10.1, we can see that this measurement categorises the piece of plastic as a mini-microplastic. Thus, we write the first part of the SCS code as follows: **MMP/0.8**

Note: If the size category of the piece of plastic was determined by a sieving method and thus, the exact measurements of each piece were not taken, then simply use the category abbreviation without specifying the size. Therefore, in that scenario, the fist part of the SCS in this example would be as follows: **MMP/**

Step 2: type

For plastics smaller than 5 mm in size, further sorting is undertaken based on their appearance. Microplastics and mini-microplastics should be sorted into their appropriate type, based on their size, shape and basic composition, by referring to Table 10.2. Each piece of plastic is then allocated the appropriate type abbreviation, which is then added to the end of the SCS code.

If further morphological categorisation is not required and it is simply wished to report the abundance of all types of plastic below 5 mm in size, then use the category abbreviation for a plasticle (PLT).

If further sorting based on colour is desired, go to **Step 3**. If not, use the colour abbreviation (ALL) and then go to **Step 4**.

Table 10.2 Categorisation of microplastics based on morphology

Abbreviation	Type	Size	Definition
PT	Pellet	<5 mm–1 mm	A small spherical piece of plastic less than 5 mm to 1 mm in diameter
MBD	Microbead	<1 mm–1 μm	A small spherical piece of plastic less than 1 mm to 1 μm in diameter
FR	Fragment	<5 mm–1 mm	An irregular shaped piece of plastic less than 5 mm to 1 mm in size along its longest dimension
MFR	Microfragment	<1 mm–1 μm	An irregular shaped piece of plastic less than 1 mm to 1 μm in size along its longest dimension
FB	Fibre	<5 mm–1 mm	A strand or filament of plastic less than 5 mm to 1 mm in size along its longest dimension
MFB	Microfibre	<1 mm–1 μm	A strand or filament of plastic less than 1 mm to 1 μm in size along its longest dimension
FI	Film	<5 mm–1 mm	A thin sheet or membrane-like piece of plastic less than 5 mm to 1 mm in size along its longest dimension

Table 10.2 **Categorisation of microplastics based on morphology—cont'd**

Abbreviation	Type	Size	Definition
MFI	Microfilm	<1 mm–1 µm	A thin sheet or membrane-like piece of plastic less than 1 mm to 1 µm in size along its longest dimension
FM	Foam	<5 mm–1 mm	A piece of sponge, foam, or foam-like plastic material less than 5 mm to 1 mm in size along its longest dimension
MFM	Microfoam	<1 mm–1 µm	A piece of sponge, foam, or foam-like plastic material less than 1 mm to 1 µm in size along its longest dimension

Example

The piece of plastic used in the example is a mini-microplastic and has a fragment-like appearance so we categorise it as a microfragment. Thus, by referring to Table 10.2, we add the abbreviation for a microfragment to the end of the SCS code as follows: **MMP/0.8/MFR**

Step 3: colour

The next step requires that all pieces of plastic are given an individual colour code by referring to Table 10.3. Combinations of colour codes can also be used and are combined using a hyphen. For example, DK-BL means dark blue and MT-GN means metallic green.

Table 10.3 **Categorisation based on colour**

Colour	Abbreviation
Any colour	ALL
All opaque	AO
All transparent	AT
Amber	AM
Beige	BG
Black	BK
Blue	BL
Brown	BN
Bronze	BZ

Continued

Table 10.3 **Categorisation based on colour—cont'd**

Colour	Abbreviation
Charcoal	CH
Clear	CL
Dark	DK
Gold	GD
Green	GN
Grey	GY
Ivory	IV
Light	LT
Metallic	MT
Olive	OL
Opaque	OP
Orange	OR
Pink	PK
Purple	PR
Red	RD
Silver	SV
Speckled	SP
Tan	TN
Transparent	TP
Turquoise	TQ
Violet	VT
White	WT
Yellow	YL

Example

The colour of the microfragment is light-blue. Thus, by referring to Table 10.3, we combine the colour abbreviation for 'light' and 'blue' with a hyphen as follows: LT-BL.

We then add this colour information to the end of the SCS code as follows: **MMP/0.8/MFR/LT-BL**

Step 4: polymer

All pieces of plastic are then analysed by either visual examination, spectroscopic methods or other analytical techniques to determine the type of polymer. Once this is determined, allocate each plastic with the appropriate polymer abbreviation from Table 10.4.

Table 10.4 **Polymer abbreviation codes**

Polymer	Abbreviation
Acrylonitrile-butadiene-styrene	ABS
Acrylate-styrene-acrylonitrile	ASA
Butadiene rubber	BR
Cellulose acetate	CA
Cellulose acetate-butyrate	CAB
Cellulose acetate propionate	CAP
Cellulose	CE
Carboxymethyl cellulose	CMC
Cellulose nitrate	CN
Cellulose propionate	CP
Polychloroprene (neoprene)	CR
Chlorosulfonated polyethylene	CSM
Ethylene chlorotrifluoroethylene	ECTFE
Ethylene-propylene rubber	EPR
Expanded polystyrene	EPS
Ethylene vinyl acetate	EVA
Ethylene vinyl alcohol	EVOH
Fluorinated ethylene propylene	FEP
High-density polyethylene	HDPE
Hydroxyethyl methacrylate	HEMA
High-impact polystyrene	HIPS
Low-density polyethylene	LDPE
Linear low-density polyethylene	LLDPE
Methacrylate butadiene styrene	MBS
Medium-density polyethylene	MDPE
Melamine formaldehyde	MF
Acrylonitrile butadiene rubber	NBR
Natural rubber	NR
Polyamide (nylon)	PA
Nylon 4,6	PA 46
Nylon 6	PA 6
Nylon 6,10	PA 610
Nylon 6,6	PA 66
Nylon 6,6/6,10 copolymer	PA 66/610
Nylon 11	PA 11
Nylon 12	PA 12
Polyarylamide	PAA
Polyamide imide	PAI

Continued

Table 10.4 **Polymer abbreviation codes—cont'd**

Polymer	Abbreviation
Polyacrylonitrile	PAN
Polybutylene	PB
Polybutylene terephthalate	PBT
Polycarbonate	PC
Polycaprolatone	PCL
Polyethylene	PE
Polyether block amide	PEBA
Polyetheretherketone	PEEK
Polyester elastomer	PEEL
Polyester imide	PEI
Polyetherketone	PEK
Polyether sulfone	PES
Polyethylene terephthalate	PET
Polyethylene terephthalate glycol-modified	PETG
Phenol formaldehyde	PF
Perfluoroalkoxy alkane	PFA
Polyhydroxybutyrate	PHB
Poly(3-hydroxybutyrate-*co*-3-hydroxyvalerate)	PHBV
Polyhydroxyvalerate	PHV
Polyimide	PI
Polyisocyanurate	PIR
Polylactic acid	PLA
Poly(methyl methacrylate)	PMA
Polymethylpentene	PMP
Polyoxymethylene	POM
Polypropylene	PP
Poly(*p*-phenylene ether)	PPE
Poly(*p*-phenylene oxide)	PPO
Polyphenylene sulphide	PPS
Polyphenylene sulphide sulfone	PPSS
Polyphenylenesulfone	PPSU
Polypropylene terephthalate	PPT
Polystyrene	PS
Polysulfone	PSU
Polytetrafluoroethylene	PTFE
Polytrimethylene terephthalate	PTT
Polyurethane	PUR
Polyvinyl acetate	PVA
Polyvinyl butytral	PVB
Polyvinyl chloride	PVC
Chlorinated polyvinyl chloride	PVCC
Polyvinylidene chloride	PVDC

Table 10.4 **Polymer abbreviation codes—cont'd**

Polymer	Abbreviation
Polyvinylidene fluoride	PVDF
Polyvinyl fluoride	PVF
Polyvinyl alcohol	PVOH
Styrene acrylonitrile	SAN
Styrene butadiene rubber	SBR
Styrene-butadiene-styrene	SBS
Styrene-ethylene-butadiene-styrene	SEBS
Styrene-isoprene-styrene	SIS
Styrene maleic anhydride	SMA
Thermoplastic polyurethane	TPUR
Urea formaldehyde	UF
Ultra-high-molecular weight polyethylene	UHMWPE
Cross-linked polyethylene	XLPE
Extruded polystyrene	XPS

Example
The microfragment has been identified to be polyethylene, which has the abbreviation PE. Thus, we add the relevant polymer abbreviation from Table 10.4 to the end of the SCS code as follows: **MMP/0.8/MFR/LT-BL/PE**

Step 5: quantity

The final step is to report the quantity of each categorised piece of plastic. Thus, the number 1 is written after the code for every unique piece of categorised plastic. However, if there is more than one piece of plastic which are identical, then the code is only written once and the number of identical pieces is written at the end of the code. Thus, for eight identical pieces of plastic, the number 8 would be written at the end of the SCS code.

Each piece of plastic should now have an SCS code which identifies the category, exact size (optional), type, colour, type of polymer and the quantity of plastic pieces.

In the case of mesoplastics and macroplastics, the category, exact size (optional), colour and type of polymer are represented in the code. In the case of nanoplastics, the category and type of polymer are represented in the code. These SCS codes can then be used to report the type and abundance of plastics in the environment.

All SCS codes which have been generated may also be entered into a tabular format, as shown by the examples in Table 10.5. Two extra columns have been added to the right of the table to provide examples of what the written code would look like and the interpretation of those codes.

Table 10.5 Examples of SCS data codes and their interpretation

Example #	Category	Optional: exact size (mm)	Type	Colour	Polymer	Quantity	Example of written code	Interpretation of code
1	MAP	38	—	BK	ABS	1	MAP/38/BK/ABS/1	A macroplastic measuring 38 mm in size which is black in colour and is composed of acrylonitrile butadiene styrene
2	MEP	16	—	LT-GY	PVC	2	MEP/16/LT-GY/PVC/2	2 mesoplastics measuring 16 mm in size which are light grey in colour and are composed of polyvinyl chloride
3	MP	1.4	PT	AM	PP	1	MP/1.4/PT/AM/PP/1	A microplastic measuring 1.4 mm in size of the pellet type which is amber in colour and composed of polypropylene
4	MP	1.9	FI	BL	PE	1	MP/1.9/FR/BL/	A microplastic measuring 1.9 mm in size of the film type which is blue in colour and composed of polyethylene
5	MP	2.0	FR	TP	PC	1	MP/2.0/FR/TP/PC/1	A microplastic measuring 2.0 mm in size of the fragment type which is transparent in colour and is composed of polycarbonate

6	MP	1.2	FM	WT	EPS	1	MP/1.2/FM/WT/EPS/1	A microplastic measuring 1.2 mm in size of the foam type which is white in colour and composed of expanded polystyrene
7	MMP	0.6	MBD	DK-GN	HDPE	1	MMP/0.7/MBD/DK-GN/HDPE/1	A mini-microplastic measuring 0.6 mm in size of the microbead type which is dark green in colour and is composed of high-density polyethylene
8	MMP	0.8	MFB	LT-BL	PE	8	MMP/0.8/MFB/LT-BL/PE/8	Eight mini-microplastics measuring 0.8 mm in size of the microfragment type which are light-blue in colour and are composed of polyethylene
9	MMP	—	MFR	GD	PET	2	MMP/MFR/GD/PET/2	Two mini-microplastics of the microfibre type which are gold in colour and are composed of polyethylene terephthalate
10	NP	—	—	ALL	PS	20	NP/ALL/PS/20	20 nanoplastics composed of polystyrene

Example
If there were eight identical pieces of plastic, then we would only write the SCS code once and put an 8 at the end of the code. Thus, the code would be written as follows: **MMP/0.8/MFR/LT-BL/PE/8**

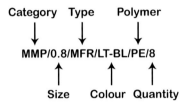

Figure 10.2 The standardised size and colour sorting (SCS) system coding method.

In the SCS system, reporting the exact size of a piece of plastic is optional. Thus, when the exact size is not reported, it is still possible to still derive from the SCS code what the size-range of the piece of plastic is. For example, in Example 9 in Table 10.5, it can be seen that the reported SCS code is MMP/MFR/GD/PET/2. From Fig. 10.2, it can be seen that the first part of the code corresponds to the category of plastic. In this case, it is reported as (MMP). By looking at Table 10.1, it can be seen that (MMP) is a mini-microplastic and the size range is (<1 mm–1 μm). Therefore, it is possible to deduce that that the piece of plastic must be (<1 mm–1 μm) in size. Thus, the standardised size and color sorting system developed for this book enables detailed information about a piece of plastic to be conveniently expressed as a simple and unique code. This facilitates entry into databases for subsequent use with analysis and comparison tools, and importantly, enables detailed standardised reporting of comparable data in the literature on the type and abundance of plastics in the environment.

Scanning electron microscope (SEM)

A scanning electron microscope (SEM) is used to create an image of a small surface of a sample by firing a high intensity beam of electrons at the samples surface and scanning it in a zig-zag type pattern (raster scanning). Since electrons are used to image the sample, levels of detail lower than 0.5 nm resolution are achievable at very high magnifications, such as 2,000,000 times. This is far higher than standard optical microscopes which are only practically useful up to around 1000 times magnification. The generation of electrons is achieved in two ways, via a hot filament source or field emission. With the filament method, a high electrical current is passed through a tungsten filament to heat the filament to around 5000°C, thereby providing the electrons with enough energy to overcome a potential energy barrier which normally confines them within the tungsten. The electrons are then emitted from the tungsten (thermionic emission) and are characterised by a Maxwell–Boltzmann distribution of energies. However, to ensure that the electrons efficiently reach the sample and are not absorbed or scattered by the air, the filament is heated in a vacuum. The field emission method exploits the quantum tunnelling of electrons through the potential energy barrier and utilises a tungsten crystal to produce electrons which possess a very narrow energy range, which is beneficial in attaining a high resolution.

The beam of electrons (incident beam) is then tightly focussed by electromagnetic lenses towards the sample. Scanning coils then move the beam over the sample surface in a raster pattern resulting in electrons in the sample near-elastically scattering the incident electrons (backscattered electrons). These electrons are then collected by a detector and converted to a signal. Elements with a high atomic number tend to produce a greater number of backscattered electrons. As such, when examining an SEM image, the brightest regions in the image contain elements with a high atomic number. However, the majority of electrons are inelastically scattered within the sample and can break molecular bonds, resulting in the emission of electrons from the sample (secondary electrons). This tends to occur at shallow depths in the sample surface, and the collection of secondary electrons is beneficial in resolving a high degree of detail pertaining to the samples surface. X-rays are also generated when the electron beam hits the samples surface and can be converted into voltage in relation to the intensity of their emission by an X-ray microanalyser via the process of energy dispersive X-ray (EDX) analysis, thereby providing detailed quantitative information as to the samples elemental composition (Fig. 10.3).

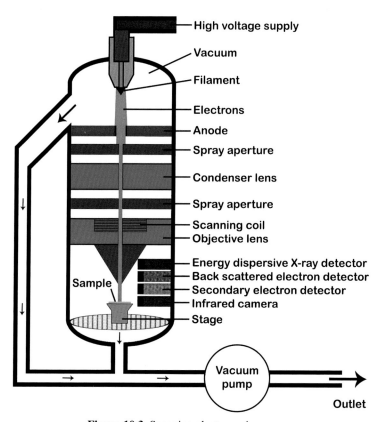

Figure 10.3 Scanning electron microscopy.

Since an electron microscope has a high depth of field, samples can be visualised in black and white with a high amount of detail in three dimensions and with a high degree of tilt, such as 45 degrees, thereby by providing topographical, morphological and compositional information. However, for samples to be visualised, their surface must be able to conduct electricity, as well as being electrically grounded to avoid the build-up of electrostatic charge. As such, the samples are required to be coated with a conductive film by way of high-vacuum evaporation or low-vacuum sputter-coating, typically with a very thin layer of gold, gold/palladium, osmium, iridium, platinum, chromium, tungsten or graphite. However, coating with metals with a high atomic number, such as gold, results in a greater abundance of emitted electrons and an improved signal to noise ratio. Thus, scanning electron microscopy is an expensive technique which requires a trained operator. Furthermore, the instrument is large and must be situated in an area free from significant electromagnetic and vibrational interface.

Scanning electron microscopy can be used to analyse the physical characteristics of microplastics recovered from environmental samples, as well as to determine their physical size and the specific dimensions of any surface features.[116,426] Thus, based upon the surface morphology, a scanning electron microscope (SEM) can help distinguish a plastic item from a non-plastic item. Unlike infrared spectroscopic techniques, SEM is not normally used to identify the type of plastic. However, if the SEM is equipped with an energy-dispersive X-ray microanalyser, energy-dispersive X-ray (EDX) analysis can be undertaken and the inorganic chemical composition of the microplastic can be examined and confirmed, as well the identity of any inorganic plastic additives that the microplastics may contain. Thus, the SEM technique has successfully been used to differentiate mini-microplastics (<1 mm–1 μm), including multi-coloured microbeads, from other materials, such as coal ash (Figs. 10.4 and 10.5).

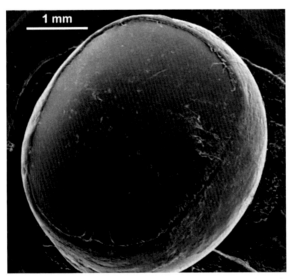

1 mm

Figure 10.4 An SEM image of a 4 mm pristine polyethylene microplastic pellet.

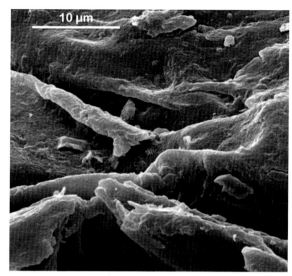

Figure 10.5 SEM images of the surface of a weathered polystyrene microplastic recovered from the marine environment.

Pyrolysis–gas chromatography–mass spectrometry (Pyr-GC–MS)

Pyrolysis–gas chromatography–mass spectrometry (Pyr-GC–MS) is a technique which thermally decomposes the large high-molecular weight molecules of a sample via heat mediated cleavage in the presence of an inert atmosphere, or a vacuum, to create a suite of smaller low molecular weight moieties. The composition of these moieties is subsequently determined by mass spectrometry (MS) and provides characteristic information as to the structural composition of the samples large high-molecular weight molecules, thereby allowing the sample composition to be identified (see Fig. 10.6).

As a destructive technique which thermally decomposes the sample, further analysis of the microplastics is precluded. Consequently, this may be a limiting factor in some cases. Nevertheless, the great advantage of Pyr-GC–MS is that the technique utilises direct introduction of the sample with minimal pre-treatment. Thus, unlike traditional GC–MS, preparation of a highly refined organic solution is not required and the sample can be analysed directly. For example, a sample of sediment suspected to contain microplastics would first be extracted with an organic solvent to remove any low molecular weight unbound compounds that may hinder the analysis by obscuring the data. The sample is then placed in a sample chamber and is heated (typically 200–600°C) in an oxygen free environment. Importantly, only a very small amount of sample is required and if the sample contains a large amount of carbon atoms, then less than 1 mg is needed, thereby facilitating trace analysis. A chromatogram is then produced which can then be compared with an electronic reference database to identify the type of microplastics present in the sample.[319] Thus, as a result of the direct introduction of a sample and the subsequent chromatographic separation, the technique is capable of providing valuable

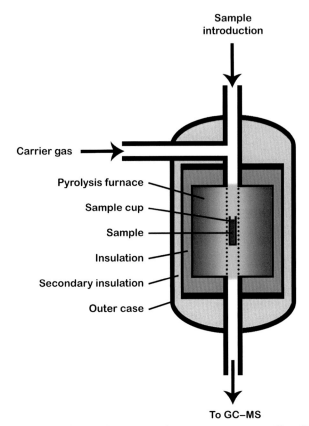

Figure 10.6 Pyrolysis–gas chromatography–mass spectrometry (Pyr-GC–MS).

and unique information that is not available with other analytical techniques. As such, the technique can be utilised for the identification of microplastics in environmental samples, as well as simultaneously identifying any plastic additives present.[142] However, since samples must be manually placed in the instrument and analysed individually, the analysis of large quantities of microplastics is limited, as well as the range of sizes of items which can be effectively handled.[263]

Nuclear magnetic resonance (NMR) spectroscopy

Nuclear magnetic resonance (NMR) spectroscopy relies upon the actuality that information about the chemical environment surrounding the nuclei of an atom can be gained by obtaining information about the behaviour of that nuclei when it is simultaneously exposed to a magnetic field and electromagnetic radiation (see Fig. 10.7).

The technique is based upon an intrinsic form of angular momentum, termed spin, of subatomic particles, such as protons and neutrons, which can be thought of as

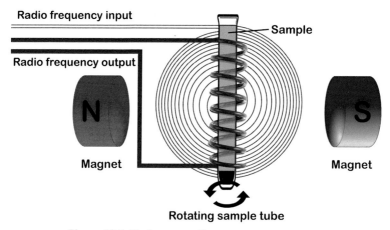

Figure 10.7 Nuclear magnetic resonance spectroscopy.

spinning on their axes. In an atom, there is a small dense region consisting of protons and neutrons, termed as the nucleus, which is surrounded by electrons. These protons and neutrons are in turn, composed of gluons and quarks. The proton is composed of a single quark with a −e/3 charge and two quarks with a +2e/3 charge, so the overall charge is positive. However, the neutron is composed of a single quark with a +2e/3 charge and two quarks with a −e/3 charge, so the overall charge is zero (see Fig. 10.8).

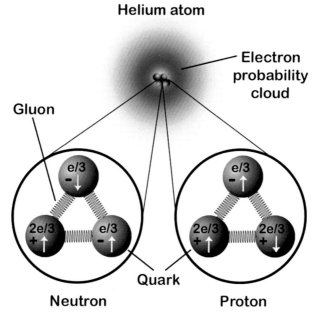

Figure 10.8 The atomic nucleus and spin of subatomic particles.

The up and down arrows in Fig. 10.8 indicate the component of spin on the z-axis with the up arrow denoting the value $+^1/_2$ and the down arrow denoting the value $-^1/_2$. Thus, the spin of protons and neutrons is equal to $^1/_2$. If the number of protons and neutrons in the nucleus are both even, then the spin of these subatomic particles will counteract one another and the overall spin of the nucleus is considered to be zero. However, if the number of protons and neutrons in the nucleus are both odd, then the nucleus is considered to possess an integer spin, such as 1 or 2. Furthermore, if the number of protons is odd and the number of neutrons in the nucleus is even, or vice versa, then the nucleus is considered to possess a half-integer spin, such as $^1/_2$ or $^5/_2$ (see Table 10.6)

Table 10.6 The nuclear spin quantum number

Number of protons	Number of neutrons	Nuclear spin	Nuclei example
Even	Even	0	12C, 16O
Odd	Odd	1, 2, etc.	2H
Odd	Even	1/2, 5/2, etc.	13C
Even	Odd	1/2, 5/2, etc.	15N

Since the nucleus is spinning and possesses a charge, it generates a small magnetic field known as the nuclear magnetic moment, which is proportional to the spin. Thus, nuclei with zero spin, such as ^{12}C and ^{16}O, have no nuclear magnetic moment, while ^1H has the largest. When an external magnetic field is applied, the nuclei will spin with precessional motion around the magnetic field. However, there are two possible spin energy states, $-^1/_2$ and $+^1/_2$. In the $+^1/_2$ spin energy state, the atomic nucleus is aligned with the externally applied magnetic field. Thus, the magnetic moment does not oppose the externally applied magnetic field and the nucleus is considered to be in a stable low energy state. On the contrary, the $-^1/_2$ spin energy state opposes the externally applied magnetic field and is considered to be in a less-stable high energy state (see Fig. 10.9).

The difference in energy between the $-^1/_2$ and $+^1/_2$ spin states is very small and is dependent upon the strength of the magnetic field. Thus, in Fig. 10.9 it can be observed that in the absence of a magnetic field, the magnetic dipoles of the nuclei are in a random orientation and consequently, there is an absence of net magnetisation. However, the application of a magnetic field results in the creation of two distinct energy levels where the small difference in energy between the two states increases as the magnetic field strength increases. Since the difference in energy is small, thermal collisions produce enough energy to flip a considerable number of nuclei into the higher energy state. Indeed, at a temperature of absolute zero, practically all nuclei would favour the low energy spin state. However, as the temperature increases, thermal collisions oppose the preferential low energy state and offset the difference by equalising the two states. The number of nuclei in low and high energy spin states

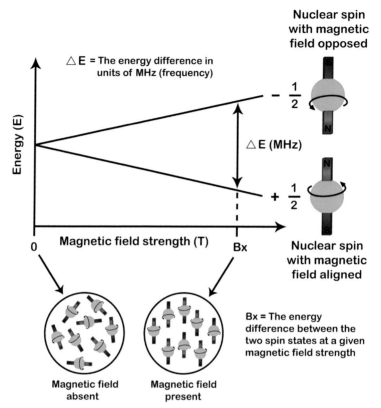

Figure 10.9 The effect of a magnetic field of increasing strength on the nuclear magnetic moment.

can be described by the Boltzmann distribution. Thus, from the Boltzmann equation, which expresses the relationship between temperature and the related energy, it can be deduced that at room temperature, there will be slightly more nuclei in the low energy state than in the high energy state. For this reason, there is always a small excess of nuclei aligned with the externally applied magnetic field, in comparison to the nuclei opposed against it.

The small difference in energy between spin states is normally expressed as a frequency (MHz) ranging from 20 to 1000 MHz. When a nucleus in the $+^1/_2$ low energy spin state is exposed to electromagnetic radiation of the same specific frequency as the energy difference (20–1000 MHz), it will absorb that electromagnetic energy and be excited into the $-^1/_2$ higher energy spin state, thereby opposing the magnetic field. The nucleus is then in resonance with the electromagnetic radiation at a specific resonance frequency. Thus, the strength of the magnetic field directly affects the amount of electromagnetic energy required to excite the nucleus to flip to the higher energy state, with a stronger magnetic field requiring more electromagnetic energy.

Table 10.7 The atomic nuclei frequently utilised in the analysis of plastic materials by nuclear magnetic resonance (NMR) spectroscopy

Name	Nuclei	Mass of atom	Chemical shift (ppm)	Nuclear spin	Natural abundance (%)	Gyromagnetic ratio (γ) (MHz/tesla)	Resonance frequency at 1.5 tesla (MHz)	Sensitivity (equal number of nuclei in a magnetic field)
Hydrogen	^1H	1.007825	15	$^1/_2$	99.9885	42.58	63.87	1.0
Deuterium	^2H	2.014102	15	1	0.0115	6.53	9.79	0.00965
Carbon	^{13}C	13.003355	220	$^1/_2$	1.07	10.71	16.07	0.0159
Nitrogen	^{14}N	14.003074	900	1	99.632	3.09	4.64	0.00101
Nitrogen	^{15}N	15.000109	900	$^1/_2$	0.368	4.34	6.51	0.00104
Oxygen	^{17}O	16.999132	800	$^5/_2$	0.038	5.81	8.71	0.0291
Fluorine	^{19}F	18.998403	800	$^1/_2$	100	40.05	60.08	0.83
Phosphorous	^{31}P	30.973762	700	$^1/_2$	100	17.23	25.85	0.0663

Traditionally, NMR techniques were undertaken by varying the frequency of the electromagnetic radiation at a constant magnetic field strength, or vice versa, and measuring the absorption of this electromagnetic radiation by different nuclei. However, by using a technique termed as pulsed Fourier transform nuclear resonance spectroscopy, improved sensitivity and resolution can be attained. In this technique, a pulse of broadband radio frequencies is emitted at the sample which excites all of the nuclei in the sample at the same time, resulting in the resonance of all the nuclei at the same time. After the pulse ceases, the excited nuclei relax and emit a superposition of all the excitation frequencies simultaneously. The way in which this signal evolves over time is recorded by the instrument, and a Fourier transform mathematical operation is carried out on the data to produce a high resolution NMR spectrum. This spectrum then provides information about the chemical environment surrounding the nucleus.

The technique is very useful in the determination of the chemical structure of the polymer chain of plastic materials and highly detailed information can be obtained relating to the sequence of monomers in a copolymer and the degree of crystallinity in a semi-crystalline plastic, as well as information relating to branching and tacticity. For example, the stereochemistry of polypropylene can be examined to ascertain the tactic form present (see Chapter 4). Furthermore, chemical changes in the plastic material can also be detected, such as oxidation states, with a high degree of sensitivity. The typical nuclei used for the NMR spectroscopy of plastic materials are listed in Table 10.7.

Importantly, the natural abundance of a particular NMR-active atomic nucleus, as well as its gyromagnetic ratio (γ), has a direct effect on the sensitivity of the technique to that nucleus. The gyromagnetic ratio is the ratio between the nuclear magnetic moment and the angular momentum of a spin. Thus, the gyromagnetic ratio is a proportionality constant unique to each nucleus that has a nuclear magnetic moment and is directly proportional to the signal strength in NMR. For this reason, abundant nuclei with a high gyromagnetic ratio, such as ^1H and ^{19}F, are the most responsive to NMR measurements and are considered as some of the most important nuclei to study in NMR spectroscopy.

Fourier-transform infrared (FTIR) spectroscopy

Fourier-transform infrared (FTIR) spectroscopy is the most popular and widely used technique for the positive identification of the type of plastic that microplastics in environmental samples are composed of.[178,188,263,406] While the reason for the popularity of the technique is in part due to its straightforwardness and reliability, the predominant reason is that FTIR is highly accurate in identifying the type of plastic present by producing highly specific infrared (IR) spectra which contain distinct band patterns,[188] thereby allowing differentiation between plastic materials and natural materials. The technique relies upon the actuality that most molecules absorb light in the infrared region of the electromagnetic spectrum (see Fig. 10.10).

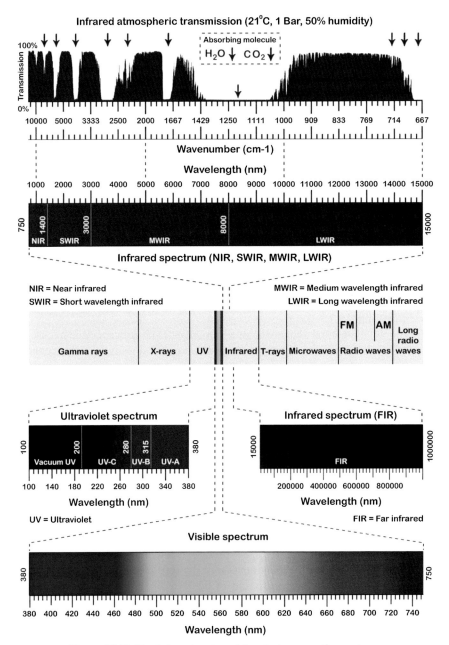

Figure 10.10 The infrared region of the electromagnetic spectrum.

With a wavelength longer than that of visible light, and just outside the red region of the visible spectrum, infrared light exhibits a wavelength of 750 nm–1 mm. If a sample is irradiated with a beam of infrared light, analysis of the elements such as carbon, hydrogen, nitrogen and oxygen can be undertaken by measuring the degree to which the molecules in the sample absorb specific wavelengths of the infrared light. The photons which make up the infrared light may be absorbed by the sample (absorption) or may not interact with the sample and pass straight through (transmittance). The molecules of the sample which absorb photons gain energy and as a consequence, the bonds of the molecules will distort more energetically by means of bending and stretching (see Fig. 10.11). Thus, infrared spectroscopy is a technique which irradiates a sample with specific wavelengths of infrared light and then examines the transmitted light to deduce the amount of energy that was absorbed by the molecules at each wavelength, thereby providing information about the molecules present in the sample.

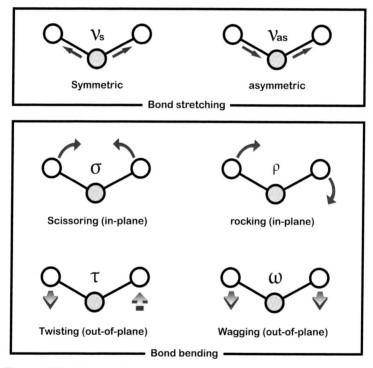

Figure 10.11 Molecules bend and stretch when exposed to infrared radiation.

Thus, by measuring the amount of absorption of infrared radiation at different frequencies, it is possible to generate an absorption spectrum that can provide insight concerning the molecular structure of the sample. The infrared spectrum contains a series of absorption peaks which correspond to the different frequencies of vibrations between the bonds of the atoms in the molecules of the sample. Since different types of plastic have unique combination of atoms, no two plastic materials will produce an identical IR spectrum. For this reason, FTIR spectrums are unique to each type of plastic and can be used to positively identify the type of plastic that a microplastic is composed of. Since FTIR collects spectral data at a high resolution over a wide range of spatial frequencies (typically 4000–600 cm⁻¹), the technique is particularly suited for identifying the different groups of specific atoms (functional groups) in a molecule (Fig. 10.12).

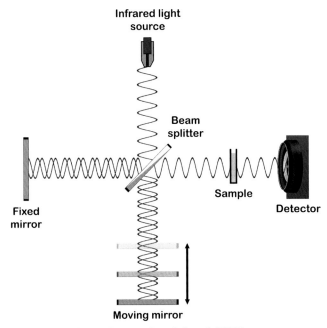

Figure 10.12 Fourier-transform infrared (FTIR) spectroscopy.

However, infrared spectroscopy can be problematic with regard to the preparation of samples. For example, the utilisation of transmission techniques requires that a sample is adequately transparent that infrared wavelengths can pass into, and transmit through, the sample. With most polymers, this is simply not attainable. Thus, suitable processing of the sample is required to allow the transmission of infrared radiation. This can be achieved in three ways:

1. Suspension of the polymer within a compressed potassium bromide (KBr) disc that is transparent to infrared light.
2. Dispersion of the polymer in mineral oil (nujol mull).
3. Dissolution of the polymer in a solvent.

The KBr method is regarded as being predominantly difficult in that the attainment of a suitably transparent disc requires a very specific set of conditions to be met. For this reason, relatively large pieces of plastic are best analysed using a reflectance technique, such as attenuated total reflection (ATR) (see Fig. 10.13).

As a technique that only analyses the surface of the plastic sample, ATR only requires that the sample is in sufficiently close proximity to a small crystal. This

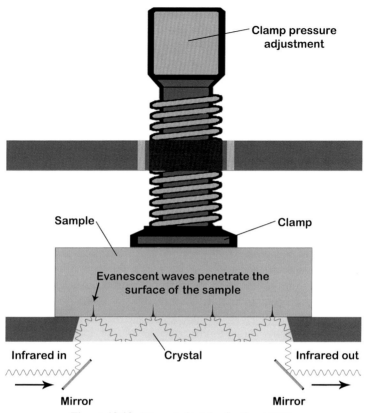

Figure 10.13 Attenuated total reflection (ATR).

crystal is typically composed of either germanium (Ge), diamond or zinc selenide (ZnSe). Once in contact with the crystal, an evanescent wave propagates 0.5–5 μm past the crystal surface and into the surface of the sample. Thus, the main requirement with ATR is that the sample is sufficiently in contact with this crystal to allow penetration of infrared radiation into the sample. This can be achieved by making use of a clamp to press the sample against the crystal.

Nevertheless, ATR is not a completely flawless technique and does exhibit some difficulties. For example, the material being analysed must have a refraction index which is lower than the crystal, otherwise infrared light will be lost to the sample. Furthermore, the degree to which a sample is in contact with the crystal directly affects the intensity of the observed bands. This is because shorter wavelengths cannot penetrate as deeply into the sample. Consequently, the effect on intensity is greatest between 2800 and 4000 cm^{-1}. Importantly, signals from C—H, O—H and N—H vibrations are represented in this region and consequently, may go unnoticed if insufficient pressure is applied. Furthermore, it is important to apply sufficient pressure to prevent air becoming trapped between the sample and the crystal, thereby ensuring that the evanescent wave will travel through the sample and not any trapped air. For this reason, the instrumental software typically displays a pressure gauge on the user interface which provides feedback as to the amount of clamp pressure being applied. The clamp can then be adjusted until the indicated pressure is within the recommended range to produce an adequate spectrum. However, handling and clamping of microplastics which are smaller than 500 μm can be difficult and consequently the resultant spectra is not always reliable for microplastics of this size.[448]

Thus, for microplastics smaller than 500 μm, an FTIR microscope can be utilised and used in various modes, such as transmittance, reflectance or ATR. This allows the collection of a spectrum from a sample, as well as mapping of the sample and even simultaneous visualisation.[355] Incidentally, ATR-FTIR microscopy was found to have an advantage over FTIR microscopy in reflectance mode because it was able to identify absorbance bands between 750 and 700 cm^{-1} in the fingerprint region (1450–600 cm^{-1}) which correspond to the scissoring of C—H bonds.[178] However, very small microfibres are commonly lost during sample preparation for analysis by ATR-FTIR microscopy,[448] where physical clamping of the samples can be difficult. Incidentally, the fingerprint region of the spectrum is an area of the spectrum between 1450 and 600 cm^{-1} in which it is difficult to assign all the absorption bands to molecular bonds due to the complex and unique nature of the peaks in this region, hence the fingerprint analogy. Thus, the identification of the scissoring of C—H bonds in this region by ATR-FTIR may be particularly useful in distinguishing between different types of microplastics, such as polyethylene and polyvinyl chloride. Nevertheless, in reflectance mode, the FTIR microscope is capable of molecular mapping analysis which is useful for detecting microplastics in sediments without the need for visual identification. However, due to refractive errors, this has proven extremely difficult with irregularly shaped microplastics.[178] Consequently, the FTIR microscope has to be used in ATR mode to identify irregularly shaped microplastics, such as microfragments (Fig. 10.14).

More recently, FTIR microscopes equipped with a focal plane array (FPA) detector have been assessed for the identification of microplastics.[263] The FPA detector allows a large sample area to be measured with a high degree of lateral resolution in a short amount of time, typically within minutes. The FPA detector typically consists of a 128 × 128 array of pixels with each pixel capable of producing an individual spectrum. Thus, over 16,000 individual spectrums of the entire sample area can be rapidly acquired simultaneously. As such, FTIR microscopy is also considered to be an imaging technique that provides information about the molecular composition of the sample area. However, FTIR imaging is performed in transmission mode and due to the attributes of a suitable filter, the FTIR microscope has generally only been capable of imaging microplastics in a limited spatial frequency range (3800–1250 cm^{-1}), thereby greatly restricting polymer identification. However, this restriction was overcome recently with the use of a novel silicon filter substrate which possessed sufficient transparency to mid-infrared spatial frequencies in the range (4000–600 cm^{-1}), thereby allowing successful analysis of samples for microplastics.[216] Importantly, when reporting the use of this technique, the number of transmission mode scans and the resolution should be quoted.

Once an IR spectrum has been obtained, it can be compared with an electronic database of reference spectrums to identify the type of plastic. However, there will often be minor differences between different spectra for the same type of plastic, which is typically the result of impurities in the plastic material or the scanning of a sample which may not be sufficiently free from water. Thus, samples need to be sufficiently dry, otherwise a very large O—H stretching signal will be apparent in the spectra at around 3300 cm^{-1}, which

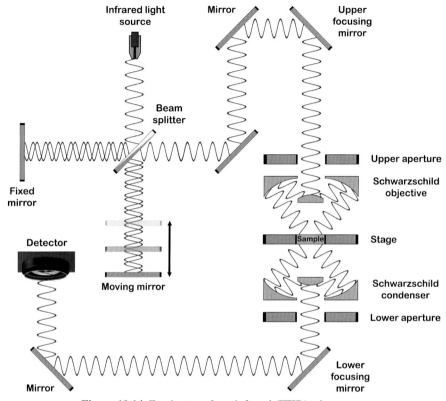

Figure 10.14 Fourier-transform infrared (FTIR) microscopy.

may mask other peaks. Furthermore, a carbon dioxide (CO_2) adsorption signal may occasionally be present at around 2200–2400 cm^{-1}, as a result of atmospheric CO_2.[140] As such, it is important to run a background scan with no sample present that the instrument can subtract any atmospheric water and CO_2 signals. Additional interferences can arise from weathered plastics which may result in poor quality signals in the spectrum and complicate identification by electronic databases.[389]

As a result of the possible inconsistencies in infrared spectra, it has been recommended that spectra compared using an electronic database should only be accepted if the match to a reference spectrum is greater than 70% similarity, with some researchers even suggesting 90% similarity.[336] However, if the similarity between the acquired spectrum and reference spectra is 60–70%, then further manual interpretation of the spectrum is required. Any spectra with less than 60% similarity should be automatically rejected.[140,147] Nevertheless, it is highly recommended to inspect each spectrum manually for clear evidence of characteristic peaks which correspond to the known functional groups of plastic materials as often, there are mismatches with spectra compared using electronic databases as a result of degradation, biofouling, laboratory contamination or the surface adsorption of natural organic or inorganic substances.[140] Thus, it is a good practice for laboratories in the field of microplastic research to consult a list of reference spectra of common plastics for manual interpretation and comparison. Consequently, a reference FTIR spectra for common plastic materials is included at the end of this section.

While the infrared spectrum for each type of plastic is different, there are characteristic absorption bands which are seen in infrared spectrums that correspond to the molecular bending and stretching of specific types of functional group. Thus, assigning these observed absorption bands to specific functional groups can help to identify the type of plastic in question. For example, in the case of polyethylene, a strong intensity double-peaked signal is present in the spectrum around $2800 \, cm^{-1}$, which represents a C—H stretch. However, in the case of polyethylene terephthalate, only a very weak intensity signal is seen in the same region but a strong intensity signal is seen around $1750 \, cm^{-1}$, which corresponds to a carbonyl stretch (C=O). Furthermore, polypropylene exhibits two medium intensity signals at around 1460 and $1500 \, cm^{-1}$. However, polyethylene only exhibits a single medium intensity signal at around $1500 \, cm^{-1}$, thereby allowing polyethylene and polypropylene to be distinguished from one another. The characteristic infrared absorption bands for different bonds are shown in Fig. 10.15, as well as a

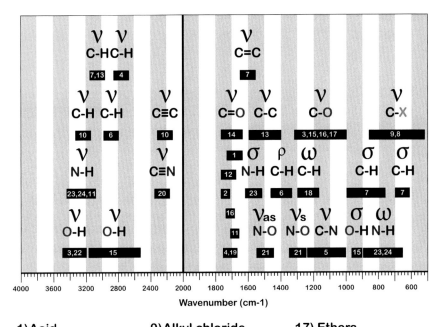

1) Acid	9) Alkyl chloride	17) Ethers
2) Acyl chloride	10) Alkyne	18) Haloalkane
3) Alcohols	11) Amide	19) Ketone
4) Aldehyde	12) Anhydride	20) Nitrile
5) Aliphatic amines	13) Aromatics	21) Nitro
6) Alkane	14) Carbonyls	22) Phenols
7) Alkene	15) Carboxylic acids	23) Primary amine
8) Alkyl bromide	16) Esters	24) Secondary amine

Figure 10.15 Characteristic infrared absorption bands.

protocol for FTIR analysis in Fig. 10.16. Ultimately, FTIR spectroscopy equipment is expensive and the procedure is time-consuming and requires a trained operator.[355] However, the technique is a suitable method for the identification of plastic materials and remains as the most popular choice for the identification of microplastics which have been separated from environmental samples.

- Prior to the analysis of a sample by ATR-FTIR spectroscopy, it is important to run a background scan with no sample present to allow normalisation whereby the instrument will subtract any background interference signals from atmospheric water and carbon dioxide (CO_2).

- The parameters of the scan are then set in the instrument menu, including the number of scans for each sample, such as 128[140] and normally no less than 20 scans. The wavenumber range to scan is generally set to 4000–600 cm^{-1}, which includes the fingerprint region (1450–600 cm^{-1}) as signals in this region are unique to the analysed sample.

- The background spectrum is then obtained.

- Following this, the sample to be analysed is placed onto the ATR crystal window and secured with the overhead clamp by applying sufficient clamping pressure. Care should be taken not to apply too much pressure as smaller microplastics may be squeezed out from underneath the clamp and lost, or even crushed and destroyed. At the same time, insufficient pressure may result in insufficient contact of the sample with the ATR crystal, thereby hindering propagation of the evanescent wave into the sample. Thus, the pressure gauge on the user interface screen should be monitored during clamping to apply the correct pressure.

- A final visual check of the sample is undertaken prior to scanning to make sure it is in sufficient contact and correctly placed over the ATR crystal to ensure sufficient exposure to the infrared light.

- The scan button is then activated and the instrument produces an ATR-FTIR spectrum of the sample. This spectrum is then compared to a selected electronic database and a percentage similarity generated. The spectrum can also be manually interpreted and compared to a consulted list of reference spectrums and while it is good practice to manually check all spectra, this is especially important if the database match is only 60–70% similarity. Any spectra with less than 60% similarity should be automatically rejected.

- For storage, the produced spectra can be stored on the instruments computer or printed out and saved for manual interpretation and comparison.

Figure 10.16 A general protocol for the analysis of microplastics by ATR-FTIR.

Fourier-transform infrared (FTIR) spectroscopy reference spectra for plastic materials

Polyethylene terephthalate (Fig. 10.17)

PET

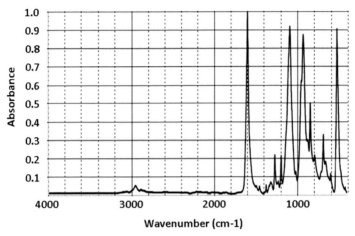

Figure 10.17 The FTIR spectrum of polyethylene terephthalate (PET).

Polyethylene (Fig. 10.18)

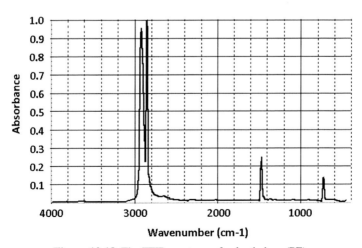

Figure 10.18 The FTIR spectrum of polyethylene (PE).

Polyvinyl chloride (Fig. 10.19)

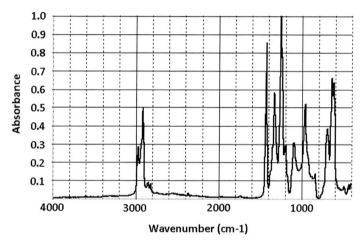

Figure 10.19 The FTIR spectrum of polyvinyl chloride (PVC).

Polypropylene (Fig. 10.20)

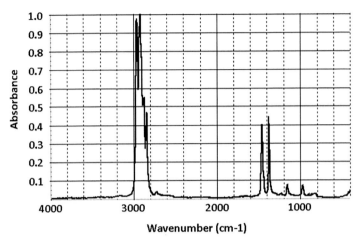

Figure 10.20 The FTIR spectrum of polypropylene (PP).

Polystyrene (Fig. 10.21)

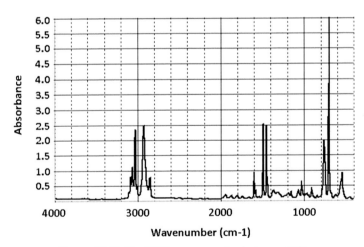

Figure 10.21 The FTIR spectrum of polystyrene (PS).

Poly(methyl methacrylate) (Fig. 10.22)

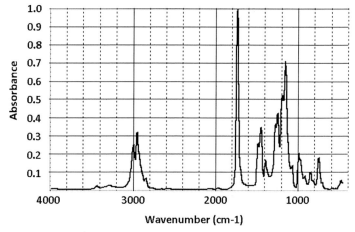

Figure 10.22 The FTIR spectrum of poly(methyl methacrylate).

Polytetrafluoroethylene (Fig. 10.23)

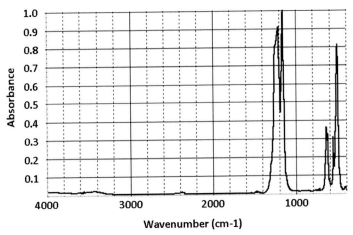

Figure 10.23 The FTIR spectrum of polytetrafluoroethylene (PTFE).

Polyamide (nylon) (Fig. 10.24)

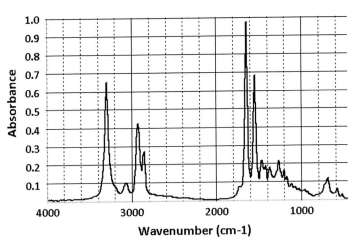

Figure 10.24 The FTIR spectrum of polyamide (nylon).

Near-infrared (NIR) and short-wavelength infrared (SWIR) spectroscopy

Plastics may also be differentiated by irradiating them with light in the near-infrared (NIR) and short-wavelength infrared (SWIR) region (750–3000 nm) of the electromagnetic spectrum (see Fig. 10.10). Upon exposure to NIR light, the constituent molecules of a plastic material absorb this electromagnetic radiation to produce molecular overtone and combination vibrations. Consequently, plastic materials can be identified by differences in the characteristic C—H, N—H and C—O bands typically observed with these types of materials. While the technique is advantageous in that NIR spectroscopy can penetrate far deeper into the plastic material than FTIR spectroscopy, the technique is not particularly sensitive. However, it is useful in examining bulk samples of plastics without having to carry out sample preparation, to rapidly identify the types of plastics present (see Fig. 10.25).

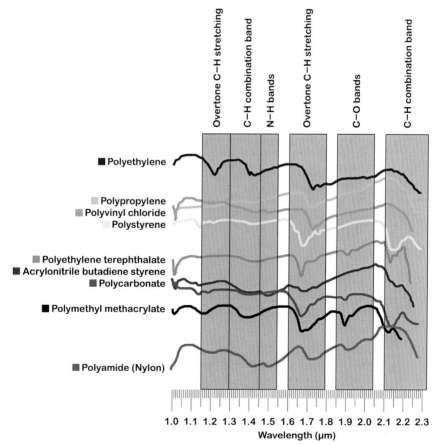

Figure 10.25 A comparison of the near-infrared (NIR) and short-wavelength infrared (SWIR) spectra between different common plastic materials.

Raman spectroscopy

The analysis of plastic materials using Raman spectroscopy was first published[385] by R. Signer and J. Weiler in 1932, in which they obtained a Raman spectrum for polystyrene. Since then, a large amount of literature has been published concerning the use of Raman spectroscopy to analyse plastics, especially since the arrival of laser Raman spectroscopy. Thus, the identification of microplastics can be undertaken by directing a beam of monochromatic laser light (incident beam) onto the sample in question which results in some of the light becoming absorbed, reflected or scattered (see Fig. 10.26).

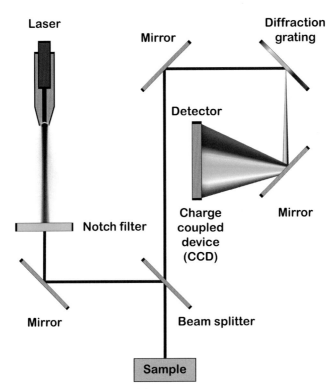

Figure 10.26 Raman spectroscopy.

It is the scattered light which is of interest in Raman spectroscopy and is the result of photons of light interacting with the molecules of the plastic material. There are two forms of scattered light, Rayleigh scattering (elastic) and the weaker Raman scattering (inelastic). The vast majority of scattered light is in the form of Rayleigh scattering and is the same frequency as the incident light, while on average, only one photon in every 30 million will be inelastically scattered. Thus, the small amount of inelastically scattered light produced means that it can easily be masked by unwanted fluorescence of the sample. The number of inelastically scattered photons is proportional to the size of the molecular bonds. For this reason, a great advantage of Raman spectroscopy over Fourier transform infrared spectroscopy is that water tends to have a negligible effect on the technique since the very small bonds present in a molecule of water will scatter very few photons. With respect to the frequency of the incident beam, Raman inelastically scattered light can have an increased frequency towards the blue end of the electromagnetic spectrum (anti-Stokes shift), or a decreased frequency towards the red end of the electromagnetic spectrum (Stokes shift). This change in frequency equates to the vibrational frequency of a molecular bond.

While Raman scattering can be explained with the classical wave interpretation of light, the quantum particle interpretation of light provides a further explanation by examining the electronic states of a molecule and the transitions between them (see Fig. 10.27). Most scattered light is elastically scattered where a photon in the incident beam is absorbed and excites a molecule in the sample to a virtual energy state. The molecule then instantly relaxes back to the ground vibrational state and emits a photon of the same energy as the photon in the incident beam (Rayleigh elastic scattering). However, occasionally the molecule will relax to a higher vibrational energy level than the ground vibrational state, and the emitted photon will have less energy than the photon in the incident beam (Stokes shifted Raman scattering). Alternatively, the molecule is already in a higher vibrational energy level and the emitted photon will have more energy than the photon in the incident beam (anti-Stokes shifted Raman scattering). However, anti-Stokes scattering is less likely to occur than Stokes scattering since most molecules will tend to be in the ground vibrational state at room temperature.

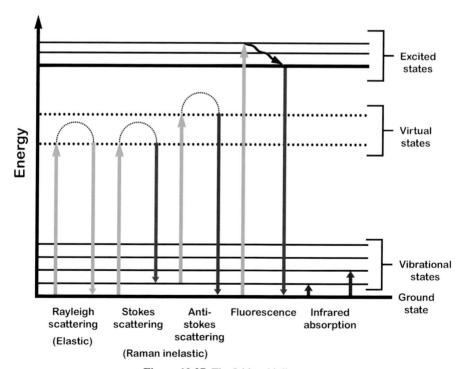

Figure 10.27 The Jablonski diagram.

The energy change in the incident photons is characteristic of the types of bonds that they are interacting with and provides precise information about the molecular structure of the sample. Thus, by collecting the Raman scattered light by a detector perpendicular to the incident beam, a Raman spectrum is produced with characteristic peaks at wavelengths associated with these different vibrations and bonds.[387] The spectrum can then be interpreted and compared to Raman reference spectrums to identify the type of plastic in the sample. Importantly, since Raman is a surface technique, heavily fouled microplastics may need some element of clean-up prior to obtaining a spectrum.

Nevertheless, the analysis of plastics is particularly suited to Raman due to the inherent sensitivity of the method to non-polar molecular species. Thus, the carbon—carbon bonds and double bonds (C—C, C=C) commonly comprising the back-bone of plastic compounds are readily detected by the technique. Indeed, Raman spectroscopy is considered to be complimentary to Fourier transform infrared (FTIR) spectroscopy since Raman exhibits better sensitivity to non-polar symmetrical bonds while FTIR provides better identification of polar groups.[252] Consequently, Raman spectroscopy permits the observation of very slight variations in:

1. the polymers molecular conformity
2. the degree of crystalline regions, with respect to amorphous regions
3. the stereoregularity of the polymer.

For these reasons, Raman is recognised as an excellent choice in the analysis of polymer morphology and provides valuable information about orientation effects and the crystalline structure of the polymer.[308] Nevertheless, Raman tends to be the second choice in polymer identification after FTIR. This is mainly a result of issues with sample fluorescence. Most plastic materials are rarely the pure polymer and are typically of impure composition as a result of the incorporation of a wide variety of additives and colouring pigments during manufacture (see Chapter 4). Occasionally, these impurities absorb light from the excitation laser thereby generating heat in the process. This can annihilate the Raman signal and induce thermal degradation of the sample. Indeed, a study[252] observed that additives incorporated into plastics during manufacture, such as fillers and colourants, can considerably alter the Raman spectrum by introducing the overlay of foreign bands and influencing absorption and fluorescence. Furthermore, Raman spectra obtained after the exposure of plastic materials to ultraviolet radiation may result in a lower intensity signal of characteristic bands. As such, this can lead to the misidentification of microplastics, as well as difficulties in obtaining satisfactory spectra. Nevertheless, the advent of 1064 nm Fourier transform (FT) excitation utilising neodymium-doped yttrium aluminium garnet (Nd:YAG) lasers, as well as ultraviolet (UV) dispersive excitation and near-infrared (NIR) visible, has enabled these difficulties to be largely overcome.

Like FTIR spectroscopy, Raman spectroscopy is a non-destructive technique that does not affect the sample. Thus, further analysis can be undertaken following identification of the microplastic, such as the extraction of any sorbed persistent organic pollutants (POPs) for identification and quantification via gas chromatography–mass spectrometry (GC–MS). However, unlike the transmission and reflectance methods utilised in infrared spectroscopy, Raman is a scattering technique. This is advantageous over infrared spectroscopy in that thicker and strongly absorbing microplastics can be analysed. Furthermore, in comparison to FTIR, microplastics of a very small size can be analysed by Raman spectroscopy and a wider range of infrared wavelengths can be utilised for analysis of the sample.[252]

Raman spectroscopy can also be coupled to microscopy (Raman microspectroscopy) to identify microplastics as small as 1 μm.[64] Thus, the great advantage of Raman spectroscopy is the ability to provide structural and chemical characterisation of samples as small as 1 μm, which other spectroscopic techniques cannot achieve. Furthermore, when combined with Raman mapping, it is possible to analyse the entire field of view at a minimum spatial resolution of 1 μm[263] to acquire spatial chemical images of the entire sample which provide information as to the distribution of components of interest in heterogeneous materials. However, the time taken to acquire such high resolution spatial chemical images may be a limiting factor with respect to high-throughput analysis. Nonetheless, the technique may be useful in the analysis of the distribution of microplastics in a sediment sample. Ultimately, Raman spectroscopy is a straightforward, efficient and reliable technique which requires minimal sample preparation and has been successfully used for the identification of microplastics that have been separated from environmental samples.[64,69,199,216,252,309,424,426]

Raman spectroscopy reference spectra for plastic materials

Polyethylene terephthalate (Fig. 10.28)

PET

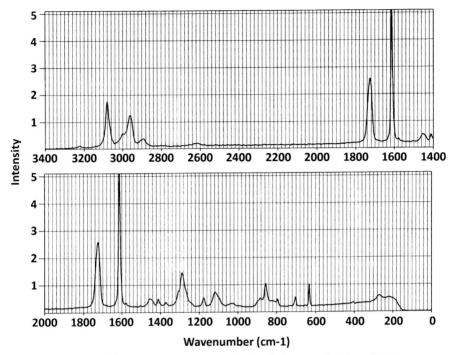

Figure 10.28 The Raman spectrum of polyethylene terephthalate (PET).

Polyethylene (Figs. 10.29 and 10.30)

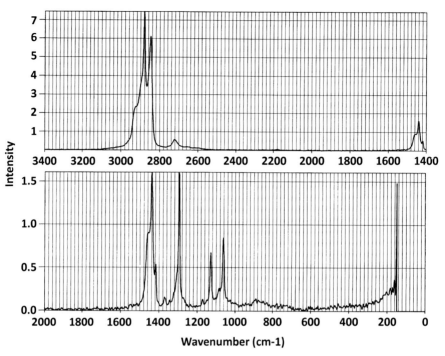

Figure 10.29 The Raman spectrum of low-density polyethylene (PE).

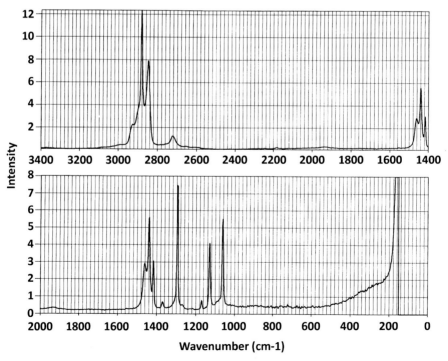

Figure 10.30 The Raman spectrum of high-density polyethylene (PE).

Polyvinyl chloride (Fig. 10.31)

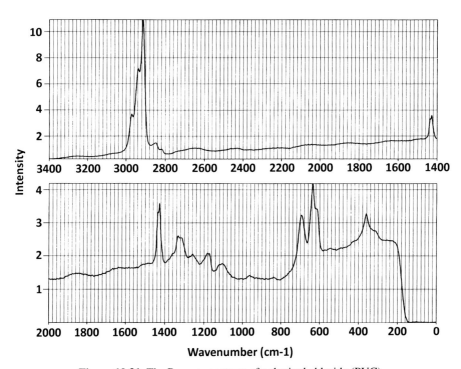

Figure 10.31 The Raman spectrum of polyvinyl chloride (PVC).

Polypropylene (Figs. 10.32 and 10.33)

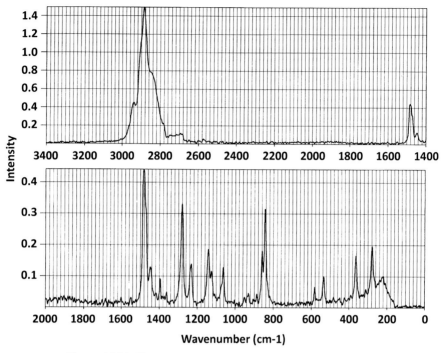

Figure 10.32 The Raman spectrum of atactic polypropylene (PP).

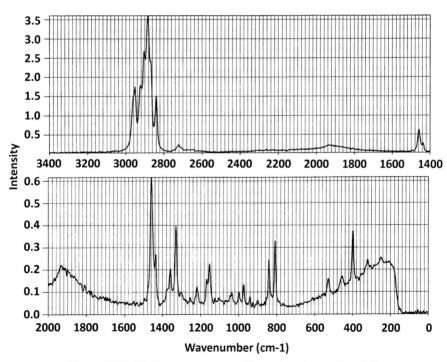

Figure 10.33 The Raman spectrum of isotactic polypropylene (PP).

Polystyrene (Fig. 10.34)

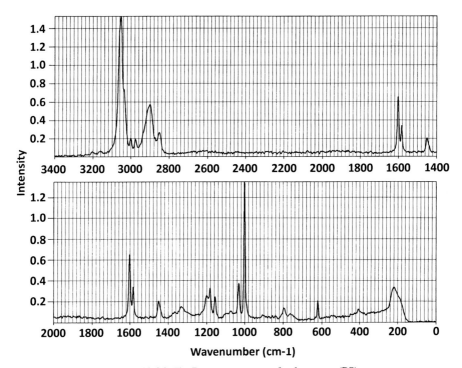

Figure 10.34 The Raman spectrum of polystyrene (PS).

Polytetrafluoroethylene (Fig. 10.35)

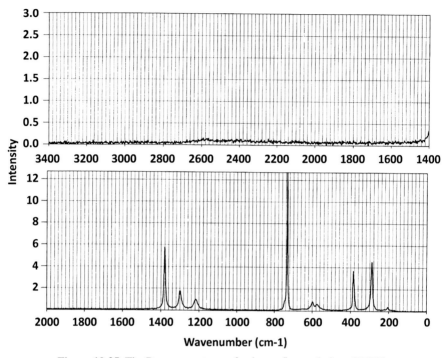

Figure 10.35 The Raman spectrum of polytetrafluoroethylene (PTFE).

Polyamide (nylon) (Figs. 10.36 and 10.37)

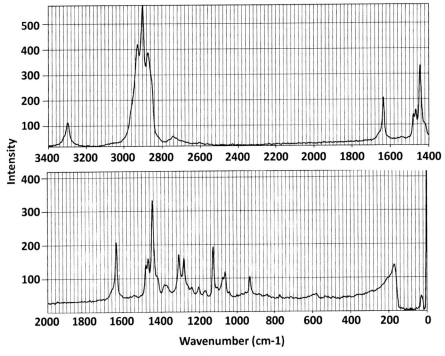

Figure 10.36 The Raman spectrum of nylon 6.

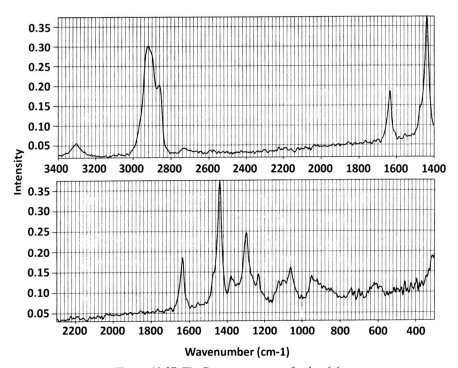

Figure 10.37 The Raman spectrum of nylon 6,6.

Future analysis

There are two predominant difficulties which require to be overcome in identifying small pieces of plastic. The first is the time and resources required to identify microplastics separated from environmental water or sediment samples, while the second is the challenge in identifying plasticles between 20 nm and 10 µm in size.[156] Thus, in terms of identifying very small plasticles, highly efficient analytical techniques need to be developed to effectively extract these small items from environmental samples. At the present time, no validated methods yet exist. However, it is likely that in due course, Raman microspectroscopy mapping techniques will be increasingly utilised, especially where speeds are sufficiently fast enough to map the entire area of a heterogeneous sample with high spatial resolution as part of a high-throughput method. Furthermore, continual spectroscopic analysis of water flowing through a portable Raman microspectrometer may facilitate the rapid detection of microplastics in water without the need for a sample filtration step. Indeed, implementation of this technique may enable the direct onsite analysis of the horizontal and vertical profiles of microplastics in marine and freshwater aquatic environments in real time. Ultimately, while microplastic research is a newly developing field, it is a rapidly expanding area of scientific research that is at the forefront in understanding the complex interactions of microplastics in the environment. Thus, through the continued study and research in the field, important and advanced breakthroughs will be made to recover, analyse, identify, monitor and study microplastic pollutants.

References

[1] Abu-Hilal AH, Al-Najjar TH. Plastic pellets on the beaches of the northern Gulf of Aqaba, Red Sea. Aquatic Ecosystem Health & Management 2009;12:461–70.

[2] Ahmed AS, Webster L, Pollard P, Davies IM, Russell M, Walsham P, Packer G, Moffat CF. The distribution and composition of hydrocarbons in sediments from the Fladen Ground, North Sea, an area of oil production. Journal of Environmental Monitoring 2006;8:307–16.

[3] Aliani S, Griffa A, Molcard A. Floating debris in the Ligurian Sea, north-western Mediterranean. Marine Pollution Bulletin 2003;46:1142–9.

[4] Alomar C, Estarellas F, Deudero S. Microplastics in the Mediterranean Sea: deposition in coastal shallow sediments, spatial variation and preferential grain size. Marine Environmental Research 2016;115:1–10.

[5] Alter H. The recovery of plastics from waste with reference to froth flotation. Resource Conservation and Recycling 2005;43:119–32.

[6] Anastasopoulou A, Mytilineou C, Smith CJ, Papadopoulou KN. Plastic debris ingested by deep-water fish of the Ionian Sea (Eastern Mediterranean). Deep Sea Research Part I: Oceanographic Research Papers 2013;74:11–3.

[7] Andrady AL. Microplastics in the marine environment. Marine Pollution Bulletin 2011;62:1596–605.

[8] Andrady AL. Plastics and the environment. New Jersey: John Wiley & Sons, Inc.; 2003.

[9] Andrady AL. Plastics and environmental sustainability. New Jersey: John Wiley & Sons, Inc.; 2015.

[10] Anthony SD, Meizhong L, Christopher EB, Robin LB, David LF. Involvement of linear plasmids in aerobic biodegradation of vinyl chloride. Applied and Environmental Microbiology 2004;70:6092–7.

[11] Apul OG, Karanfil T. Adsorption of synthetic organic contaminants by carbon nanotubes: a critical review. Water Research 2015;68:34–55.

[12] Arthur C, Baker J, Bamford H. In: Proceedings of the international research workshop on the occurrence, effects and fate of microplastic marine debris. NOAA Technical Memorandum NOS-OR&R-30; 2009. p. 49.

[13] Ascer L, Custódio M, Turra A. Morphological changes in polyethylene abrasives of Brazilian cosmetics caused by mechanical stress. In: Fate and impact of microplastics in marine ecosystems. Micro International Workshop 2014. Plouzané, France. 2014.

[14] Avio CG, Gorbi S, Milan M, Benedetti M, Fattorini D, d'Errico G, Pauletto M, Bargelloni L, Regoli F. Pollutants bioavailability and toxicological risk from microplastics to marine mussels. Environmental Pollution 2015;198:211–22.

[15] Avio CG, Gorbi S, Regoli F. Experimental development of a new protocol for extraction and characterization of microplastics in fish tissues: first observations in commercial species from Adriatic Sea. Marine Environmental Research 2015;111:18–26.

[16] Bakir A, Rowland SJ, Thompson RC. Competitive sorption of persistent organic pollutants onto microplastics in the marine environment. Marine Pollution Bulletin 2012;64:2782–9.

[17] Bakir A, Rowland SJ, Thompson RC. Transport of persistent organic pollutants by microplastics in estuarine conditions. Estuarine, Coastal and Shelf Science 2014;140:14–21.

[18] Ballachey BE, Bodkin JL. Challenges to sea otter recovery and conservation. In: Larson SE, Bodkin JL, VanBlaricom GR, editors. Sea otter conservation. 2014 Dec 23. p. 63–88.

[19] Barbes L, Rădulescu C, Stihi C. ATR-FTIR spectrometry characterisation of polymeric materials. Romanian Reports in Physics 2014;66:765–77.

[20] Barnes DK. Biodiversity: invasions by marine life on plastic debris. Nature 2002; 416(6883):808–9.

[21] Barnes DKA, Milner P. Drifting plastic and its consequences for sessile organism dispersal in the Atlantic Ocean. Marine Biology 2005;146(4):815–25.

[22] Bejgarn S, MacLeod M, Bogdal C, Breitholtz M. Toxicity of leachate from weathering plastics: an exploratory screening study with *Nitocra spinipes*. Chemosphere 2015;132:114–9.

[23] Béland P, Deguise S, Girard C, Lagace A, Martineau D, Michaud R, Muir DCG, Norstrom RJ, Pelletier E, Ray S, Shugart LR. Toxic compounds and health and reproductive effects in St. Lawrence beluga whales. Journal of Great Lakes Research 1993;19:766–75.

[24] Bendell LI. Favored use of anti-predator netting (APN) applied for the farming of clams leads to little benefits to industry while increasing nearshore impacts and plastics pollution. Marine Pollution Bulletin 2015;91(1):22–8.

[25] Bendell-Young LI, Arifin Z. Application of a kinetic model to demonstrate how selective feeding could alter the amount of cadmium accumulated by the blue mussel (*Mytilus trossulus*). Journal of Experimental Marine Biology and Ecology 2004;298:21–33.

[26] Bergmann M, Klages M. Increase of litter at the Arctic deep-sea observatory HAUSGARTEN. Marine Pollution Bulletin 2012;64(12):2734–41.

[27] Besseling E, Foekema EM, Van Franeker JA, Leopold MF, Kühn S, Rebolledo EB, Heße E, Mielke L, IJzer J, Kamminga P, Koelmans AA. Microplastic in a macro filter feeder: humpback whale *Megaptera novaeangliae*. Marine Pollution Bulletin 2015;95(1):248–52.

[28] Besseling E, Wegner A, Foekema EM, Martine JH, Koelmans AA. Effects of microplastic on fitness and PCB bioaccumulation by the lugworm *Arenicola marina* (L.). Environmental Science and Technology 2013;47:593–600.

[29] Bergami E, Bocci E, Vannuccini ML, Monopoli M, Salvati A, Dawson KA, Corsi I. Nano-sized polystyrene affects feeding, behavior and physiology of brine shrimp Artemia franciscana larvae. Ecotoxicology and Environmental Safety 2016 Jan 31;123:18–25.

[30] Boerger CM, Lattin GL, Moore SL, Moore CJ. Plastic ingestion by planktivorous fishes in the North Pacific Central Gyre. Marine Pollution Bulletin 2010;60:2275–8.

[31] Bornehag CG, Sundell J, Weschler CJ, Sigsgaard T, Lundgren B, Hasselgren M, Hägerhed-Engman L. The association between asthma and allergic symptoms in children and phthalates in house dust: a nested case-control study. Environmental Health Perspectives 2004:1393–7.

[32] Boucher C, Morin M, Bendell LI. The influence of cosmetic microbeads on the sorptive behaviour of cadmium and lead within intertidal sediments: a laboratory study. Regional Studies in Marine Science 2016;3:1–7.

[33] Bouwman H, Evans SW, Cole N, Yive NSCK. The flip-or-flop boutique: Marine debris on the shores of St Brandon's rock, an isolated tropical atoll in the Indian Ocean. Marine Environmental Research 2016;114:58–64.

[34] Breivik K, Alcock R, Li Y, Bailey RE, Fiedler H, Pacyna JM. Primary sources of selected POPs: regional and global scale emission inventories. Environmental Pollution 2004;128:3–16.

[35] Breivik K, Sweetman A, Pacyna JM, Jones KC. Towards a global historical emission inventory for selected PCB congeners — a mass balance approach: 2. Emissions. The Science of the Total Environment 2002;290:199–224.

[36] Brennecke D, Ferreira EC, Costa TM, Appel D, da Gama BA, Lenz M. Ingested microplastics (>100 µm) are translocated to organs of the tropical fiddler crab *Uca rapax*. Marine Pollution Bulletin 2015;96(1):491–5.

[37] Brooks SJ, Farmen E, Heier LS, Blanco-Rayón E, Izagirre U. Differences in copper bioaccumulation and biological responses in three *Mytilus* species. Aquatic Toxicology 2015;160:1–12.

[38] Browne MA, Dissanayake A, Galloway TS, Lowe DM, Thompson RC. Accumulations of microplastic on shorelines worldwide: sources and sinks. Environmental Science and Technology 2011;45:9175–9.

[39] Browne MA, Dissanayake A, Galloway TS, Lowe DM, Thompson RC. Ingested microscopic plastic translocates to the circulatory system of the mussel, *Mytilus edulis* (L.). Environmental Science and Technology 2008;42(13):5026–31.

[40] Browne MA, Galloway TS, Thompson RC. Microplastic – an emerging contaminant of potential concern. Integrated Environmental Assessment and Management 2007;3:559–66.

[41] Browne MA, Galloway TS, Thompson RC. Spatial patterns of plastic debris along estuarine shorelines. Environmental Science and Technology 2010;44:3404–9.

[42] Browne MA, Niven SJ, Galloway TS, Rowland SJ, Thompson RC. Microplastic moves pollutants and additives to worms, reducing functions linked to health and biodiversity. Current Biology 2013;23(23):2388–92.

[43] Cacciari I, Quatrini P, Zirletta G, Mincione E, Vinciguerra V, Lupattelli P, Sermanni GG. Isotactic polypropylene biodegradation by a microbial community: physicochemical characterization of metabolites produced. Applied and Environmental Microbiology 1993;59(11):3695–700.

[44] Campagna C, Piola AR, Marin MR, Lewis M, Zajaczkovski U, Fernández T. Deep divers in shallow seas: southern elephant seals on the Patagonian shelf. Deep Sea Research Part 1: Oceanographic Research Papers 2007;54:1792–814.

[45] Carlos de Sá L, Luís LG, Guilhermino L. Effects of microplastics on juveniles of the common goby (*Pomatoschistus microps*): confusion with prey, reduction of the predatory performance and efficiency, and possible influence of developmental conditions. Environmental Pollution 2015;196:359–62.

[46] Carman VG, Machain N, Campagna C. Legal and institutional tools to mitigate plastic pollution affecting marine species: Argentina as a case study. Marine Pollution Bulletin 2015;92(1):125–33.

[47] Carpenter EJ, Anderson SJ, Harvey GR, Miklas HP, Peck BB. Polystyrene spherules in coastal waters. Science 1972;178:749–50.

[48] Carpenter EJ, Smith KL. Plastics on the Sargasso Sea surface. Science 1972;175:1240–1.

[49] Carr SA, Liu J, Tesoro AG. Transport and fate of microplastic particles in wastewater treatment plants. Water Research 2016;91:174–82.

[50] Carson HS. The incidence of plastic ingestion by fishes: from the prey's perspective. Marine Pollution Bulletin 2013;74:170–4.

[51] Carson HS, Colbert SL, Kaylor MJ, Mcdermid KJ. Small plastic debris changes water movement and heat transfer through beach sediments. Marine Pollution Bulletin 2011;62:1708–13.

[52] Castañeda RA, Avlijas S, Simard MA, Ricciardi A. Microplastic pollution in St. Lawrence river sediments. Canadian Journal of Fisheries and Aquatic Sciences 2014;71(12):1767–71.

[53] Cauwenberghe LV, Claessens M, Vandegehuchte MB, Janssen CR. Microplastics are taken up by mussels (*Mytilus edulis*) and lugworms (*Arenicola marina*) living in natural habitats. Environmental Pollution 2015;199:10–7.

[54] Cauwenberghe LV, Janssen CR. Microplastics in bivalves cultured for human consumption. Environmental Pollution 2014;193:65–70.

[55] Chapman PM, Wang F. Assessing sediment contamination in estuaries. Environmental Toxicology & Chemistry 2001;20:3–22.

[56] Chen CL. Regulation and management of marine litter. In: Bergmann M, Gutow L, Klages M, editors. Marine anthropogenic litter. Berlin: Springer; 2015.

[57] Cheng X, Shi H, Adams CD, Ma Y. Assessment of metal contaminations leaching out from recycling plastic bottles upon treatments. Environmental Science and Pollution Research 2010;17(7):1323–30.

[58] Cherbut C, Ruckebusch Y. The effect of indigestible particles on digestive transit time and colonic motility in dogs and pigs. British Journal of Nutrition 1985;53:549–57.

[59] Cherel Y, Guinet C, Tremblay Y. Fish prey of Antarctic fur seals *Arctocephalus gazelle* at IIe de Croy, Kerguelen. Polar Biology 1997;17:87–90.

[60] Chua EM, Shimenta J, Nugegoda D, Morrison PD, Clarke BO. Assimilation of polybrominated diphenyl ethers from microplastics by the marine amphipod, *Allorchestes Compressa*. Environmental Science and Technology 2014;48:8127–34.

[61] Claessens M, Meester SD, Landuyt LV, Clerck KD, Janssen CR. Occurrence and distribution of microplastics in marine sediments along the Belgian coast. Marine Pollution Bulletin 2011;62:2199–204.

[62] Claessens M, Cauwenberghe LV, Vandegehuchte MB, Janssen CR. New techniques for the detection of microplastics in sediments and field collected organisms. Marine Pollution Bulletin 2013;70:227–33.

[63] Cole G, Sherrington C. Study to quantify pellet emissions in the UK. 2016. Report for Fidra.

[64] Cole M, Lindeque P, Fileman E, Halsband C, Goodhead R, Moger J, Galloway TS. Microplastic ingestion by zooplankton. Environmental Science and Technology 2013;47:6646–55.

[65] Cole M, Lindeque P, Halsband C, Galloway TS. Microplastics as contaminants in the marine environment: a review. Marine Pollution Bulletin 2011;62:2588–97.

[66] Cole M, Webb H, Lindeque PK, Fileman ES, Halsband C, Galloway TS. Isolation of microplastics in biota-rich seawater samples and marine organisms. Scientific Reports 2014;4(4528):1–8.

[67] Coleman FC, Wehle DHS. Plastic pollution: a worldwide problem. Parks 1984;9:9–12.

[68] Colborn RE, Buckley DJ, Adams ME. Acrylonitrile-butadiene-styrene. Shropshire, United Kingdom: Rapra Technology Ltd; 1997.

[69] Collard F, Gilbert B, Eppe G, Parmentier E, Das K. Detection of anthropogenic particles in fish stomachs: an isolation method adapted to identification by Raman spectroscopy. Archives of Environmental Contamination and Toxicology 2015;69:331–9.

[70] Collignon A, Hecq J, Galgani F, Collard F, Goffart A. Annual variation in neustonic micro- and meso-plastic particles and zooplankton in the Bay of Calvi (Mediterranean–Corsica). Marine Pollution Bulletin 2014;79:293–8.

[71] Collignon A, Hecq JH, Glagani F, Voisin P, Collard F, Goffart A. Neustonic microplastic and zooplankton in the North Western Mediterranean Sea. Marine Pollution Bulletin 2012;64:861–4.

[72] Colton JB, Knapp FD, Burns BR. Plastic particles in surface waters of the Northwestern Atlantic. Science 1974;185:491–7.

[73] Cooper DA, Corcoran PL. Effects of mechanical and chemical processes on the degradation of plastic beach debris on the island of Kauai, Hawaii. Marine Pollution Bulletin 2010;60:650–4.

[74] Corcoran PL, Biesinger MC, Grifi M. Plastics and beaches: a degrading relationship. Marine Pollution Bulletin 2009;58:80–4.

[75] Corcoran PL, Norris T, Ceccanese T, Walzak MJ, Helm PA, Marvin CH. Hidden plastics of Lake Ontario, Canada and their potential preservation in the sediment record. Environmental Pollution 2015;204:17–25.

[76] Costa DP, Kuhn CE, Weise MJ, Shaffer SA, Arnould JPY. When does physiology limit the foraging behaviour of freely diving mammals? International Congress Series 2004;1275:359–66.

[77] Cózar A, Echevarría F, González-Gordillo JI, Irigoien X, Ubeda B, Hernández-León S, et al. Plastic debris in the open ocean. Proceedings of the National Academy of Sciences 2014;111(28):10239–44.

[78] Ferreira P, Fonte E, Soares ME, Carvalho F, Guilhermino L. Effects of multi-stressors on juveniles of the marine fish Pomatoschistus microps: Gold nanoparticles, microplastics and temperature. Aquatic Toxicology 2016 Jan 31;170:89–103.

[79] Cózar A, Sanz-Martín M, Martí E, González-Gordillo JI, Ubeda B, Gálvez JÁ, et al. Plastic accumulation in the Mediterranean Sea. PLoS One 2015;10(4):e0121762. http://dx.doi.org/10.1371/journal.pone.0121762.

[80] Frias JP, Sobral P, Ferreira AM. Organic pollutants in microplastics from two beaches of the Portuguese coast. Marine Pollution Bulletin 2010 Nov 30;60(11):1988–92.

[81] Crise A, Kaberi H, Ruiz J, Zatsepin A, Arashkevich E, Giani M, Karageorgis AP, Prieto L, Pantazi M, Gonzalez-Fernandez D, d'Alcalà MR. A MSFD complementary approach for the assessment of pressures, knowledge and data gaps in Southern European Seas: the PERSEUS experience. Marine Pollution Bulletin 2015;95(1):28–39.

[82] Green DS, Boots B, Sigwart J, Jiang S, Rocha C. Effects of conventional and biodegradable microplastics on a marine ecosystem engineer (Arenicola marina) and sediment nutrient cycling. Environmental Pollution 2016 Jan 31;208:426–34.

[83] Daly GL, Wania F. Organic contaminants in mountains. Environmental Science and Technology 2005;39:385–98.

[84] Daniels CA. Polymers: structure and properties. Technomic Publishing Company, Inc.; 1989.

[85] Davis A, Sims D. Weathering of polymers. United Kingdom: Applied Science Publishers Ltd; 1983 Nov 30.

[86] Davis ME, Zuckerman JE, Choi CHJ, Seligson D, Tolcher A, Alabi CA, Yen Y, Heidel JD, Ribas A. Evidence of RNAi in humans from systemically administered siRNA via targeted nanoparticles. Nature 2010;464(7291):1067–70.

[87] Davison P, Asch RG. Plastic ingestion by mesopelagic fishes in the North Pacific Subtropical Gyre. Marine Ecology Progress Series 2011;432:173–80.

[88] Day RH, Shaw DG, Ignell SE. The quantitative distribution and characteristics of neuston plastic in the North Pacific Ocean, 1984–1988. In: Proceedings of the second international conference on marine debris, April 2–7, 1989. Honolulu, Hawaii. NOAA Technical Memorandum, NOAA-TM-NMFS-SWFSC-154; 1990. p. 182–211.

[89] Karapanagioti HK, Klontza I. Testing phenanthrene distribution properties of virgin plastic pellets and plastic eroded pellets found on Lesvos island beaches (Greece). Marine Environmental Research 2008 May 31;65(4):283–90.

[90] De Guise S, Martineau D, Be'land P, Fournier M. Possible mechanisms of action of environmental contaminants on St. Lawrence beluga whales (Delphinapterus leucas). Environmental Health Perspectives 1995;103:73–7.

[91] De Lucia GA, Caliani I, Marra S, Camedda A, Coppa S, Alcaro L, et al. Amount and distribution of neustonic micro-plastic off the western Sardinian coast (Central-Western Mediterranean Sea). Marine Environmental Research 2014;100:10–6.

[92] Laurier F, Mason R. Mercury concentration and speciation in the coastal and open ocean boundary layer. Journal of Geophysical Research: Atmospheres 2007 Mar 27:112(D6).

[93] De Stephanis R, Giménez J, Carpinelli E, Gutierrez-Exposito C, Cañadas A. As main meal for sperm whales: plastics debris. Marine Pollution Bulletin 2013;69(1):206–14.

[94] De Wael K, Gason FG, Baes CA. Selection of an adhesive tape suitable for forensic fiber sampling. Journal of Forensic Sciences 2008;53(1):168–71.

[95] Deguchi T, Kitaoka Y, Kakezawa M, Nishida T. Purification and characterization of a nylon-degrading enzyme. Applied and Environmental Microbiology 1998;64(4):1366–71.

[96] Dekiff JH, Remy D, Klasmeier J, Fries E. Occurrence and spatial distribution of microplastics in sediments from Norderney. Environmental Pollution 2014; 186:248–56.

[97] Lutz PL. Studies on the ingestion of plastic and latex by sea turtles. In: Proceedings of the Workshop on the Fate and Impact of Marine Debris, Honolulu; 1990. p. 719–35.

[98] Desforges J-PW, Galbraith M, Dangerfield N, Ross PS. Widespread distribution of microplastics in subsurface seawater in the NE Pacific Ocean. Marine Pollution Bulletin 2014;79:94–9.

[99] Dhananjayan V, Muralidharan S. Polycyclic aromatic hydrocarbons in various species of fishes from Mumbai Harbour, India, and their dietary intake concentration to human. International Journal of Oceanography 2012;2012.

[100] Dimarogana M, Nikolaivits E, Kanelli M, Christakopoulos P, Sandgren M, Topakas E. Structural and functional studies of a *Fusarium oxysporum* cutinase with polyethylene terephthalate modification potential. Biochimica et Biophysica Acta (BBA) – General Subjects 2015;1850(11):2308–17.

[101] Dodiuk H, Goodman S. Handbook of thermoset plastics. 3rd ed. Massachusetts, USA: Elsevier Inc.; 2014.

[102] DouAbul AA, Heba HM, Fareed KH. Polynuclear aromatic hydrocarbons (PAHs) in fish from the Red Sea Coast of Yemen. In: Asia-Pacific conference on science and management of coastal environment. Netherlands: Springer; 1997. p. 251–62.

[103] Doyle MJ, Watson W, Bowlin NM, Sheavly SB. Plastic particles in coastal pelagic ecosystems of the Northeast Pacific Ocean. Marine Environmental Research 2011;71:41–52.

[104] Driedger AG, Dürr HH, Mitchell K, Van Cappellen P. Plastic debris in the Laurentian Great Lakes: a review. Journal of Great Lakes Research 2015;41(1):9–19.

[105] Dris R, Gasperi J, Saad M, Mirande C, Tassin B. Synthetic fibres in atmospheric fallout: a source of microplastics in the environment? Marine Pollution Bulletin 2016;104:290–3.

[106] Dubaish F, Liebezeit G. Suspended microplastics and black carbon particles in the Jade system, Southern North sea. Water, Air, & Soil Pollution 2013;224:1–8.

[107] Duis K, Coors A. Microplastics in the aquatic and terrestrial environment: sources (with a specific focus on personal care products), fate and effects. Environmental Sciences Europe 2016;28:1–25.

[108] EC 2013. Green Paper on a European strategy on plastic waste in the environment. Brussels, 7.3.2013; COM(2013) 123 final. 2013. p. 20.

[109] ECE. State of knowledge report of the UN ECE Task Force on persistent organic pollutants: for the convention on long-range transboundary air pollution: for presentation to the meetings of the working groups on technology and on effects 28th June–1st July. Geneva: Department of Indian Affairs and Northern Development Canada. Environmental Services and Research Division; 1994.

[110] Eerkes-Medrano D, Thompson RC, Aldridge DC. Microplastics in freshwater systems: a review of the emerging threats, identification of knowledge gaps and prioritisation of research needs. Water Research 2015. http://dx.doi.org/10.1016/j.watres.2015.02.012.

[111] El-Shahawi MS, Hamza A, Bashammakh AS, Al-Saggaf WT. An overview on the accumulation, distribution, transformations, toxicity and analytical methods for the monitoring of persistent organic pollutants. Talanta 2010;80:1587–97.

[112] Endo S, Takizawa R, Okuda K, Takada H, Chiba K, Kanehiro H, Ogi H, Yamashita R, Date T. Concentration of polychlorinated biphenyls (PCBs) in beached resin pellets: variability among individual particles and regional differences. Marine Pollution Bulletin 2005;50:1103–14.

[113] McCauley SJ, Bjorndal KA. Conservation implications of dietary dilution from debris ingestion: sublethal effects in post-hatchling loggerhead sea turtles. Conservation Biology 1999 Aug 1;13(4):925–9.

[114] Erdman Jr JW, MacDonald IA, Zeisel SH, editors. Present knowledge in nutrition. John Wiley & Sons; 2012.

[115] Eriksen M, Lebreton LCM, Carson HS, Thiel M, Moore CJ, Borerro JC, Galgani F, Ryan PG, Reisser J. Plastic pollution in the world's oceans: more than 5 trillion plastic pieces weighing over 250,000 tons afloat at sea. PLoS One 2014;9:e111913.

[116] Eriksen M, Mason S, Wilson S, Box C, Zellers A, Edwards W, Farley H, Amato S. Microplastic pollution in the surface waters of the Laurentian Great Lakes. Marine Pollution Bulletin 2013;77:177–82.

[117] Eriksen M, Maximenko N, Thiel M, Cummins A, Lattin G, Wilson S, Hafner J, Zellers A, Rifman S. Plastic pollution in the South Pacific Subtropical Gyre. Marine Pollution Bulletin 2013;68:71–6.

[118] Eriksson C, Burton H. Origins and biological accumulation of small plastic particles in fur-seal scats from Macquarie Island. Ambio 2003;32:380–4.

[119] EU. Directive 2008/56/EC of the European Parliament and of the Council of 17 June 2008 establishing a framework for community action in the field of marine environmental policy (Marine Strategy Framework Directive). Brussels: EU; 2008.

[120] European Commission. Guidance on monitoring of marine litter in European Seas. A guidance document within the common implementation strategy for the Marine Strategy Framework Directive. Ispra: European Commission, Joint Research Centre, MSFD Technical Subgroup on Marine Litter; 2013.

[121] European Food Safety Authority (EFSA). Perfluoroalkylated substances in food: occurrence and dietary exposure. EFSA Journal 2012;10:2743–98.

[122] European Food Safety Authority. Results of the monitoring of non-dioxin-like PCBs in food and feed. EFSA Journal 2010;8(7):1701.

[123] Fähnrich KA, Pravda M, Guilbault GG. Immunochemical detection of polycyclic aromatic hydrocarbons (PAHs). Analytical Letters 2002;35(8):1269–300.

[124] Farrel P, Nelson K. Trophic level transfer of microplastic: *Mytilus edulis* (L.) to *Carcinus maenas* (L.). Environmental Pollution 2013;177:1–3.

[125] Faure F, Saini C, Potter G, Galgani F, Alencastro LF, Hagmann P. An evaluation of surface micro- and mesoplastic pollution in pelagic ecosystems of the Western Mediterranean Sea. Environmental Science and Pollution Research 2015;22:12190–7.

[126] Fendall LS, Sewell MA. Contributing to marine pollution by washing your face. Microplastics in facial cleansers. Marine Pollution Bulletin 2009;58:1225–8.

[127] Fiedler H. Polychlorinated biphenyls (PCBs): uses and environmental releases. In: Proceedings of the subregional awareness raising workshop on Persistent Organic Pollutants (POPs). Abu Dhabi: United Arab Emirates; 1998.

[128] Fischer V, Elsner NO, Brenke N, Schwabe E, Brandt A. Plastic pollution of the Kuril–Kamchatka trench area (NW pacific). Deep Sea Research Part II: Topical Studies in Oceanography 2015;111:399–405.

[129] Fisk A, Hobson K, Nortsrom R. Factors on trophic transfer of persistent organic pollutants in the Northwater Polyna marine food web. Environmental Science and technology 2001;35:732–8.

[130] Fisner M, Taniguchi S, Moreira F, Bícego MC, Turra A. Polycyclic aromatic hydrocarbons (PAHs) in plastic pellets: variability in the concentration and composition at different sediment depths in a sandy beach. Marine Pollution Bulletin 2013;70(1):219–26.

[131] Fisner M, Taniguchi S, Majer AP, Bícego MC, Turra A. Concentration and composition of polycyclic aromatic hydrocarbons (PAHs) in plastic pellets: implications for small-scale diagnostic and environmental monitoring. Marine Pollution Bulletin 2013;76(1):349–54.

[132] Foekema EM, De Gruijter C, Mergia MT, Van Franeker JA, Murk AJ, Koelmans AA. Plastic in North Sea fish. Environmental Science & Technology 2013;47:8818–24.

[133] Fok L, Cheung PK. Hong Kong at the Pearl River Estuary: a hotspot of microplastic pollution. Marine Pollution Bulletin 2015;99:112–8.

[134] Fossi MC, Coppola D, Baini M, Giannetti M, Guerranti C, Marsili L, Panti C, Sabtata E, Clò S. Large filter feeding marine organisms as indicators of microplastic in the pelagic environment: the case studies of the Mediterranean basking shark (*Cetorhinus maximus*) and fin whale (*Balaenoptera physalus*). Marine Environmental Research 2014;100:17–24.

[135] Fossi MC, Marsili L, Baini M, Giannetti M, Coppola D, Guerranti C, Caliani I, Minutoli R, Lauriano G, Finoia MG, Rubegni F, Panigada S, Bérubé M, Urbán Ramírez J, Panti C. Fin whales and microplastics: the Mediterranean Sea and the Sea of Cortez scenarios. Environmental Pollution 2016;209:68–78.

[136] Fossi MC, Panti C, Guerranti C, Coppola D, Giannetti M, Marsili L, Minutoli R. Are baleen whales exposed to the threat of microplastics? A case study of the Mediterranean fin whale (*Balaenoptera physalus*). Marine Pollution Bulletin 2012;64:2374–9.

[137] Fraunholcz N. Separation of waste plastics by froth flotation—a review, Part I. Minerals Engineering 2004;17:261–8.

[138] Free CM, Jensen OP, Mason SA, Eriksen M, Williamson NJ, Boldgiv B. High-levels of microplastic pollution in a large, remote, mountain lake. Marine Pollution Bulletin 2014;85:156–63.

[139] Freinkel S. Plastic: a toxic love story. New York: Houghton Mifflin Harcourt; 2011.

[140] Frias JPGL, Gago J, Otero V, Sobral P. Microplastics in coastal sediments from Southern Portuguese shelf waters. Marine Environmental Research 2016;114:24–30.

[141] Frias JPGL, Otero V, Sobral P. Evidence of microplastics in samples of zooplankton from Portuguese coastal waters. Marine Environmental Research 2014;95:89–95.

[142] Fries E, Dekiff JH, Willmeyer J, Nuelle MT, Ebert M, Remy D. Identification of polymer types and additives in marine microplastic particles using pyrolysis-GC/MS and scanning electron microscopy. Environmental Science: Processes & Impacts 2013;15:1949–56.

[143] Fries E, Zarfl C. Sorption of polycyclic aromatic hydrocarbons (PAHs) to low and high density polyethylene (PE). Environmental Science and Pollution Research 2012;19:1296–304.

[144] Fromme H, Küchler T, Otto T, Pilz K, Müller J, Wenzel A. Occurrence of phthalates and bisphenol A and F in the environment. Water Research 2002;36(6):1429–38.

[145] Fukuzaki H, Yoshida M, Asano M, Kumakura M. Synthesis of copoly(D,L-lactic acid) with relative low molecular weight and in vitro degradation. European Polymer Journal 1989;25:1019–26.

[146] Galgani F, Burgeot T, Bocquéné G, Vincent F, Leauté JP, Labastie J, Forest A, Guichet R. Distribution and abundance of debris on the continental shelf of the Bay of Biscay and in Seine Bay. Marine Pollution Bulletin 1995;30:58–62.

[147] Galgani F, Hanke G, Werner S, De Vrees L. Marine litter within the European Marine Strategy Framework Directive. ICES Journal of Marine Science: Journal du Conseil 2013;70:1055–64.

[148] Galgani F, Jaunet S, Campillot A, Guenegen X, His E. Distribution and abundance of debris on the continental shelf of the north-western Mediterranean Sea. Marine Pollution Bulletin 1995;31:713–7.

[149] Galgani F, Leaute JP, Moguedet P, Souplet A, Verin Y, Carpentier A, Goraguer H, Latrouite D, Andral B, Cadiou Y, Mahe JC, Poulard JC, Nerisson P. Litter on the sea floor along European coasts. Marine Pollution Bulletin 2000;40:516–27.

[150] Galgani F, Souplet A, Cadiou Y. Accumulation of debris on the deep sea floor off the French Mediterranean coast. Marine Ecology Progress Series 1996;142:225–34.

[151] Miranda DD, de Carvalho-Souza GF. Are we eating plastic-ingesting fish? Marine Pollution Bulletin 2016 Feb 15;103(1):109–14.

[152] Gallagher A, Rees A, Rowe R, Stevens J, Wright P. Microplastics in the Solent estuarine complex, UK: an initial assessment. Marine Pollution Bulletin 2016;102(2):243–9.

[153] Gedde U. Polymer physics. Dordrecht (The Netherlands): Kluwer Academic Publishers; 2001.

[154] Gehringer JW, Aron W. Field techniques. Zooplankton sampling. UNESCO monographs on oceanographic methodology, vol. 2. 1968. p. 87–104.

[155] Geissen V, Mol H, Klumpp E, Umlauf G, Nadal M, van der Ploeg M, van de Zee SE, Ritsema CJ. Emerging pollutants in the environment: a challenge for water resource management. International Soil and Water Conservation Research 2015;3(1):57–65.

[156] Gesamp. Sources, fate and effects of microplastics in the marine environment: a global assessment. In: Kershaw PJ, editor. IMO/FAO/UNESCO-IOC/UNIDO/WMO/IAEA/UN/UNEP/UNDP Joint Group of Experts on the Scientific Aspects of Marine Environmental Protection. Rep. Stud. GESAMP; 2015.

[157] Giari L, Guerranti C, Perra G, Lanzoni M, Fano EA, Castaldelli G. Occurrence of perfluorooctanesulfonate and perfluorooctanoic acid and histopathology in eels from north Italian waters. Chemosphere 2015;118:117–23.

[158] Giesy JP, Kannan K. Global distribution of perfluorooctane sulfonate in wildlife. Environmental Science & Technology 2001;35(7):1339–42.

[159] Gilan I, Hadar Y, Sivan A. Colonization, biofilm formation and biodegradation of polyethylene by a strain of Rhodococcus ruber. Applied Microbiology and Biotechnology 2004;65:97–104.

[160] Gilfillan LR, Ohman MD, Doyle MJ, Watson W. Occurrence of plastic micro-debris in the Southern California Current System. California Cooperative Oceanic Fisheries Investigations Reports 2009;50:123–33.

[161] Goldstein MC, Rosenberg M, Cheng L. Increased oceanic microplastic debris enhances oviposition in an endemic pelagic insect. Biol Letters 2012;8(5):817–20.

[162] Goldstein MC, Titmus AJ, Ford M. Scales of spatial heterogeneity of plastic marine debris in the northeast Pacific Ocean. PLoS One 2013;8(11):e80020. http://dx.doi.org/10.1371/journal.pone.0080020.

[163] González-Gaya B, Dachs J, Roscales JL, Caballero G, Jiménez B. Perfluoroalkylated substances in the global tropical and subtropical surface oceans. Environmental Science and Technology 2014;48:13076–84.

[164] Gordon G. Eliminating land-based discharges of marine debris in California: a plan of action from the plastic debris project. California, USA: California State Water Resources Control Board; 2006.

[165] Gouin T, Roche N, Lohmann R, Hodges G. A thermodynamic approach for assessing the environmental exposure of chemicals absorbed to microplastic. Environmental Science and Technology 2011;45:1466–72.

[166] Gregory MR. Plastic 'scrubbers' in hand cleansers: a further (and minor) source for marine pollution identified. Marine Pollution Bulletin 1996;32:867–71.

[167] Nicolau L, Marçalo A, Ferreira M, Sá S, Vingada J, Eira C. Ingestion of marine litter by loggerhead sea turtles, Caretta caretta, in Portuguese continental waters. Marine Pollution Bulletin 2016 Feb 15;103(1):179–85.

[168] Güven O, Gülyavuz H, Deva MC. Benthic debris accumulation in Bathyal Grounds in the Antalya Bay, eastern Mediterranean. Turkish Journal of Fisheries and Aquatic Sciences 2013;13:43–9.

[169] Habib D, Locke DC, Cannone LJ. Synthetic fibers as indicators of municipal sewage sludge, sludge products, and sewage treatment plant effluents. Water, Air, and Soil Pollution 1998;103:1–8.

[170] Hadad D, Geresh S, Sivan A. Biodegradation of polyethylene by the thermophilic bacterium *Brevibacillus borstelensis*. Journal of Applied Microbiology 2005;98:1093–100.

[171] Halden RU. Epistemology of contaminants of emerging concern and literature meta-analysis. Journal of Hazardous Materials 2015;282:2–9.

[172] Halden RU. Plastics and health risks. Annual Review of Public Health 2010;31:179–94.

[173] Hammer S, Nager RG, Johnson PCD, Furness RW, Provencher JF. Plastic debris in great skua (*Stercorarius skua*) pellets corresponds to seabird prey species. Marine Pollution Bulletin 2016;103:206–10.

[174] Hanke G, Galgani F, Werner S, Oosterbaan L, Nilsson P, Fleet D, Al E. MSFD GES technical subgroup on marine litter. Guidance on monitoring of marine litter in European Seas. Luxembourg: Joint Research Centre – Institute for Environment and Sustainability, Publications Office of the European Union; 2013.

[175] Harper CA. Handbook of plastics technologies. New York, USA: The McGraw-Hill Companies, Inc.; 2006.

[176] Harris ME, Walker B. A novel, simplified scheme for plastics identification. Journal of Chemical Education 2010;87:147–9.

[177] Harrison RM. Pollution: causes and effects. 5th ed. Cambridge (United Kingdom): Royal Society of Chemistry Publishing; 2014.

[178] Harrison JP, Ojeda JJ, Romero-González ME. The applicability of reflectance micro-Fourier-transform infrared spectroscopy for the detection of synthetic microplastics in marine sediments. Science of the Total Environment 2012;416:455–63.

[179] Harshvardhan K, Jha B. Biodegradation of low-density polyethylene by marine bacteria from pelagic waters, Arabian Sea, India. Marine Pollution Bulletin 2013;77:100–6.

[180] Hart H, Craine LE, Hart DJ, Hadad CM. Organic chemistry. Boston (USA): Houghton Mifflin Company; 2007.

[181] Hauser R, Calafat AM. Phthalates and human health. Occupational and Environmental Medicine 2005;62(11):806–18.

[182] Hays H, Cormons G. Plastic particles found in tern pellets, on coastal beaches and at factory sites. Marine Pollution Bulletin 1974;5:44–6.

[183] Obbard RW, Sadri S, Wong YQ, Khitun AA, Baker I, Thompson RC. Global warming releases microplastic legacy frozen in Arctic Sea ice. Earth's Future 2014 Jun 1;2(6):315–20.

[184] Helcom. Regional action plan for marine litter in the Baltic Sea. 2015. Helsinki.

[185] Heo NW, Hong SH, Han GM, Hong S, Lee J, Song YK, Jang M, Shim WJ. Distribution of small plastic debris in cross-section and high strandline on Heungnam beach, South Korea. Ocean Science Journal 2013;48:225–33.

[186] Heskett M, Takada H, Yamashita R, Yuyama M, Ito M, Geok YB, Ogata Y, Kwan C, Heckhausen A, Taylor H, Powell T, Morishige C, Young D, Patterson H, Robertson B, Bailey E, Mermoz J. Measurement of persistent organic pollutants (POPs) in plastic resin pellets from remote islands: toward establishment of background concentrations for International Pellet Watch. Marine Pollution Bulletin 2012;64:445–8.

[187] HHS. United States Department of Health and Human Services. Toxicological profile for alpha-, beta-, gamma-, and delta-hexachlorocyclohexane. Agency for Toxic Substances and Disease Registry; 2005. p. 1–377.

[188] Hidalgo-Ruz V, Gutow L, Thompson RC, Thiel M. Microplastics in the marine environment: a review of the methods used for identification and quantification. Environmental Science and Technology 2012;46:3060–75.

[189] Hidalgo-Ruz V, Thiel M. Distribution and abundance of small plastic debris on beaches in the SE Pacific (Chile): a study supported by a citizen science project. Marine Environmental Research 2013;87–88:12–8.

[190] Hinojosa IA, Thiel M. Floating marine debris in fjords, gulfs and channels of southern Chile. Marine Pollution Bulletin 2009;58:341–50.

[191] Hirai H, Takada H, Ogata Y, Yamashita R, Mizukawa K, Saha M, Kwan C, Moore C, Gray H, Laursen D, Zettler ER, Farrington JW, Reddy CM, Peacock EE, Ward MW. Organic micropollutants in marine plastics debris from the open ocean and remote and urban beaches. Marine Pollution Bulletin 2011;62:1683–92.

[192] Holmes LA, Turner A, Thompson RC. Adsorption of trace metals to plastic resin pellets in the marine environment. Environmental Pollution 2012;160:42–8.

[193] Holmes LA, Turner A, Thompson RC. Interactions between trace metals and plastic production pellets under estuarine conditions. Marine Chemistry 2014;167:25–32.

[194] Howard GT. Biodegradation of polyurethane. International Biodeterioration and Biodegradation 2002;49:213.

[195] Huerta Lwanga E, Gertsen H, Gooren H, Peters P, Salánki T, van der Ploeg M, Besseling E, Koelmans AA, Geissen V. Microplastics in the terrestrial ecosystem: implications for *Lumbricus terrestris* (Oligochaeta, Lumbricidae). Environmental Science and Technology 2016;50(5):2685–91.

[196] Hui YH. Handbook of Food Science, Technology and Engineering, vol. 3. Florida, USA: CRC Press; 2006.

[197] ICES. OSPAR request on development of a common monitoring protocol for plastic particles in fish stomachs and selected shellfish on the basis of existing fish disease surveys. 2015. ICES Special Request Advice.

[198] Imhof H, Jschmid J, Niessner R, Ivleva NP, Laforsch C. A novel, highly efficient method for the quantification of plastic particles in sediments of aquatic environments. Limnology and Oceanography: Methods 2012;10:524–37.

[199] Imhof HK, Ivleva NP, Schmid J, Niessner R, Laforsch C. Contamination of beach sediments of a subalpine lake with microplastic particles. Current Biology 2013;23:R867–8.

[200] Ingram AG, Hoskins JH, Sovik JH, Maringer RE, Holden FC. Study of microplastic properties and dimensional stability of materials. Technical Report AFML-TR-67-232, Part II. United States Air Force Materials Laboratory; 1968.

[201] Pedà C, Caccamo L, Fossi MC, Gai F, Andaloro F, Genovese L, Perdichizzi A, Romeo T, Maricchiolo G. Intestinal alterations in European sea bass Dicentrarchus labrax (Linnaeus, 1758) exposed to microplastics: Preliminary results. Environmental Pollution 2016 May 31;212:251–6.

[202] Ivar do Sul JA, Costa MF. The present and future of microplastic pollution in the marine environment. Environmental Pollution 2014;185:352–64.

[203] Jacobsen JK, Massey L, Gulland F. Fatal ingestion of floating net debris by two sperm whales (*Physeter macrocephalus*). Marine Pollution Bulletin 2010;60:765–7.

[204] Jambeck JR, Geyer R, Wilcox C, Siegler TR, Perryman M, Andrady A, Narayan R, Law KL. Plastic waste inputs from land into the ocean. Science 2015;347:768–71.

[205] Jayasiri HB, Purushothaman CS, Vennila A. Plastic litter accumulation on high-water strandline of urban beaches in Mumbai, India. Environmental Monitoring and Assessment 2013;185:7709–19.

[206] Jovíc M, Stankovíc S. Human exposure to trace metals and possible public health risks via consumption of mussels *Mytilus galloprovincialis* from the Adriatic coastal area. Food and Chemical Toxicology 2014;70:241–51.

[207] Jones KC, Voogt P. Persistent organic pollutants (POPs): state of the science. Environmental Pollution 1999;100:209–21.

[208] JRC (Joint Research Centre). MSFD GES technical subgroup on marine litter, and technical recommendations for the implementation of MSFD requirements. 2011. Luxembourg.

[209] Kaberi H, Tsangaris C, Zeri C, Mousdis G, Papadopoulos A, Streftaris N. Microplastics along the shoreline of a Greek island (Kea Isl., Aegean Sea): types and densities in relation to beach orientation, characteristics and proximity to sources. In: 4th international conference on environmental management, engineering, planning and economics (CEMEPE) and SECOTOX conference. Mykonos Island, Greece. 2013.

[210] Kalantzi OI, Hall AJ, Thomas GO, Jones KC. Polybrominated diphenyl ethers and selected organochlorine chemicals in grey seals (*Halichoerus grypus*) in the North Sea. Chemosphere 2005;58:345–54.

[211] Kalogerakis N, Arff J, Banat IM, Broch OJ, Daffonchio D, Edvardsen T, Eguiraun H, Giuliano L, Handå A, López-de-Ipiña K, Marigomez I. The role of environmental biotechnology in exploring, exploiting, monitoring, preserving, protecting and decontaminating the marine environment. New Biotechnology 2015;32(1):157–67.

[212] Kampire E, Rubidge G, Adams JB. Distribution of polychlorinated biphenyl residues in sediments and blue mussels (*Mytilus galloprovincialis*) from Port Elizabeth Harbour, South Africa. Marine Pollution Bulletin 2015;91:173–9.

[213] Kannan K, Choi JW, Iseki N, Senthilkumar K, Kim DH, Masunaga S, Giesy JP. Concentrations of perfluorinated acids in livers of birds from Japan and Korea. Chemosphere 2002;49(3):225–31.

[214] Kannan K, Tao L, Sinclair E, Pastva SD, Jude DJ, Giesy JP. Perfluorinated compounds in aquatic organisms at various trophic levels in a Great Lakes food chain. Archives of Environmental Contamination and Toxicology 2005;48(4):559–66.

[215] Kang JH, Kwon OY, Lee KW, Song YK, Shim WJ. Marine neustonic microplastics around the southeastern coast of Korea. Marine Pollution Bulletin 2015;96(1):304–12.

[216] Käppler A, Windrich F, Löder MGJ, Malanin M, Fischer D, Labrenz M, Eichhorn K-J, Voit B. Identification of microplastics by FTIR and Raman microscopy: a novel silicon filter substrate opens the important spectral range below 1300 cm^{-1} for FTIR transmission measurements. Analytical and Bioanalytical Chemistry 2015;407:6791–801.

[217] Karapanagioti HK, Endo S, Ogata Y, Takada H. Diffuse pollution by persistent organic pollutants as measured in plastic pellets sampled from various beaches in Greece. Marine Pollution Bulletin 2011;62:312–7.

[218] Karger-Kocsis J. Polypropylene structure, blends and composites: copolymers and blends. London: Chapman & Hall; 1995. p. 6.

[219] Kataoka T, Hinata H. Evaluation of beach cleanup effects using linear system analysis. Marine Pollution Bulletin 2015;91(1):73–81.

[220] Kataoka T, Hinata H, Kato S. Analysis of a beach as a time-invariant linear input/output system of marine litter. Marine Pollution Bulletin 2013;77(1):266–73.

[221] Kee YL, Mukherjee S, Pariatamby A. Effective remediation of phenol, 2,4-bis (1,1-dimethylethyl) and bis(2-ethylhexyl) phthalate in farm effluent using Guar gum – a plant based biopolymer. Chemosphere 2015;136:111–7.

[222] Keller AA, Fruh EL, Johnson MM, Simon V, McGourty C. Distribution and abundance of anthropogenic marine debris along the shelf and slope of the US West Coast. Marine Pollution Bulletin 2010;60(5):692–700.

[223] Kelly BC, Ikonomou MG, Blair JD, Morin AE, Gobas FAPC. Food web – specific biomagnification of persistent organic pollutants. Science 2007;317:236–9.

[224] Kleeberg I, Hetz C, Kroppenstedt RM, Muller RJ, Deckwer WD. Biodegradation of aliphatic-aromatic copolyesters by *Thermomonospora fusca* and other thermophilic compost isolates. Applied and Environmental Microbiology 1998;64:1731–5.

[225] Klein E, Lukeš V, Cibulková Z. On the energetics of phenol antioxidants activity. Petroleum and Coal 2005;47:33–9.

[226] Koch HM, Calafat AM. Human body burdens of chemicals used in plastic manufacture. Philosophical Transactions of the Royal Society B: Biological Sciences 2009;364(1526):2063–78.

[227] Koch HM, Drexler H, Angerer J. An estimation of the daily intake of di(2-ethylhexyl) phthalate (DEHP) and other phthalates in the general population. International Journal of Hygiene and Environmental Health 2003;206(2):77–83.

[228] Koelmans AA. ET&C perspectives. Environmental Toxicology and Chemistry 2014; 33:5–10.

[229] Koelmans AA, Besseling E, Wegner A, Foekma EM. Plastic as a carrier of POPs to aquatic organisms: a model analysis. Environmental Science and Technology 2013;47:7812–20.

[230] Korherr C, Roth R, Holler E. Poly(β-L-malate) hydrolase from plasmodia of *Physarum polycephalum*. Canadian Journal of Microbiology 1995;41(Suppl. 1):192–9.

[231] Koutsodendris A, Papatheodorou G, Kougiourouki O, Georgiadis M. Benthic marine litter in four Gulfs in Greece, Eastern Mediterranean; abundance, composition and source identification Estuarine. Coastal and Shelf Science 2008;77(3):501–12.

[232] Kukulka T, Proskurowski G, Morét-Ferguson S, Meyer DW, Law KL. The effect of wind mixing on the vertical distribution of buoyant plastic debris. Geophysical Research Letters 2012;39.

[233] Kutz FW, Wood PH, Bottimore DP. Organochlorine pesticides and polychlorinated biphenyls in human adipose tissue. Reviews of Environmental Contamination and Toxicology 1991;20:1–82.

[234] Laglbauer BJL, Franco-Santos RM, Andreu-Cazenave M, Brunelli L, Papadatou M, Palatinus A, Grego M, Deprez T. Macrodebris and microplastics from beaches in Slovenia. Marine Pollution Bulletin 2014;89:356–66.

[235] Lambert S, Sinclair C, Boxall A. Occurrence, degradation, and effect of polymer-based materials in the environment. Reviews of Environmental Contamination and Toxicology 2014;227:1–53. Springer International Publishing.

[236] Lambert S, Wagner M. Characterisation of nanoplastics during the degradation of polystyrene. Chemosphere 2016;145:265–8.

[237] Land MF, Osorio DC. Marine optics: dark disguise. Current Biology 2011;21:918–20.

[238] Lang IA, Galloway TS, Scarlett A, Henley WE, Depledge M, Wallace RB, Melzer D. Association of urinary bisphenol A concentration with medical disorders and laboratory abnormalities in adults. JAMA 2008;300(11):1303–10.

[239] Quinn B, Gagné F, Costello M, McKenzie C, Wilson J, Mothersill C. The endocrine disrupting effect of municipal effluent on the zebra mussel (Dreissena polymorpha). Aquatic Toxicology 2004 Feb 25;66(3):279–92.

[240] Lattin GL, Moore CJ, Zellers AF, Moore SL, Weisberg SB. A comparison of neustonic plastic and zooplankton at different depths near the southern California shore. Marine Pollution Bulletin 2004;49:291–4.

[241] Rehse S, Kloas W, Zarfl C. Short-term exposure with high concentrations of pristine microplastic particles leads to immobilisation of Daphnia magna. Chemosphere 2016 Jun 30;153:91–9.

[242] Law KL, Morét-Ferguson SE, Goodwin DS, Zettler ER, DeForce E, Kukulka T, et al. Distribution of surface plastic debris in the eastern Pacific Ocean from an 11-year data set. Environmental Science and Technology 2014;48(9):4732–8.

[243] Law KL, Moret-Ferguson S, Maximenko NA, Proskurowski G, Peacock E, Hafner J, Reddy CM. Plastic accumulation in the North Atlantic Subtropical Gyre. Science 2010;329:1185–8.

[244] Law KL, Thompson RC. Microplastics in the seas. Science 2014;345:144–5.

[245] Lebreton LCM, Greer SD, Borrero JC. Numerical modelling of floating debris in the world's oceans. Marine Pollution Bulletin 2012;64:653–61.

[246] Lechner A, Keckeis H, Lumesberger-Loisl F, Zens B, Krusch R, Tritthart M, Glas M, Schludermann E. The Danube so colourful: a potpourri of plastic litter outnumbers fish larvae in Europe's second largest river. Environmental Pollution 2014;188:177–81.

[247] Lechner A, Ramler D. The discharge of certain amounts of industrial microplastic from a production plant into the River Danube is permitted by the Austrian legislation. Environmental Pollution 2015;200:159–60.

[248] Lee J, Pedersen AB, Thomsen M. The influence of resource strategies on childhood phthalate exposure—the role of REACH in a zero waste society. Environment International 2014;73:312–22.

[249] Lee RF, Sanders DP. The amount and accumulation rate of plastic debris on marshes and beaches on the Georgia coast. Marine Pollution Bulletin 2015;91(1):113–9.

[250] Lee H, Shim WJ, Kwon J. Sorption capacity of plastic debris for hydrophobic organic chemicals. Science of the Total Environment 2014;470–471:1545–52.

[251] Lee J, Hong S, Song YK, Hong SH, Jang YC, Jang M, Heo NW, Han GM, Lee MJ, Kang D, Shim WJ. Relationships among the abundances of plastic debris in different size classes on beaches in South Korea. Marine Pollution Bulletin 2013;77:349–54.

[252] Lenz R, Enders K, Stedmon CA, Mackenzie DMA, Nielsen TG. A critical assessment of visual identification of marine microplastic using Raman spectroscopy for analysis improvement. Marine Pollution Bulletin 2015;100:82–91.

[253] Leslie HA. Review of microplastics in cosmetics. IVM Institute for Environmental Studies; 2014. R14/29.

[254] Leslie HA, Meulen MD, Kleissen FM, Vethaak AD. Microplastic litter in the Dutch marine environment. Dutch Ministry of Infrastructure and Environmnet; 2011. 1203772-000.

[255] Leslie HA, Van Velzen MJM, Vethaak AD. Microplastic survey of the Dutch environment. Novel data set of microplastics in North Sea sediments, treated wastewater effluents and marine biota. Amsterdam: Institute for Environmental Studies, VU University Amsterdam; 2013.

[256] Li L, Stramski D, Reynolds RA. Characterization of the solar light field within the ocean mesopelagic zone based on radiative transfer simulations. Deep Sea Research Part I 2014;87:53–69.

[257] Liebezeit G, Dubaish F. Microplastics in beaches of the East Frisian islands Spiekeroog and Kachelotplate. Bulletin of Environmental Contamination and Toxicology 2012;89:213–7.
[258] Lima ARA, Barletta M, Costa MF. Seasonal distribution and interactions between plankton and microplastics in a tropical estuary. Estuarine, Coastal and Shelf Science 2015;165:213–25.
[259] Lima ARA, Costa MF, Barletta M. Distribution patterns of microplastics within the plankton of a tropical estuary. Environmental Research 2014;132:146–55.
[260] Lithner D, Larsson A, Dave G. Environmental and health hazard ranking and assessment of plastic polymers based on chemical composition. Science of the Total Environment 2011;409:3309–24.
[261] Liu Y, Li J, Zhao Y, Wen S, Huang F, Wu Y. Polybrominated diphenyl ethers (PBDEs) and indicator polychlorinated biphenyls (PCBs) in marine fish from four areas of China. Chemosphere 2011;83:168–74.
[262] Lobelle D, Cunliffe M. Early microbial biofilm formation on marine plastic debris. Marine Pollution Bulletin 2011;62(1):197–200.
[263] Löder M, Gerdts G. Methodology used for the detection and identification of microplastics—a critical appraisal. In: Bergmann M, Gutow L, Klages M, editors. Marine anthropogenic litter. Berlin: Springer; 2015.
[264] Van A, Rochman CM, Flores EM, Hill KL, Vargas E, Vargas SA, Hoh E. Persistent organic pollutants in plastic marine debris found on beaches in San Diego, California. Chemosphere 2012 Jan 31;86(3):258–63.
[265] Long M, Moriceau B, Gallinari M, Lambert C, Huvet A, Raffray J, Soudant P. Interactions between microplastics and phytoplankton aggregates: impact on their respective fates. Marine Chemistry 2015;175:39–46.
[266] Luís LG, Ferreira P, Fonte E, Oliveira M, Guilhermino L. Does the presence of microplastics influence the acute toxicity of chromium(VI) to early juveniles of the common goby (*Pomatoschistus microps*)? A study with juveniles from two wild estuarine populations. Aquatic Toxicology 2015;164:163–74.
[267] Lusher AL, Burke A, O'connor I, Officer R. Microplastic pollution in the Northeast Atlantic Ocean: validated and opportunistic sampling. Marine Pollution Bulletin 2014;88:325–33.
[268] Lusher AL, Hernandez-Milian G, O'Brien J, Berrow S, O'Connor I, Officer R. Microplastic and macroplastic ingestion by a deep diving, oceanic cetacean: the True's beaked whale *Mesoplodon mirus*. Environmental Pollution 2015;199:185–91.
[269] Lusher AL, McHugh M, Thompson RC. Occurrence of microplastics in the gastrointestinal tract of pelagic and demersal fish from the English Channel. Marine Pollution Bulletin 2013;67:94–9.
[270] Mansui J, Molcard A, Ourmieres Y. Modelling the transport and accumulation of floating marine debris in the Mediterranean basin. Marine Pollution Bulletin 2015;91(1):249–57.
[271] Mariussen E. Neurotoxic effects of perfluoroalkylated compounds: mechanisms of action and environmental relevance. Archives of Toxicology 2012;86(9):1349–67.
[272] Marques GA, Tenorio JAS. Use of froth flotation to separate PVC/PET mixtures. Waste Management 2000;20:265–9.
[273] Martineau D, Béland P, Desjardins C, Lagacé A. Levels of organochlorine chemicals in tissues of beluga whales (*Delphinapterus leucas*) from the St. Lawrence Estuary, Quebec, Canada. Archives of Environmental Contamination and Toxicology 1987;16:137–47.
[274] Martins J, Sobral P. Plastic marine debris on the Portuguese coastline: a matter of size? Marine Pollution Bulletin 2011;62:2649–53.

[275] Martins PLG, Marques LG, Colepicolo P. Antioxidant enzymes are induced by phenol in the marine microalga *Lingulodinium polyedrum*. Ecotoxicology and Environmental safety 2015;116:84–9.

[276] Masura J, Baker J, Foster G, Arthur C. Laboratory methods for the analysis of microplastics in the marine environment: recommendations for quantifying synthetic particles in waters and sediments. NOAA Technical Memorandum, NOS-OR&R-48; 2015. p. 1–39.

[277] Matarese AC, Blood DM, Picquelle SJ, Benson JL. Atlas of the abundance and distribution patterns of ichthyoplankton from the Northeast Pacific Ocean and Bering Sea ecosystems based on research conducted by the Alaska Fisheries Science Centre (1972–1996). NOAA Professional Paper NMFS 1. 2003. p. 281.

[278] Mathalon A, Hill P. Microplastic fibres in the intertidal ecosystem surrounding Halifax Harbour, Nova Scotia. Marine Pollution Bulletin 2014;81:69–79.

[279] Mato Y, Isobe T, Takada H, Kanehiro H, Ohtake C, Kaminuma T. Plastic resin pellets as a transport medium for toxic chemicals in the marine environment. Environmental Science and Technology 2001;35:318–24.

[280] Maximenko N, Hafner J, Niiler P. Pathways of marine debris derived from trajectories of Lagrangian drifters. Marine Pollution Bulletin 2012;65(1–3):51–62.

[281] McCormick A, Hoellein TJ, Mason SA, Schluep J, Kelly JJ. Microplastic is an abundant and distinct microbial habitat in an urban river. Environmental Science and Technology 2014;48:11863–71.

[282] Mcdermid KJ, Mcmullen TL. Quantitative analysis of small-plastic debris on beaches in the Hawaiian Archipelago. Marine Pollution Bulletin 2004;48:790–4.

[283] McIntyre A, Vincent RM, Perkins AC, Spiller RC. Effect of bran, ispaghula, and inert plastic particles on gastric emptying and small bowel transit in humans: the role of physical factors. Gut 1997;40:223–7.

[284] Wang J, Tan Z, Peng J, Qiu Q, Li M. The behaviors of microplastics in the marine environment. Marine Environmental Research 2016 Feb 29;113:7–17.

[285] McKeen L. The effect of creep and other time related factors on plastics and elastomers. 3rd ed. Massachusetts, USA: Elsevier; 2015.

[286] Zarfl C, Matthies M. Are marine plastic particles transport vectors for organic pollutants to the Arctic? Marine Pollution Bulletin 2010;60:1810–4.

[287] Meeker JD, Sathyanarayana S, Swan SH. Phthalates and other additives in plastics: human exposure and associated health outcomes. Philosophical Transactions of the Royal Society B: Biological Sciences 2009;364(1526):2097–113.

[288] Megharaj M, Ramakrishnan B, Venkateswarlu K, Sethunathan N, Naidu R. Bioremediation approaches for organic pollutants: a critical perspective. Environment International 2011;37:1362–75.

[289] Melzer D, Rice NE, Lewis C, Henley WE, Galloway TS. Association of urinary bisphenol a concentration with heart disease: evidence from NHANES 2003/06. PLoS One 2010;5(1):e8673.

[290] Méndez-Fernandez P, Webster L, Chouvelon T, Bustamante P, Ferreira M, González AF, López A, Moffat CF, Pierce GJ, Read FL, Russell M, Santos MB, Spitz J, Vingada JV, Caurant F. An assessment of contaminant concentrations in toothed whale species of the NW Iberian Peninsula: Part I. Persistent organic pollutants. Science of the Total Environment 2014;484:196–205.

[291] Meng XY, Li YS, Zhou Y, Zhang YY, Yang L, Qiao B, Wang NN, Hu P, Lu SY, Ren HL, Liu ZS. An enzyme-linked immunosorbent assay for detection of pyrene and related polycyclic aromatic hydrocarbons. Analytical Biochemistry 2015;473:1–6.

[292] Mills OH, Kligman AM. Evaluation of abrasives in acne therapy. Cutis 1979;23:704–5.
[293] Miranda-Urbina D, Thiel M, Luna-Jorquera G. Litter and seabirds found across a longitudinal gradient in the South Pacific Ocean. Marine Pollution Bulletin 2015;96(1):235–44.
[294] Mitchell A. Thinking without the 'circle': marine plastic and global ethics. Political Geography 2015;47:77–85.
[295] Mogil'nitskii GM, Sagatelyan RT, Kutishcheva TN, Zhukova SV, Kerimov SI, Parfenova TB. Disruption of the protective properties of the polyvinyl chloride coating under the effect of microorganisms. Protection of Metals 1987;23(Engl. Transl.):173–5.
[296] Moore CJ. Synthetic polymers in the marine environment: a rapidly increasing, long-term threat. Environmental Research 2008;108:131–9.
[297] Moore CJ, Lattin GL, Zellers AF. Quantity and type of plastic debris flowing from two urban rivers to coastal waters and beaches of Southern California. Journal of Integrated Coastal Zone Management 2011;11:65–73.
[298] Moore CJ, Moore SL, Leecaster MK, Weisberg SB. A comparison of plastic and plankton in the North Pacific Central Gyre. Marine Pollution Bulletin 2001;42:1297–300.
[299] Moore CJ, Moore SL, Weisberg SB, Lattin GL, Zellers AF. A comparison of neustonic plastic and zooplankton abundance in southern California's coastal waters. Marine Pollution Bulletin 2002;44(10):1035–8.
[300] Moos NV, Burkhardt-Holm P, Köhler A. Uptake and effects of microplastics on cells and tissue of the blue mussel Mytilus edulis L. after an experimental exposure. Environmental Science and Technology 2012;46:11327–35.
[301] Mordecai G, Tyler P, Masson DG, Huvenne VAI. Litter in submarine canyons off the west coast of Portugal. Deep Sea Research Part II 2011;58:2489–96.
[302] Moreira FT, Prantoni AL, Martini B, De Abreu MA, Stoiev SB, Turra A. Small-scale temporal and spatial variability in the abundance of plastic pellets on sandy beaches: methodological considerations for estimating the input of microplastics. Marine Pollution Bulletin 2016;102:114–21.
[303] Morét-Ferguson S, Law KL, Proskurowski G, Murphy EK, Peacock EE, Reddy CM. The size, mass, and composition of plastic debris in the western North Atlantic Ocean. Marine Pollution Bulletin 2010;60:1873–8.
[304] Zetsche EM, Ploug H. Marine chemistry special issue: particles in aquatic environments: from invisible exopolymers to sinking aggregates. Marine Chemistry 2015;175:1–4.
[305] MSFD-TSGML. Guidance on monitoring of marine litter in European Seas. 2013. A guidance document within the common implementation strategy for the Marine Strategy Framework Directive. EUR-26113 EN JRC Scientific and Policy Reports JRC83985.
[306] Muenmee S, Chiemchaisri W, Chiemchaisri C. Microbial consortium involving biological methane oxidation in relation to the biodegradation of waste plastics in a solid waste disposal open dump site. International Biodeterioration & Biodegradation 2015;102:172–81.
[307] Murphy F, Ewins C, Carbonnier F, Quinn B. Wastewater Treatment Works (WwTW) as a source of microplastics in the aquatic environment. Environmental Science and Technology 2016;50:5800–8.
[308] Murray F, Cowie PR. Plastic contamination in the decapod crustacean Nephrops norvegicus (Linnaeus, 1758). Marine Pollution Bulletin 2011;62:1207–17.
[309] Zettler ER, Mincer TJ, Amaral-Zettler A. Life in the 'Plastisphere': microbial communities on plastic marine debris. Environmental Science and Technology 2013;47:7137–46.
[310] Nagata M, Kiyotsukuri T, Minami S, Tsutsumi N, Sakai W. Enzymatic degradation of poly(ethylene terephthalate) copolymers with aliphatic dicarboxylic acids and/or poly(ethylene glycol). European Polymer Journal 1997;33:1701–5.

[311] Ocean Studies Board, Marine Board. Oil in the sea III: inputs, fates, and effects. National Academies Press; 2003.

[312] Neufeld L, Stassen F, Sheppard R, Gilman T. In: The new plastics economy: rethinking the future of plastics. World Economic Forum; 2016.

[313] Newman S, Watkins E, Farmer A, Ten Brink P, Schweitzer JP. The economics of marine litter. In: Bergmann M, Gutow L, Klages M, editors. Marine anthropogenic litter. Berlin: Springer; 2015.

[314] Ng KL, Obbard JP. Prevalence of microplastics in Singapore's coastal marine environment. Marine Pollution Bulletin 2006;52:761–7.

[315] Nobre CR, Santana MFM, Maluf A, Cortez FS, Cesar A, Pereira CDS, Turra A. Assessment of microplastic toxicity to embryonic development of the sea urchin *Lytechinus variegatus* (Echinodermata: Echinoidea). Marine Pollution Bulletin 2015;92(1):99–104.

[316] Noren F. Small plastic particles in Coastal Swedish waters. 2007. N-Research report, commissioned by KIMO, Sweden.

[317] Noren F, Naustvoll F. Survey of microscopic anthropogenic particles in Skagerrak. 2010. Commissioned by Klima-og Forurensningsdirektoratet, Norway.

[318] NTP. Technical report on the toxicology and carcinogenesis studies of 2,3,7,8-tetrachlorodibenzo-p-dioxin (TCDD) (CAS No. 1746-01-6) in female Harlan Sprague–Dawley rats (Gavage Studies). National Toxicology Program Technical Report Series 2006;521:4–232.

[319] Nuelle M, Dekiff JH, Remy D, Fries E. A new analytical approach for monitoring microplastics in marine sediments. Environmental Pollution 2014;184:161–9.

[320] Ogata Y, Takada H, Mizukawa K, Hirai H, Iwasa S, Endo S, Mato Y, Saha M, Okuda K, Nakashima A, Murakami M, Zurcher N, Booyatumanondo R, Zakaria MP, Dung LQ, Gordon M, Miguez C, Suzuki S, Moore C, Karapanagioti HK, Weerts S, McClurg T, Burres E, Smith W, Van Velkenburg M, Lang JS, Lang RC, Laursen D, Danner B, Stewardson N, Thompson RC. International Pellet Watch: global monitoring of persistent organic pollutants (POPs) in coastal waters. 1. Initial phase data on PCBs, DDTs, and HCHs. Marine Pollution Bulletin 2009;58:1437–46.

[321] Olencycz M, Sokołowski A, Niewińska A, Wołowicz M, Namieśnik J, Hummel H, Jansen J. Comparison of PCBs and PAHs levels in European coastal waters using mussels from the *Mytilus edulis* complex as biomonitors. Oceanologia 2015 Jun 30;57(2):196–211.

[322] Oliveira M, Ribeiro A, Hylland K, Guilhermino L. Single and combined effects of microplastics and pyrene on juveniles (0+ group) of the common goby *Pomatoschistus microps* (Teleostei, Gobiidae). Ecological Indicators 2013;34:641–7.

[323] O'Shea OR, Hamann M, Smith W, Taylor H. Predictable pollution: an assessment of weather balloons and associated impacts on the marine environment – an example for the Great Barrier Reef, Australia. Marine Pollution Bulletin 2014;79(1):61–8.

[324] OSPAR. Guideline for monitoring marine litter on the beaches in the OSPAR maritime area. 2010. London.

[325] OSPAR. Quality status report 2010. 2010. London.

[326] Panyala NR, Peña-Méndez EM, Havel J. Silver or silver nanoparticles: a hazardous threat to the environment and human health. Journal of Applied Biomedicine 2008;6(3):117–29.

[327] Patchell J. What's choking our sewer lines? Plumbing Connection 2012.

[328] Patrick GL. Medicinal chemistry. 5th ed. Oxford (United Kingdom): Oxford University Press; 2013.

[329] Peacock A. Handbook of polyethylene: structures, properties and applications. Switzerland: Marcel Dekker, Inc.; 2000.

[330] Perelo LW. Review: in situ and bioremediation of organic pollutants in aquatic sediments. Journal of Hazardous Materials 2010;177:81–9.

[331] Peters CA, Bratton SP. Urbanization is a major influence on microplastic ingestion by sunfish in the Brazos River Basin, Central Texas, USA. Environmental Pollution 2016;210:380–7.

[332] Pham CK, Ramirez-Llodra E, Alt CHS, Amaro T, Bergmann M, Canals M, Company JB, Davies J, Duineveld G, Galgani F, Howell KL, Huvenne VAI, Isidro E, Jones DOB, Lastras G, Morato T, Gomes-Pereira JN, Purser A, Stewart H, Tojeira I, Tubau X, Van Rooij D, Tyler PA. Marine litter distribution and density in European Seas, from the shelves to deep basins. PLoS One 2014;9:e95839.

[333] Phillips MB, Bonner TH. Occurrence and amount of microplastic ingested by fishes in watersheds of the Gulf of Mexico. Marine Pollution Bulletin 2015:264–9.

[334] Phuong NN, Zalouk-Vergnoux A, Poirier L, Kamari A, Châtel A, Mouneyrac C, Lagarde F. Is there any consistency between the microplastics found in the field and those used in laboratory experiments? Environmental Pollution 2016;211:111–23.

[335] PlasticsEurope. Plastics – the facts 2013. An analysis of European latest plastics production, demand and waste data. Brussels: Plastics Europe: Association of Plastic Manufacturers; 2013. p. 1–38.

[336] Zhang K, Gong W, Lv J, Xiong X, Wu C. Accumulation of floating microplastics behind the Three Gorges Dam. Environmental Pollution 2015;204:117–23.

[337] PlasticsEurope. Plastics – the facts 2015. An analysis of European latest plastics production, demand and waste data. Brussels: Plastics Europe: Association of Plastic Manufacturers; 2015. p. 1–30.

[338] Possatta FE, Barletta M, Costa MF, Ivar do Sul JA, Dantas DV. Plastic debris ingestion by marine catfish: an unexpected fisheries impact. Marine Pollution Bulletin 2011;62:1098–102.

[339] Quinn B. Chronic toxicity of pharmaceutical compounds on bivalve mollusc primary cultures. P11-35. Abstracts Toxicology Letters 2012;211S:S43–216.

[340] Quinn B. Preparation and maintenance of live tissues and primary cultures for toxicity studies. In: Gagné B, editor. Biochemical ecotoxicology: principles and methods. San Diego (California, USA): Elsevier Inc.; 2014.

[341] Quinn B, Gagné F, Blaise C. An investigation into the acute and chronic toxicity of eleven pharmaceuticals (and their solvents) found in wastewater effluent on the cnidarian, *Hydra attenuate*. Science of the Total Environment 2008;389:306–14.

[342] Zhang W, Ma X, Zhang Z, Wang Y, Wang J, Wang J, Ma D. Persistent organic pollutants carried on plastic resin pellets from two beaches in China. Marine Pollution Bulletin 2015;99:28–34.

[343] Quinn B, Gagné F, Blaise C. Evaluation of the acute, chronic and teratogenic effects of a mixture of eleven pharmaceuticals on the cnidarian, *Hydra attenuate*. Science of the Total Environment 2009;407:1072–9.

[344] Zhao S, Zhu L, Li D. Characterization of small plastic debris on tourism beaches around the South China Sea. Regional Studies in Marine Science 2015;1:55–62.

[345] Zhao S, Zhu L, Li D. Microplastic in three urban estuaries, China. Environmental Pollution 2015;206:597–604.

[346] Zhao S, Zhu L, Wang T, Li D. Suspended microplastics in the surface water of the Yangtze Estuary System, China: first observations on occurrence, distribution. Marine Pollution Bulletin 2014;86:562–8.

[347] Ramirez-Llodra E, De Mol B, Company JB, Coll M, Sardà F. Effects of natural and anthropogenic processes in the distribution of marine litter in the deep Mediterranean Sea. Progress in Oceanography 2013;118:273–87.

[348] Rawn DF, Forsyth DS, Ryan JJ, Breakell K, Verigin V, Nicolidakis H, Hayward S, Laffey P, Conacher HB. PCB, PCDD and PCDF residues in fin and non-fin fish products from the Canadian retail market 2002. Science of the Total Environment 2006;359(1):101–10.

[349] Reisser J, Shaw J, Hallegraeff G, Proietti M, Barnes DKA, Thums M, Wilcox C, Hardesty BD, Pattiaratchi C. Millimetre-sized marine plastics: a new pelagic habitat for microorganisms and invertebrates. PLoS One 2014;9:e100289.

[350] Reisser J, Shaw J, Wilcox C, Hardesty BD, Proietti M, Thums M, et al. Marine plastic pollution in waters around Australia: characteristics, concentrations, and pathways. PLoS One 2013;8(11):e80466. http://dx.doi.org/10.1371/journal.pone.0080466.

[351] Rice MR, Gold HS. Polypropylene as an adsorbent for trace organics in water. Analytical Chemistry 1984;56:1436–40.

[352] Rillig MC. Microplastic in terrestrial ecosystems and the soil? Environmental Science and Technology 2012;46:6453–4.

[353] Rios LM, Jones PR, Moore C, Narayan UV. Quantitation of persistent organic pollutants adsorbed on plastic debris from the Northern Pacific Gyre's "eastern garbage patch". Journal of Environmental Monitoring 2010;12:2189–312.

[354] Rios LM, Moore C, Jones PR. Persistent organic pollutants carried by synthetic polymers in the ocean environment. Marine Pollution Bulletin 2007;54:1230–7.

[355] Rocha-Santos T, Duarte AC. A critical overview of the analytical approaches to the occurrence, the fate and the behaviour of microplastics in the environment. Trends in Analytical Chemistry 2015;65:47–53.

[356] Rochman CM, Browne MA, Halpern BS, Hentschel BT, Hoh E, Karapanagioti HK, Rios-Mendoza LM, Takada H, Teh S, Thompson RC. Classify plastic waste as hazardous. Nature 2013;494:169–71.

[357] Rochman CM, Hentschel BT, Teh SJ. Long-term sorption of metals is similar among plastic types: implications for plastic debris in aquatic environments. PLoS One 2014;9(1):e85433.

[358] Rochman CM, Manzano C, Hentschel BT, Simonich SLM. Polystyrene plastic: a source and sink for polycyclic aromatic hydrocarbons in the marine environment. Environmental Science and Technology 2013;47:13976–84.

[359] Romeo T, Pietro B, Pedà C, Consoli P, Andaloro F, Fossi MC. First evidence of presence of plastic debris in stomach of large pelagic fish in the Mediterranean Sea. Marine Pollution Bulletin 2015;95(1):358–61.

[360] Rowthorn C, Bartlett R, Bender A, Clark M. Japan, lonely planet. 2007.

[361] Rummel CD, Löder MGJ, Fricke NF, Lang T, Griebeler E-M, Janke M, Gerdts G. Plastic ingestion by pelagic and demersal fish from the North Sea and Baltic Sea. Marine Pollution Bulletin 2016;102:134–41.

[362] Russell JR, Huang J, Anand P, Kucera K, Sandoval AG, Dantzler KW, Hickman D, Jee J, Kimovec FM, Koppstein D, Marks DH. Biodegradation of polyester polyurethane by endophytic fungi. Applied and Environmental Microbiology 2011;77(17):6076–84.

[363] Russell M, Webster L, Walsham P, Packer G, Dalgarno EJ, McIntosh AD, Fryer RJ, Moffat CF. Composition and concentration of hydrocarbons in sediment samples from the oil producing area of the East Shetland Basin, Scotland. Journal of Environmental Monitoring 2008;10:559–69.

[364] Russell M, Webster L, Walsham P, Packer G, Dalgarno EJ, McIntosh AD, Moffat CF. The effects of oil exploration and production in the Fladen Ground: composition and concentration of hydrocarbons in sediment samples collected during 2001 and their comparison with sediment samples collected in 1989. Marine Pollution Bulletin 2005;50:638–51.

[365] Ryan PG. A simple technique for counting marine debris at sea reveals steep litter gradients between the Straits of Malacca and the Bay of Bengal. Marine Pollution Bulletin 2013;69:128–36.

[366] Ryan PG, Connell AD, Gardner BD. Plastic ingestion and PCBs in seabirds: is there a relationship? Marine Pollution Bulletin 1988;19:174–6.

[367] Ryan PG, Moore CJ, Franeker JA, Moloney CL. Monitoring the abundance of plastic debris in the marine environment. Philosophical Transactions of the Royal Society B 2009;364:1999–2012.

[368] Sadri SS, Thompson RC. On the quantity and composition of floating plastic debris entering and leaving the Tamar Estuary, Southwest England. Marine Pollution Bulletin 2014;81:55–60.

[369] Sammons C, Yarwood J, Everall N. An FT-IR study of the effect of hydrolytic degradation on the structure of thin PET films. Journal of Polymer Degradation and Stability 2000;67:149–58.

[370] Samsonek J, Puype F. Occurrence of brominated flame retardants in black thermo cups and selected kitchen utensils purchased on the European market. Food Additives & Contaminants: Part A 2013;30(11):1976–86.

[371] Sanchez W, Bender C, Porcher J. Wild gudgeons (*Gobio gobio*) from French rivers are contaminated by microplastics: preliminary study and first evidence. Environmental Research 2014;128:98–100.

[372] Santos RG, Andrades R, Boldrini MA, Martins AS. Debris ingestion by juvenile marine turtles: an underestimated problem. Marine Pollution Bulletin 2015;93(1):37–43.

[373] Schettler TED. Human exposure to phthalates via consumer products. International Journal of Andrology 2006;29(1):134–9.

[374] Schmidt W, O'Shea T, Quinn B. The effect of shore location on biomarker expression in wild *Mytilus* spp. and its comparison with long line cultivated mussels. Marine Environmental Research 2012;80:70–6.

[375] Schmidt W, Power E, Quinn B. Seasonal variations of biomarker responses in the marine blue mussel (*Mytilus* spp.). Marine Pollution Bulletin 2013;74:50–5.

[376] Schultz I. In: Summary of Expert Discussion Forum on possible human health risks from microplastics in the marine environment. U.S. Environmental Protection Agency; February 6, 2015.

[377] Zheng Y, Yanful EK, Bassi AS. A review of plastic waste biodegradation. Critical Reviews in Biotechnology 2005;25:243–50.

[378] Secchi E, Zarzur S. Plastic debris ingested by a Blainville's beaked whale, *Mesoplodon densirostris*, washed ashore in Brazil. Aquatic Mammals 1999;25:21–4.

[379] Setälä O, Fleming-Lehtinen V, Lehtiniemi M. Ingestion and transfer of microplastics in the planktonic food web. Environmental Pollution 2014;185:77–83.

[380] Shashoua Y. Conservation of plastics. Burlington, USA: Elsevier Ltd; 2008.

[381] Zhu X. Optimization of elutriation device for filtration of microplastic particles from sediment. Marine Pollution Bulletin 2015;92:69–72.

[382] Sherrington C, Darrah C, Hann S, Cole G, Corbin M. Study to support the development of measures to combat a range of marine litter sources. 2016. Report for the European Commission DG Environment.

[383] Shen L, Haufe J, Patel MK. Product overview and market projection of emerging bio-based plastics. PRO-BIP. Universiteit Utrecht; 2009.

[384] Shumway SE, Cucci TL. The effects of the toxic dinoflagellate *Protogonyaulax tamarensis* on the feeding and behaviour of bivalve molluscs. Aquatic Toxicology 1987;10:9–27.

[385] Signer R, Weiler J. Raman spectra and constitution of compounds of high molecular weight. LXII. Formation of higher polymers. Helvetica Chimica Acta 1932;15(1):649–57.

[386] Sivan A, Szanto M, Pavlov V. Biofilm development of the polyethylene-degrading bacterium *Rhodococcus ruber*. Applied Microbiology and Biotechnology 2006;72(2):346–52.

[387] Smith WE, Dent G. Modern Raman spectroscopy – a Practical approach. John Wiley & Sons, Ltd; 2005.

[388] Song YK, Hong SH, Jang M, Kang J, Kwon OY, Han GM, et al. Large accumulation of micro-sized synthetic polymer particles in the sea surface microlayer. Environmental Science and Technology 2014;48:9014–21.

[389] Song YK, Hong SH, Jang M, Han GM, Rani M, Lee J, et al. A comparison of microscopic and spectroscopic identification methods for analysis of microplastics in environmental samples. Marine Pollution Bulletin 2015;93:202–9.

[390] Spengler A, Costa MF. Methods applied in studies of benthic marine debris. Marine Pollution Bulletin 2008;56:226–30.

[391] Zitco V, Hanlon M. Another source of pollution by plastics: skin cleaners with plastic scrubbers. Marine Pollution Bulletin 1991;22:41–2.

[392] Strafella P, Fabi G, Spagnolo A, Grati F, Polidori P, Punzo E, Fortibuoni T, Marceta B, Raicevich S, Cvitkovic I, Despalatovic M. Spatial pattern and weight of seabed marine litter in the northern and central Adriatic Sea. Marine Pollution Bulletin 2015;91(1):120–7.

[393] Streb J. Shaping the national system of inter-industry knowledge exchange vertical integration, licensing and repeated knowledge transfer in the German plastics industry. Research Policy 2003;32:1125–40.

[394] Strychar KB, MacDonald BA. Impacts of suspended peat particles on feeding retention rates in cultured eastern oysters (Crassostrea virginica, Gmelin). Journal of Shellfish Research 1999;18:437–44.

[395] Suhrhoff TJ, Scholz-Böttcher BM. Qualitative impact of salinity, UV radiation and turbulence on leaching of organic plastic additives from four common plastics – a lab experiment. Marine Pollution Bulletin 2016;102:84–94.

[396] Sun H, Gu Q, Liao Y, Sun C. Research of amoxicillin microcapsules preparation playing micro-jetting technology. The Open Biomedical Engineering Journal 2015;9:115.

[397] Takada H, Mato Y, Endo S, Yamashita R, Zakaria MP. Pellet Watch: global monitoring of Persistent Organic Pollutants (POPs) using beached plastic resin pellets. Unpublished. 2005.

[398] Talsness CE, Andrade AJ, Kuriyama SN, Taylor JA, vom Saal FS. Components of plastic: experimental studies in animals and relevance for human health. Philosophical Transactions of the Royal Society of London B: Biological Sciences 2009;364(1526):2079–96.

[399] Tanaka K, Takada H, Yamashita R, Mizukawa K, Fukuwaka M, Watanuki Y. Accumulation of plastic-derived chemicals in tissues of seabirds ingesting marine plastics. Marine Pollution Bulletin 2013;69:219–22.

[400] Tanabe S. PCB problems in the future: foresight from current knowledge. Environmental Pollution 1988;50(1–2):5–28.

[401] Tappin AD, Millward GE. The English Channel: contamination status of its transitional and coastal waters. Marine Pollution Bulletin 2015;95(2):529–50.

[402] Tarpley RJ, Marwitz S. Plastic debris ingestion by cetaceans along the Texas coast: two case reports. Aquatic Mammals 1993;19:93–8.

[403] Teuten EL, Rowland SJ, Galloway TS, Thompson RC. Potential for plastics to transport hydrophobic contaminants. Environmental Science and Technology 2007;41:7759–64.

[404] Thiel M, Hinojosa I, Vasquez N, Macaya E. Floating marine debris in coastal waters of the SE-Pacific (Chile). Marine Pollution Bulletin 2003;46(2):224–31.

[405] Thompson RC, Moore CJ, Vom Saal FS, Swan SH. Plastics, the environment and human health: current consensus and future trends. Philosophical Transactions of the Royal Society B: Biological Sciences 2009;364(1526):2153–66.

[406] Thompson RC, Olsen Y, Mitchell RP, Davis A, Rowland SJ, John AWG, McGonigle D, Russell AE. Lost at sea: where is all the plastic? Science 2004;304:838.

[407] Titmus AJ, Hyrenbach KD. Habitat associations of floating debris and marine birds in the North East Pacific Ocean at coarse and meso spatial scales. Marine Pollution Bulletin 2011;62(11):2496–506.

[408] Titow WV. PVC technology. 4th ed. United Kingdom: Elsevier Applied Science Publishers Ltd; 1986.

[409] Tomlin J, Read NW. Laxative properties of indigestible plastic particles. British Medical Journal 1988;297:1175–6.

[410] Tongesayi T, Tongesayi S. Contaminated irrigation water and the associated public health risks. Food, Energy and Water 2015:349–81.

[411] Trenkel VM, Hintzen NT, Farnsworth KD, Olesen C, Reid D, Rindorf A, Shephard S, Dickey-Collas M. Identifying marine pelagic ecosystem management objectives and indicators. Marine Policy 2015;55:23–32.

[412] Tubau X, Canals M, Lastras G, Rayo X, Rivera J, Amblas D. Marine litter on the floor of deep submarine canyons of the Northwestern Mediterranean Sea: the role of hydrodynamic processes. Progress in Oceanography 2015;134:379–403.

[413] Turner A, Holmes L. Occurrence, distribution and characteristics of beached plastic production pellets on the island of Malta (Central Mediterranean). Marine Pollution Bulletin 2011;62:377–81.

[414] Turra A, Manzano AB, Dias RJS, Mahiques MM, Barbosa L, Balthazar-Silva D, Moreira FT. Three-dimensional distribution of plastic pellets in sandy beaches: shifting paradigms. Scientific Reports 2014;4:4435.

[415] Zubris KAV, Richards BK. Synthetic fibres as an indicator of land application of sludge. Environmental Pollution 2005;138:201–11.

[416] UNEP. Marine litter: an analytical overview. 2005. Nairobi.

[417] UNEP. Marine litter: a global challenge. 2009. Nairobi.

[418] UNEP. Plastic in cosmetics. 2015. Nairobi.

[419] UNEP/NOAA. The Honolulu strategy: a global framework for prevention and management of marine debris. 2011. Nairobi/Silver Spring, MD.

[420] US EPA (Unites States Environmental Protection Agency). Guidance for the reregistration of pesticide products containing lindane as the active ingredient. EPARS85027. 1985. p. 4–5.

[421] US EPA (Unites States Environmental Protection Agency). Summary of Expert Discussion Forum on possible human health risks from microplastics in the marine environment. Marine Pollution Control Branch; 2014. p. 1–31.

[422] Vallak HW, Bakker DJ, Brandt I, Broström-Lundén E, Brouwer A, Bull KR, Gough C, Guardens R, Holoubek I, Jansson B, Koch R, Kuylenstierna J, Lecloux A, Mackray D, McCutcheon P, Mocarelli P, Taalman RDF. Controlling persistent organic pollutants – what next? Environmental Toxicology and Pharmacology 1998;6:143–75.

[423] Zylinski S, Johnsen S. Mesopelagic cephalopods switch between transparency and pigmentation to optimize camouflage in the deep. Current Biology 2011;21:1937–41.

[424] Van Cauwenberghe L, Claessens M, Vandegehuchte MB, Mees J, Janssen CR. Assessment of marine debris on the Belgian Continental Shelf. Marine Pollution Bulletin 2013;73:161–9.

[425] Van Cauwenberghe L, Devriese L, Galgani F, Robbens J, Janssen CR. Microplastics in sediments: a review of techniques, occurrence and effects. Marine Environmental Research 2015;111:5–17.

[426] Van Cauwenberghe L, Vanreusel A, Mees J, Janssen CR. Microplastic pollution in deep-sea sediments. Environmental Pollution 2013;182:495–9.

[427] Van Franeker JA, Blaize C, Danielsen J, Fairclough K, Gollan J, Guse N, Hansen P-L, Heubeck M, Jensen J-K, Le Guillou G, Olsen B, Olsen K-O, Pedersen J, Stienen EWM, Turner DM. Monitoring plastic ingestion by the northern fulmar *Fulmarus glacialis* in the North Sea. Environmental Pollution 2011;159:2609–15.

[428] Vandermeersch G, Van Cauwenberghe L, Janssen CR, Marques A, Granby K, Fait G, Kotterman MJJ, Diogène J, Bekaert K, Robbens J, Devriese L. A critical view on microplastic quantification in aquatic organisms. Environmental Research 2015;143(Part B):46–55.

[429] Velzeboer I, Kwadijk CJAF, Koelmans AA. Strong sorption of PCBs to nanoplastics, microplastics, carbon nanotubes, and fullerenes. Environmental Science and Technology 2014;48:4869–76.

[430] Vesilend P, editor. Wastewater treatment plant design. IWA Publishing; 2003.

[431] Vianello A, Boldrin A, Guerriero P, Moschino V, Rella R, Sturaro A, Da Ros L. Microplastic particles in sediments of Lagoon of Venice, Italy: first observations on occurrence, spatial patterns and identification. Estuarine, Coastal and Shelf Science 2013;130:54–61.

[432] Vijgen J, Abhilash PC, Li YF, Lal R, Forter M, Torres J, Singh N, Yunus M, Tian C, Schäffer A, Weber R. Hexachlorocyclohexane (HCH) as new Stockholm Convention POPs – a global perspective on the management of Lindane and its waste isomers. Environmental Science and Pollution Research International 2011;18:152–62.

[433] Walter T, Augusta J, Muller R-J, Widdecke H, Klein J. Enzymatic degradation of a model polyester by lipase from *Rhizophus delemar*. Enzyme and Microbial Technology 1995;17:218–24.

[434] Wan Y, Wiseman S, Chang H, Zhang X, Jones PD, Hecker M, Kannan K, Tanabe S, Hu J, Lam MHW, Giesy JP. Origin of hydroxylated brominated diphenyl ethers: natural compounds or man-made flame retardants? Environmental Science and Technology 2009;43:7536–42.

[435] Wang F, Shih KM, Li XY. The partition behavior of perfluorooctanesulfonate (PFOS) and perfluorooctanesulfonamide (FOSA) on microplastics. Chemosphere 2015;119:841–7.

[436] Wang H, Wang C-Q, Fu J-G, Gu G-H. Floatability and flotation separation of polymer materials modulated by wetting agents. Waste Management 2014;34:309–15.

[437] Ward J, Shumway S. Separating the grain from the chaff: particle selection in suspension- and deposit-feeding bivalves. Journal of Experimental Marine Biology and Ecology 2004;300:83–130.

[438] Watters DL, Yoklavich MM, Love MS, Schroeder DM. Assessing marine debris in deep seafloor habitats off California. Marine Pollution Bulletin 2010;60:131–8.

[439] Webster L, Phillips L, Russell M, Dalgarno E, Moffat C. Organic contaminants in the Firth of Clyde following the cessation of sewage sludge dumping. Journal of Environmental Monitoring 2005;7:1378–87.

[440] Webster L, Russell M, Adefehinti F, Dalgarno EJ, Moffat CF. Preliminary assessment of polybrominated diphenyl ethers (PBDEs) in the Scottish aquatic environment, including the Firth of Clyde. Journal of Environmental Monitoring 2008;10:463–73.

[441] Webster L, Russell M, Phillips LA, McIntosh A, Walsham P, Packer G, Dalgarno E, McKenzie M, Moffat C. Measurement of organic contaminants and biological effects in Scottish waters between 1999 and 2005. Journal of Environmental Monitoring 2007;9:616–29.

[442] Webster L, Russell M, Walsham P, Phillips LA, Hussy I, Packer G, Dalgarno EJ, Moffat CF. An assessment of persistent organic pollutants in Scottish coastal and offshore marine environments. Journal of Environmental Monitoring 2011;13:1288–307.

[443] Webster L, Russell M, Walsham P, Hussy I, Lacaze J, Phillips L, Dalgarno E, Packer G, Neat F, Moffat CF. Halogenated persistent organic pollutants in relation to trophic level in deep sea fish. Marine Pollution Bulletin 2014;88:14–27.

[444] Webster L, Russell M, Walsham P, Phillips LA, Packer G, Hussy I, Scurfield JA, Dalgarno EJ, Moffat CF. An assessment of persistent organic pollutants (POPs) in wild and rope grown blue mussels (*Mytilius edulis*) from Scottish coastal waters. Journal of Environmental Monitoring 2009;11:1169–84.

[445] Webster L, Walsham P, Russell M, Hussy I, Neat F, Dalgarno E, Packer G, Scurfield JA, Moffat CF. Halogenated persistent organic pollutants in deep water fish from waters to the west of Scotland. Chemosphere 2011;83:839–50.

[446] Webster L, Walsham P, Russell M, Neat F, Phillips L, Dalgarno E, Packer G, Scurfield JA, Moffat CF. Halogenated persistent organic pollutants in Scottish deep water fish. Journal of Environmental Monitoring 2009;11:406–17.

[447] Wei CL, Rowe GT, Nunnally C, Wicksten MK. Anthropogenic "litter" and macrophyte detritus in the deep Northern Gulf of Mexico. Marine Pollution Bulletin 2012;64:966–73.

[448] Wesch C, Barthel A-K, Braun U, Klein R, Paulus M. No microplastics in benthic eelpout (*Zoarces viviparus*): an urgent need for spectroscopic analyses in microplastic detection. Environmental Research 2016;148:36–8.

[449] Westerhoff P, Prapaipong P, Shock E, Hillaireau A. Antimony leaching from polyethylene terephthalate (PET) plastic used for bottled drinking water. Water Research 2008;42(3):551–6.

[450] Wheatley L, Levendis YA, Vouros P. Exploratory study on the combustion and PAH emissions of selected municipal waste plastics. Environmental Science and Technology 1993;27:2885–95.

[451] Wheeler J, Stancliffe J. Comparison of methods for monitoring solid particulate surface contamination in the workplace. Annals of Occupational Hygiene 1998;42:477–88.

[452] Whitt M, Vorst K, Brown W, Baker S, Gorman L. Survey of heavy metal contamination in recycled polyethylene terephthalate used for food packaging. Journal of Plastic Film and Sheeting 2013;29(2):163–73.

[453] WHO. Polychlorinated biphenyls: human health aspects. Concise International Chemical Assessment Document 55. Geneva: World Health Organization; 2003.

[454] Whyte ALH, Hook GR, Greening GE, Gibbs-Smith E, Gardner JPA. Human dietary exposure to heavy metals via the consumption of greenshell mussels (*Perna canaliculus* Gmelin 1791) from the Bay of Islands, northern New Zealand. Science of the Total Environment 2009;407:4348–55.

[455] Willem JK, Wollent I. Inventories of obsolete pesticide stocks in central and Eastern Europe. In: Proceedings of the 7th International HCH and Pesticides Forum, Kiev, Ukraine. 2005. p. 37–9.

[456] Witkowski PJ, Smith JA, Fusillo TV, Chiou CT. A review of surface-water sediment fractions and their interactions with persistent manmade organic compounds. U.S. Geological Survey Circular 993USA: United States Government Printing Office; 1987.

[457] Woodall LC, Robinson LF, Rogers AD, Narayanaswamy BE, Paterson GLJ. Deep sea litter: a comparison of seamounts, banks and a ridge in the Atlantic and Indian Oceans reveals both environmental and anthropogenic factors impact accumulation and composition. Frontiers in Marine Science 2015;2.

[458] Woodall LC, Sanchez-Vidal A, Canals M, Paterson GLJ, Coppock R, Sleight V, Calafat A, Rogers AD, Narayanaswamy BE, Thompson RC. The deep sea is a major sink for microplastic debris. The Royal Society Open Science; 2014.

[459] Woodall LC, Gwinnett C, Packer M, Thompson RC, Robinson LF, Paterson GLJ. Using a forensic science approach to minimize environmental contamination and to identify microfibres in marine sediments. Marine Pollution Bulletin 2015;95:40–6.

[460] Wright SL, Thompson RC, Galloway TS. The physical impacts of microplastics on marine organisms: a review. Environmental Pollution 2013;178:483–92.

[461] Wu Z, Liu G, Song S, Pan S. Regeneration and recycling of waste thermosetting plastics based on mechanical thermal coupling fields. International Journal of Precision Engineering and Manufacturing 2014;15:2639–47.

[462] Xanthos M. Recycling of the #5 polymer. Science 2012;337(6095):700–2.

[463] Xu S, Zhang L, Lin Y, Li R. Layered double hydroxides used as flame retardant for engineering plastic acrylonitrile-butadiene-styrene (ABS). Journal of Physics and Chemistry of Solids 2012;73:1514–7.

[464] Yamada-Onodera K, Mukumoto H, Katsuyaya Y, Saiganji A, Tani Y. Degradation of polyethylene by a fungus, *Penicillium simplicissimum* YK. Polymer Degradation and Stability 2001;72:323–7.

[465] Yamashita R, Tanimura A. Floating plastic in the Kuroshio Current area, western North Pacific Ocean. Marine Pollution Bulletin 2007;54(4):485–8.

[466] Yang J, Yang Y, Wu W, Zhao J, Jiang L. Evidence of polyethylene biodegradation by bacterial strains from the guts of plastic-eating waxworms. Environmental Science and Technology 2014;48(23):13776–84.

[467] Yang R, Jing C, Zhang Q, Wang Z, Wang Y, Li Y, Jiang G. Polybrominated diphenyl ethers (PBDEs) and mercury in fish from lakes of the Tibetan Plateau. Chemosphere 2011;83:862–7.

[468] Yonkos LT, Friedel EA, Perez-Reyes AC, Ghosal S, Arthur CD. Microplastics in four estuarine rivers in the Chesapeake Bay, U.S.A. Environmental Science & Technology 2014;48:14195–202.

[469] Yoshida S, Hiraga K, Takehana T, Taniguchi I, Yamaji H, Maeda Y, Toyohara K, Miyamoto K, Kimura Y, Oda K. A bacterium that degrades and assimilates poly(ethylene terephthalate). Science 2016;351(6278):1196–9.

[470] Schulz M, Krone R, Dederer G, Wätjen K, Matthies M. Comparative analysis of time series of marine litter surveyed on beaches and the seafloor in the southeastern North Sea. Marine Environmental Research 2015;106:61–7.

Glossary and list of abbreviations

ABS Acrylonitrile butadiene styrene

AG (Aktiengesellschaft) A German term for a company which is publicly traded.

ASA Acrylonitrile styrene acrylate

AO All opaque

ALL Any colour

AM Amber

Amorphous polymer A polymer which does not exhibit any form of crystallinity, such as brittle, glassy and ductile plastics.

AT All transparent

Attenuated total reflectance A form of infrared spectroscopy in which infrared light passes into a crystal at a greater angle than the critical angle for internal reflection within the crystal. This results in the production of an evanescent wave which penetrates into the surface of a sample in contact with the crystal.

ATM Atmosphere. A unit of pressure which is based upon the average atmospheric pressure at sea level and is defined as 1.01325 bar (equivalent to 14.696 psi).

ATR Attenuated total reflectance

BG Beige

BFR Brominated flame retardant

Bioaccumulation The accumulation of a substance, such as a toxic chemical pollutant, within the tissues of an organism over time. Bioaccumulation results when the speed at which the organism absorbs and stores the substance from the surrounding environment occurs at a faster rate than the speed at which the organism can effectively break down and remove the substance from its body.

Bioconcentration The bioaccumulative process in which the concentration of a substance in the tissues of an organism becomes higher than the concentration of the substance in the air or water surrounding the organism.

Biobased polymer A plastic produced from renewable biomass sources.

Biodegradable plastic A plant-based or oil-based plastic which naturally decomposes in the environment as a result of the activity of living organisms, such as bacteria and fungi.

Biomagnification The accumulation of a substance in the tissues of an organism at higher levels than the concentration of the substance which is found in its food. Thus, the next organism in the food chain will accumulate a greater amount of the substance from consuming the organisms at lower levels. As such, organisms in the top regions of the food chain exhibit a much greater concentration of the substance in their tissues than those in the bottom regions. Therefore, biomagnification is the tendency of a substance to concentrate to a greater extent in organisms at successively higher trophic levels.

Biomass Organic material used as fuel which is derived from living organisms, especially from plants.

Bivalve (Bivalvia) A class of marine and freshwater molluscs which have no head or radula and which have a two-part hinged shell enclosing a laterally compressed body. Bivalves typically feed by filtering the water and characteristic species are mussels, oysters, clams, scallops and cockles.

BK Black

BL Blue

BN Brown

BR Butadiene rubber

BZ Bronze

CA Cellulose acetate

CAB Cellulose acetate butyrate

CAP Cellulose acetate propionate

Carbon-fibre-reinforced polymers A plastic combined with carbon fibres to increase its rigidity and strength-to-weight ratio.

CCD Charge coupled device

CE Cellulose

CH Charcoal

CL Clear

Chromophore The part of a molecule containing a specific atom or, or group of atoms, which absorb light and is responsible for the colour of that molecule.

CMC Carboxymethyl cellulose

CN Cellulose nitrate

Commodity plastic A low-cost plastic that is mass-produced and has wide-ranging applications, where mechanical properties and operating environment are not critical.

Common plastic A plastic produced from fossil resources, such as unprocessed crude oil, petroleum products and natural gas.

Compound A chemical substance created from the combination of two or more atoms, of two or more elements, in fixed proportions by way of chemical bonds.

Contaminant A chemical, radiological, physical or biological substance which pollutes the environment, and which is normally absent in the environment. Some contaminants bioaccumulate and induce toxicological effects or mortality in organisms.

Contaminated microplastics Microplastics which have accumulated a layer of contaminants on their surface via adsorptive processes, or which contain contaminants that have diffused into their bulk via absorptive processes.

CP Cellulose propionate

CR Polychloroprene (neoprene)

CSM Chlorosulfonated polyethylene

DDD Dichlorodiphenyldichloroethane

DDE Dichlorodiphenyldichloroethylene

DDT Dichlorodiphenyltrichloroethane

DEHP Bis(2-ethylhexl) phthalate

DK Dark

EC European Commission

ECTFE Ethylene chlorotrifluoroethylene

EFSA European Food Safety Authority

Elastomer A viscoelastic polymer in which the inter-molecular forces are weak, and the substance can undergo high amount of elongation before failure; such as rubber or neoprene. If the deforming force is discontinued, the substance will revert back to its original shape.

Elastomeric Possessing or exhibiting the properties of an elastomer.

Elongation at break (%) The percentage of the initial length of a material by which the material can be elongated in length, until fracture of the material occurs. The elongation at break expresses the ability of the material to resist changes in shape without cracking.

Engineering plastic A plastic with superior thermal and/or mechanical properties, in comparison to a commodity plastic.

EPA Environmental Protection Agency

EPR Ethylene propylene rubber

EPS Expanded polystsyrene

ePTFE Expanded polytetrafluoroethylene

EVA Ethylene vinyl acetate

EVOH Ethylene vinyl alcohol

FAO Food Agriculture Organization

FB Fibre

FEP Fluorinated ethylene propylene

FI Film

Fibre A strand or filament of plastic (<5 mm–1 mm) in size along its longest dimension.

Film A thin sheet or membrane-like piece of plastic (<5 mm–1 mm) in size along its longest dimension.

Filtrate The fluid which has passed through a filter.

Filtration A technique used to separate solids from fluids (liquids or gases).

FM Foam

Foam A piece of sponge, foam, or foam-like plastic material (<5 mm–1 mm) in size along its longest dimension.

Food chain A linear series of organisms in a food web which are linked by the transfer of energy and nutrients, and which starts with a producer and ends with an apex predator. Each successive organism in the food chain is dependent upon the organism below as a source of food.

Food web A network of interconnected and mutually dependent food chains which can be graphically represented as a web.

Fourier transform A mathematical tool which deconstructs a waveform by exploiting the actuality that any waveform can be rewritten as the sum of sine and cosine functions.

Fourier transform infrared spectroscopy An analytical chemistry technique which irradiates a sample with specific wavelengths of infrared red light and then examines the transmitted light to deduce the amount of energy that was absorbed by the molecules at each wavelength, thereby providing information about the molecules present in the sample.

FR Fragment

Fragment An irregular shaped piece of plastic (<5 mm–1 mm) in size along its longest dimension.

FTIR Fourier transform infrared spectroscopy

GD Gold

GESAMP Joint Group of Experts on the Scientific Aspects of Marine Environmental Protection

Glass transition temperature (T_g) The temperature region at which a plastic material transforms from a hard glassy state to a soft flexible state.

GN Green

GPA Global Programme of Action

GY Grey

HCH Hexachlorohexane

HDPE High-density polyethylene

HELCOM Helsinki Convention on the Protection of the Marine Environment in the Baltic Sea Area

HEMA Hydroxyethyl methacrylate

HHS United States Department of Health and Human Services

HIPS High-impact polystyrene

HOC Hydrophobic organic compounds

Hydrolysis The splitting of chemical bonds by the addition of water.

Hydrolytic Pertaining to, or resulting in, hydrolysis.

Hydrophilic Having an affinity for water or readily mixing with, or dissolving in, water.

Hydrophobic Water repelling or failing to readily mix with, or dissolve in, water.

Hydrothermal vent A fissure on the ocean floor out of which mineral-rich water flows, which has been superheated by underlying molten or semi-molten rock (magma) to temperatures of up to 464°C.

Hygroscopic A substance which has the ability to absorb and retain water from the surrounding atmosphere.

ICES International Council for the Exploration of the Sea

ICI Imperial Chemical Industries

IMO International Maritime Organization

Impact test A test which indicates a plastic material's toughness by measuring the amount of energy absorbed by the plastic material during fracture.

Ingestion The process of consumption of material, such as food, by an organism, which is typically accomplished by the intake of the material through the mouth into the gastrointestinal tract.

Invasive species Animals, plants, fungi or pathogens which occur in an area outside their normal accepted distribution and which harm or pose a threat to that area.

Invertebrate Any animal which does not possess a vertebral column (backbone).

IR Infrared

IV Ivory

K_{ow} Octanol/Water partition coefficient

LDPE Low-density polyethylene

Lipophilic Having an affinity for lipids or fats or readily mixing with, or dissolving in, lipids or fats.

LLDPE Linear low-density polyethylene

LT Light

Limiting Oxygen Index An indication of flammability in which the minimum amount of oxygen required to allow sustained combustion for 3 min is expressed as a percentage.

LOD Limit of detection

Longshore drift The movement of material along a coast as a result of the prevailing wind blowing waves into the coast at an angle (swash), which pushes material up the beach at the same angle. The waves then recede directly away from the coast at right angles (backwash) and pull the material back again, thereby resulting in transportation of the material along the coast in a zig-zag fashion.

Macroplastic Any piece of plastic equal to or larger than 25 mm in size along its longest dimension.

MAP Macroplastic

MARPOL Marine pollution (International Convention for the Prevention of Pollution from Ships)

MBD Microbead

MBS Methacrylate butadiene styrene

MDP Marine Debris Program

MDPE Medium-density polyethylene

MEHP Mono-(2-ethylhexyl) phthalate

Melting point The temperature at which a material undergoes a phase transition from a solid state to a liquid state at atmospheric pressure.

MEP Mesoplastic

MF Melamine formaldehyde

MFB Microfibre

MFI Microfilm

MFM Microfoam

MFR Microfragment

Mesoplastic Any piece of plastic (<25 mm–5 mm) in size along its longest dimension.

Microbead A small spherical piece of plastic (<1 mm–1 μm) in diameter.

Microfoam A piece of sponge, foam, or foam-like plastic material (<1 mm–1 μm) in size along its longest dimension.

Microfibre A strand or filament of plastic (<1 mm–1 μm) in size along its longest dimension.

Microfilm A thin sheet or membrane-like piece of plastic (<1 mm–1 μm) in size along its longest dimension.

Microfragment An irregular shaped piece of plastic (<1 mm–1 μm) in size along its longest dimension.

Microplastic Any piece of plastic (<5 mm–1 mm) in size along its longest dimension.

Mini-microplastic Any piece of plastic (<1 mm–1 μm) in size along its longest dimension.

Mono- A prefix which means single or containing only one.

Monomer A low molecular weight molecule which can chemically bond to other low molecular weight molecules in a repeating fashion to form polymer. A subunit of a polymer.

MMP Mini-microplastic

MP Microplastic

MPPRCA Marine Plastic Pollution Research and Control Act

MSFD Marine Strategy Framework Directive

MT Metallic

Nanoplastic Any piece of plastic less than 1 μm in size along its longest dimension.

NASA National Aeronautics and Space Administration

NBR Acrylonitrile butadiene rubber

NGO Non-governmental organisation

NGT National Green Tribunal

NMDMP National Marine Debris Monitoring Program

NMR Nuclear magnetic resonance

NOAA National Oceanic and Atmospheric Administration

NP Nanoplastic

NR Natural rubber

Nuclear magnetic resonance An analytical chemistry technique which can be used to study the molecular structure of a sample by exploiting the physical phenomena that nuclei with a magnetic moment in the presence of a magnetic field will absorb and re-emit electromagnetic radiation.

OL Olive

OP Opaque

OR Orange

OSPAR Oslo/Paris convention (for the Protection of the Marine Environment of the North-East Atlantic)

Oxide A chemical compound which contains at least one oxygen atom in combination with one other chemical element in its molecular formula, such as carbon monoxide (CO), nitrous oxide (N_2O) or iron(III)oxide (Fe_2O_3).

Ozone Considered to be beneficial in the upper atmosphere, but a serious pollutant in the lower atmosphere, ozone is an unstable highly reactive and powerfully oxidizing form of oxygen gas (O_3) which has three oxygen atoms, as opposed to normal oxygen gas (O_2) which has two oxygen atoms. Ozone has a pungent odour and pale blue colour and is formed by ultraviolet light, lightning or other electrical discharges in the atmosphere.

PA Polyamide (Nylon)

PA 46 Nylon 4,6

PA 6 Nylon 6

PA 610 Nylon 6,10

PA 66 Nylon 6,6

PA 66/610 Nylon 6,6/6,10 copolymer

PA 11 Nylon 11

PA 12 Nylon 12

PAA Polyarylamide

PAES Polyarylethersulfone [usually known as polyphenylsulfone (PPSU)]

PAH Polycyclic aromatic hydrocarbon

PAI Polyamide imide

PAN Polyacrylonitrile

PB Polybutylene

PBDE Polybrominated diphenyl ether

PBT Polybutylene terephthalate

PC Polycarbonate

PCB Polychlorinated biphenyl

PCDD Polychlorinated dibenzodioxin

PCL Polycaprolactone

PCPs Personal care products

PE Polyethylene

PEBA Polyether block amide

PEEK Polyetheretherketone

PEEL Polyester elastomer

PEI Polyester imide

PEK Polyetherketone

Pellet A small spherical piece of plastic (<5 mm–1 mm) in diameter.

Persistent organic pollutant A hazardous organic chemical which exhibits substantial resistance to biodegradation and persists in the environment for a considerable period of time, while typically tending to bioaccumulate in the bodies of organisms, as well as inducing toxicological effects.

PES Polyethersulfone

PET Polyethylene terephthalate

PETG Polyethylene terephthalate glycol-modified

PF Phenol formaldehyde

PFA Perfluoroalkoxy alkane

PHB Polyhydroxybutyrate

PHBV Poly(3-hydroxybutyrate-co-3-hydroxyvalerate)

PHV Polyhydroxyvalerate

PI Polyimide

PIR Polyisocyanurate

PK Pink

PLA Polylactic acid

Plastic A synthetic material composed of macromolecules which, in turn, consists of many recurring small molecules, termed monomers, which are connected together in sequence by covalent bonds.

Plasticle An all-inclusive term, first coined by Christopher Blair Crawford and introduced with this book in 2016, which is used to describe any piece of plastic smaller than 5 mm in size along its longest dimension; such as a microplastic, a mini-microplastic and a nanoplastic.

PLT Plasticle

PMA Poly(methyl methacrylate)

PMP Polymethylpentene

Pollutant A substance, typically a waste material, which contaminates the environment, such as water, air or soil, and which has undesired effects or adversely affects the value or practicality of a resource.

Poly- A prefix which means many.

Polymer A complex high molecular weight molecule formed from many repeating monomer subunits.

Polymerization The process of connecting monomers together.

POM Polyoxymethylene

POP Persistent organic pollutant

Primary microplastic Typically, small spherical microbeads which are intentionally manufactured by the plastics industry for use in cosmetics, personal care products, dermal exfoliators, cleaning agents and sand blasting shot.

PP Polypropylene

PPE Poly(*p*-phenylene ether)

PPO Poly(*p*-phenylene oxide)

PPSS Polyphenylene sulphide sulfone

PPS Polyphenylene sulphide

PPSF Polyphenylethersulfone [usually known as polyethersulfone (PES)]

PPSU Polyphenylsulfone

PPT Polypropylene terephthalate

PPy Polypyrole

PR Purple

PS Polystyrene

PSU Polysulfone

PT Pellet

PTFE Polytetrafluoroethylene

PTT Polytrimethylene terephthalate

PUR Polyurethane

PVA Polyvinyl acetate

PVB Polyvinyl butytral

PVC Polyvinyl chloride

PVCC Chlorinated polyvinyl chloride

PVDC Polyvinylidene chloride

PVDF Polyvinylidene fluoride

PVF Polyvinyl fluoride

PVOH Polyvinyl alcohol

Radula A minutely toothed ribbon-like anatomical structure, similar to a tongue, which is unique to molluscs (except bivalves) and is used to feed by cutting food or scraping food from surfaces.

Raman spectroscopy An analytical chemistry technique which examines the interaction of specific wavelengths of light with matter. The technique is used to derive information from scattered light about the vibrational motions of molecules comprising a sample, typically to identify the samples chemical composition.

RD Red

SAN Styrene acrylonitrile

SBR Styrene butadiene rubber

SBS Styrene-butadiene-styrene

Scanning electron microscopy A technique which scans the surface of a sample with a focused beam of high-energy electrons to form a three-dimensional image of the sample from scattered electrons and the emission of secondary electrons.

SCS Size and colour sorting. A standardised system that was designed and created by Christopher Blair Crawford and first introduced with this book in 2016. The system is used for the effective categorising of any piece of plastic, based upon its size, colour and appearance.

SEBS Styrene-ethylene-butadiene-styrene

Secondary microplastic Irregular pieces of plastic that have been unintentionally produced as a result of the degradation of larger pieces of plastic.

Sediment Naturally occurring material, such as sand and silt, which settles to the bottom of a liquid.

Semi-crystalline polymer A polymer where some fraction of the material is crystallised.

SIS Styrene-isoprene-styrene

Size and colour sorting (SCS) system A standardised system that was designed and created by Christopher Blair Crawford and first introduced with this book in 2016. The system is used for the effective categorising of any piece of plastic, based upon its size, colour and appearance.

SMA Styrene maleic anhydride

Solute A chemical substance (solid, liquid or gas) which dissolves in a solvent.

Solution The mixture formed once a solute has dissolved in a solvent.

Solvent A liquid chemical substance which dissolves another solid, gas or liquid chemical substance (solute) to form a solution.

SP Speckled

Spin A fundamental quantum-mechanical property which is characterised as the intrinsic or total angular momentum carried by composite and elementary particles, as well as atomic nuclei.

Strength-to-weight ratio A measure of the specific strength of a material obtained by dividing the force applied perpendicular to the surface of the material per unit area at failure by the density of the material.

Surface-area-to-volume ratio In terms of plastic materials, the quantity of surface of a piece of plastic per unit of volume of that piece of plastic.

SV Silver

TCDD 2,3,7,8-tetrachlorodibenzo-p-dioxin

Thermoplastic A plastic material which can be heated until melted and then cooled and formed into new shapes.

Thermosetting plastic A plastic material which is generally permanently set and cannot be melted and reformed.

T_g Glass transition temperature

TN Tan

TP Transparent

TPUR Thermoplastic polyurethane

TQ Turquoise

Translocation A change in location from one place to another.

Trophic level The position at which an organism is situated within the food chain and is measured by the number of steps at which the organism is situated from the start of a food chain. Trophic level 1 is considered as the bottom of the food chain where producers reside, such as plants and algae. Trophic level 2 represents primary consumers, which are plant-eating organisms (herbivores). Trophic level 3 represents secondary consumers, which are meat-eaters (carnivores) and thus, predators of herbivores. Trophic level 4 represents tertiary consumers, which are carnivores which eat other carnivores. Trophic level 5 represents apex predators, which are defined as organisms which have no natural predator in their ecosystem. Thus, the highest trophic level is at the end of a food chain, and high levels of contaminants are often found in organisms residing at high trophic levels as a result of biomagnification.

Trophic transfer The movement of a substance from one trophic level to the next trophic level as a result of dietary bioaccumulation.

UF Urea formaldehyde

UHMWPE Ultra-high molecular weight polyethylene

UNCLOS United Nations Convention on the Law of the Sea

UNEP United Nations Environment Programme

USEPA United States Environmental Protection Agency

UV Ultraviolet

Vertebrate Any animal which possesses a vertebral column (backbone).

VT Violet

Viscoelastic A substance possessing both viscous and elastic properties, such as an elastomer.

Water absorption The amount of water absorbed by a material during specific test conditions, and which is typically expressed as a percentage weight gain of the material following submersion in water over a specified period.

Water pollution The contamination of a body of water with chemical, radiological, physical or biological material.

WFD Water Framework Directive

WH White

XLPE Cross-linked polyethylene

XPS Extruded polystyrene

YL Yellow

Zone (intertidal) The area between the low tide mark and the high tide mark, which is also referred to as the foreshore, and which is subject to the rigours of the sea, as well as the land, and where sediments are often interspersed with microplastic pollutants.

Index

Printed in the United States
By Bookmasters